Karl Schlagenhauf   Fanji Gu

The Brain and AI
Correspondence between a German Engineer and a Chinese Scientist

Volume II

Mystery of Consciousness and Myth of Mind Uploading

上海教育出版社
SHANGHAI EDUCATIONAL
PUBLISHING HOUSE

The Authors

Karl Schlagenhauf

Karl Schlagenhauf is a serial entrepreneur in the field of new technologies and has founded numerous start-ups in the high-tech sector in Europe and the USA. He was CEO of IFAO (Institute for applied organizational research) and ADI Software, a company with a focus on RDBMS, Multimedia and Internet-Applications in industry and banking, both of which he spun off from the University of Karlsruhe in the 1980s.

He stepped down as CEO in 2003, and is now chairman of the board of ADI Innovation and runs a Family Office. He also serves on the board of companies where he is invested and has served on the boards of several technology companies such as AP Automation + Productivity (now Asseco Solutions), Brandmaker, CAS Software, JPK Instruments and Web.de.

As an inventor he holds patents in the field of secure remote control via the internet and Nano-robots for protein analysis based on atomic force spectroscopy.

He is an advisor to private equity firms and governmental bodies and also a coach to young entrepreneurs.

While his focus has always been on leading edge technologies and their impact on social systems and the delicate orchestration of human, intellectual und financial resources, he has extended his interest from software, electronics and manufacturing to the life sciences and artificial intelligence.

He holds a Master's degree in Economics/Industrial Engineering, a PhD in Philosophy and a "Venia Legendi" for Sociology and Theory of Science from University of Karlsruhe (now KIT).

Fanji Gu

Prof. Fanji Gu is an emeritus professor of computational neuroscience at the School of Life Sciences, Fudan University, Shanghai, China.

He graduated in mathematics from Fudan University in 1961, worked in Department of Biophysics, Science & Technology University of China from 1961 to 1979, then worked in School of Life Sciences, Fudan University since 1979. He worked as a visiting research associate in the Department of Physiology & Biophysics, University of Illinois at Urbana-Champaign from 1983 to 1985.

His major is computational neuroscience. He published 3 monographs and about 100 papers. After his retirement in 2004, he was the managing editor of the journal *Cognitive Neurodynamics* from 2006 to 2011. He edited three proceedings of international conferences.

After that he became a popular science writer on brain science. He wrote 6 popular science books on brain science and translated seven books, including Walter Freeman's *Neurodynamics*, Christof Koch's *Quest for Consciousness*, Gerald Edelman's *A Universe of Consciousness* and V. S. Ramachandran's *Phantoms in the Brain*. His popular science books won seven awards, including the 2017 best seller in China, the 2016 Shanghai Municipal Award of Science and Technology (science popular book, third grade), the 2015 Shanghai Science Popularization Education Innovation Award (science popular book, second grade) etc. He himself also won a Merit Award conferred by the 2013 Fourth International Conference on Cognitive Neurodynamics held in Sweden, the 2017 Shanghai Science Popularization Education Innovation Award (individual contribution, second grade), and the 2018 Shanghai Award for Science & Technology (third grade).

He is now the honorary president of Shanghai Chapter of University of Illinois Alumni Association and a member of the editorial board of WeChat's subscription channel "Fanpu".

This book is a special treat. It has something to offer to everyone who is curious about the nature of human insight and how we can make progress on its understanding.

    ——Matthias Bethge (Chair of the Bernstein Center for Computational Neuroscience, Professor, University Tübingen and cofounder of deepart.io)

An unusual book which clearly stands out from many other neuroscience publications. It should be recommended to all IT experts and neuroscientists who wish to escape well-trodden paths of Main Stream research.

    ——Hans Albert Braun (Professor of Philipps University of Marburg, Head of the Neurodynamics Group, Institute of Physiology and Pathophysiology)

A fascinating dialogue of a German engineer and a Chinese biophysicist. We should thank the authors for publishing their in-depth discussion on AI, brain activity and consciousness.

    ——Yizhang Chen (Academician of Chinese Academy of Sciences, Professor of The Second Military Medicine University)

I have read "The Brain & AI" completely spellbound and would be on fire to film it one day.

    ——Serdar Dogan (German Director and Filmmaker)

An intellectual delight like very few I have encountered. It is treasure trove of ideas and insights, sometimes playful, sometimes deeply serious, yet always optimistic.

    ——Dietmar Harhoff (Director at the Max Planck Institute for Innovation and Competition in Munich)

Chinese venerable wisdom and expertise in neurophysiology meets Western social science plus information technology. "Do we need such a book?" My answer is definitely yes.

——Gert Hauske (Retired Professor at Technische Universität München, former Editor-in-chief of *Biological Cybernetics*)

I really hope Fanji Gu's and Karl Schlagenhauf's book, so full of original thoughts and insights, sparks new ways of collaboration between science, politics and business, to create the next wave of "real" A(G)I to the benefit of all human beings.

——Rafael Laguna (CEO of Open-Xchange, Director of the German Agency for Disruptive Innovation)

With wisdom and vision the authors bring readers together to discuss the leap and myth behind the age of intelligence.

——Peiji Liang (Director of the Chinese Neuroscience Society, Professor of Shanghai Jiaotong University)

Perhaps you are just interested in artificial intelligence and the human brain, or you have delved into a specific related area and would like to see the bigger picture, or you just want to learn something fun in your spare time. No matter what the case is, I believe that this book will bring you unique and fun experiences.

——Amanda Song (PhD candidate, University of California, San Diego)

A book to arouse thinking. All readers interested in brain science and artificial intelligence deserve to read this.

——Xiaowei Tang (Academician of Chinese Academy of Sciences, Professor of Zhejiang University)

A model of how to conduct scientific discussions. All interested readers in brain science, cognitive science, artificial intelligence and the philosophy of science will love this collision of views and ideas. I am convinced that this dialog is one of the most profound books on human brain and artificial intelligence published in recent years.

——Pei Wang (Associate Professor of Temple University, Vice Chair of Artificial General Intelligence Society, Chief Executive Editor, *Journal of Artificial General Intelligence,* one of the Translators of *Gödel, Escher, Bach: An Eternal Golden Braid*)

A science book as thrilling as an Agatha Christie Crime Novel.
The style of the "old debate culture of scientists", i.e. respectfully exchanging arguments in letters and developing knowledge based on solid information of arguments, is very refreshing. We should not unlearn these skills, they are extremely important in an era of twitter and short messages.

——Ehrenfried Zschech (Department Head, Fraunhofer Institute for Ceramic Technologies and Systems IKTS, Professor, Member of the Senate, European Materials Research Society)

Introduction 1

Would it be better to forget about brain mechanisms in the development of artificial intelligence? Airplanes do not copy the methods of birds, so why should artificial intelligence copy the methods of the brain? Are there principle differences between information processing in computers and human brains? Was there, at any time, a comparable progress in neuroscience to that in physics at the beginning of the last century (Planck, Einstein, Heisenberg etc.)? Is there something wrong with contemporary education and young researchers' promotion, preferring those who are faithfully following established authorities instead of promoting critical thinking? Are political skills more important than scientific concepts for winning a big grant of hundreds of millions of US dollars or euros? Do successful applicants of such big grants really believe in their promises, e.g. that the human brain can be fully simulated by digital computers? How can we talk about conscious AI without yet knowing how human consciousness is related to physical mechanisms in the brain? Are there principle limitations for brains to understand brains? Could artificial intelligence be more successful?

These are some of the topics and types of questions that are discussed in this truly exceptional book. It is written in a unique, highly interesting and entertaining style, and not just because it is in the form of the authors' e-mail correspondence. It especially provides completely new points-of-view on science and the business behind it. Even though the reader will be thoroughly informed about the current state of brain research and AI activities, discussing the particular success in these fields during recent years is not the focus of the book. Instead, it points out the many still open questions, unveils popular myths about brains and artificial intelligence, and discusses current research concepts rather skeptically, thereby critically shining on some of "The Emperor's New Clothes".

This e-mail exchange began when my two friends asked me almost simultaneously but independently of one another for my opinion on the EU "Human Brain" project which back then, in 2013, was close to its approval. I had to detach myself from this exchange of thoughts due to numerous other commitments while my friends continued and extended their discussions. When I now see the result, of course I find it quite unfortunate that I didn't continue my participation in our correspondence, and not only because I envy my friends for publishing this outstanding book. I think that I, as an engineer and neurophysiologist, could have been able to put forth additional arguments. I don't agree with the authors on all points just as the authors, despite the expressively friendly style of their debate, are far away from fully agreeing with each other.

But this is precisely one of the strengths of this book — it opens new perspectives and inspires critical thinking. It is particularly interesting, even amusing, to see how the two authors who are not only coming from different research areas but also from different cultural circles, are overtaking each other's arguments and bringing them back in new forms under different perspectives. The authors possess, in addition to their intellectual capacity, what artificial systems (despite Siri and Alexa) are still missing, namely, a subtle sense of humor.

My German friend Karl often likes to emphasize his arguments with his incomparably broad and profound knowledge of the history and theory of science. It is impressive to see how my Chinese friend Fanji, not so familiar with this background of mainly Western research, eagerly takes up all these notes and integrates them, often with surprising new results, into his own worldview, and then subtly informs Karl of his new insights from the perspective of an Eastern neurophysiologist.

On the other hand, my Chinese friend Fanji does his best to feed Karl all kinds of information about the brain which Karl, in turn, reinterprets in context with the design of artificial intelligence. For Karl, the engineer, it is difficult to accept the often imprecise use of scientific terms in neurophysiology and that the boundaries between knowledge and speculation are often blurred. He focuses on the enormous technological progress in recent years, especially in computer science and IT with promising new developments towards artificial intelligence. Nevertheless, they both have to confess that so far nobody knows whether and how an artificial system of comparable human flexibility and creativity, including consciousness, could be designed, and whether this should be desired at all. This goes along with intense discussions about similarities and differences of information processing in computers and human brains, including the old brain-mind problem, i.e. asking how mental functions and consciousness can arise from physical mechanisms. This leads to an equally interesting question for both of them, namely, to what extent the development of artificial intelligence should be based on brain-like mechanisms, or whether it would be better to go ahead with mind-like simulation strategies, i.e. searching for the most appropriate algorithms to simulate the desired functions instead of trying to copy the physiological mechanisms in all details.

Interestingly, Fanji and Karl have never met in person. I have met Fanji at various conferences, specifically at the "International Conferences on Cognitive Neurodynamics" which he has initiated. I was deeply impressed by his scientific exactness, his unpretentious nature and especially by his profound sense of humor, which is evident throughout this book making it particularly pleasant to read. I have known Karl since my high school days. We met again as students at the University of Karlsruhe, now KIT. I am still grateful today that I was allowed to belong to his Karlsruhe circle of friends. This book immediately reminded me of our student days when we used to get together for an evening drink for example at the "Pschorr-Fässle" or "Bürgerstüble". This generally happened in an interdisciplinary composition and Karl usually opened our meetings with his preferred question: "Who has a problem for today?" The more intense the discussions became, the more satisfied, also intellectually, we left the pub. The same search for stimulating discussions can still be recognized in this book of two congenial partners.

One thing is certain: this is an unusual book which clearly stands out from many other neuroscience publications. It should be recommended to all IT experts and neuroscientists who wish to escape well-trodden paths of Main Stream research. Particularly, it should be given to every young scientist who is interested in neuroscience and artificial intelligence. They will learn a lot about the current state of brain and AI research, about science as a big business, about the history of science and many more topics. Most importantly, the insights form this book will strengthen critical thinking and prevent blindly following scientific gurus and their dogmas. For this reason alone I can expressly recommend this book, quite apart from the exceptionally interesting and entertaining reading material it offers.

Hans Albert Braun [1]

Ph D, Dipl.Ing.

Professor

Head of the Neurodynamics Group

Institute of Physiology and Pathophysiology

Philipps University of Marburg

Germany

---

[1] Dr. Hans Albert Braun is a professor of physiology and Head of the Neurodynamics Group at the Institute of Physiology and Pathophysiology at Philips University of Marburg, Germany. Hans Braun's many outstanding contributions to physiology and neuroscience are documented in a focus issue of the journal *Chaos* that resulted from a scientific meeting held to honor his lifetime's work in June 2017.

See: "Introduction to Focus Issue: Nonlinear science of living systems: From cellular mechanisms to functions"
Epaminondas Rosa, Svetlana Postnova, Martin Huber, Alexander Neiman, and Sonya Bahar
Chaos 28, 106201 (2018); Published by the American Institute of Physics, doi: 10.1063/1.5065367
View online: https://doi.org/10.1063/1.5065367

Introduction 2

The rise of mankind is based on the one hand on his ability to think and his manual skills. The prerequisites for this are a large, complex brain and freely movable arms and hands made possible by the upright gait. Other essential points are conceptual and social aspects of intelligent human behavior. The authors of this book are deeply concerned with the above mentioned topics. They say: "My impression is that the majority of neurobiologists and psychologists concentrate too much on the single isolated brain. They tend to misjudge the fact that a human brain can only work meaningfully as part of a network of brains with enormous content." This is an essential and valuable message. It becomes evident that present Brain Research asks so deep and fundamental questions for which metrically operating natural sciences alone can give no answer.

So, a very attractive aspect for this book is that both authors come from very different professional fields and different cultural experiences: a Chinese expert in neurophysiology and mathematics meets a Western expert in social science and information technology. I know Fanji for a long time. He is a critical thinker, a successful academic teacher, author of many articles in relevant journals, editor of scientific journals, and author of a number of popular science books. He graduated in Mathematics and devoted his energies to Biophysics with a specialization in Neural Information Processing. He describes himself as a Computational Neuroscientist. It is a lucky circumstance that he met Karl, a German sociologist and Information Processing expert, to exchange their ideas. They march now from neurophysiological entities to society and from single cell activity to cooperation thus broadening their individual views. The interdisciplinary background is an essential condition to reach an adequate understanding of the abundance of viewpoints in this book.

The chosen format as letters leads to a more personal kind of argumentation during which the authors present themselves in some more individuality than in usual books. It is fascinating how the reader experiences creation and exchange of ideas in a very personalized manner. An excellent choice of citations of relevant literature presents a valuable possibility for the reader to go deeper into the respective topics. The presentation thus possesses the character of a lecture with exercises and this is a nice pedagogical aspect. We learn that science and scientific development are not static and abstract but intimately related to human beings and their communication. There is much wisdom in their presentation with respect to brain science, about the development of our society, but also concerning the question how and under which circumstances good science can be carried out. The authors legitimately criticize some of the contemporary attempts to intensify Brain Research, not only because these consume enormous amounts of money. Some inconsistencies in the concepts of the prestigious and ambitious attempts of this kind served as the starting point of this correspondence. The authors say that there is often an enormous amount of biological data, but these tell us nothing about how the brain really works.

Under such a wide perspective it is very exciting to follow a review of the development of existing attempts to understand the function of the nervous system under the viewpoint different concepts from Logical Networks, Cybernetics, Perceptrons, Neural Networks to Artificial Intelligence (AI). AI plays a major role, of course, and the authors are relatively optimistic about it. Some of the earlier promises of AI, however, are criticized by the authors as unrealistic. Some AI enthusiasts made their readers believe that all problems can be overcome by more computer-power and applying sufficient mathematics, but what they really need is a bit more of understanding. They wanted readers to believe that everything in the world can be calculated if you only have the right algorithm. But according to the authors' statement there is a relevant class of phenomena which are resistant to algorithms and also to numerical calculations.

Intelligence rests upon thinking processes which can be logical, problem solving or inductive. Inductive thinking with the ability to recognize regularities is assumed to be one of the main components of human intelligence. The authors ask whether Artificial Intelligence is possible after all and the answer depends on how one defines it. Obviously this is not a simple task, because 70 definitions for intelligence are possible. Problems in connection with consciousness are also discussed with the speculation whether artificial consciousness appears possible, but according to the authors' opinion it is still too early to study it. This should not mean that artificial consciousness is absolutely impossible.

It becomes clear that brains, unlike technical systems, are capable of generating intentions, goals and meanings (Freeman) or predicting the future (Hawkins). In order to solve a specific problem, brains can not only handle huge amounts of data in a short time, as required by chess playing, for example, but they also contain the means to develop adequate methods to solve problems (Wang). So they can not only improve abilities, but also create new ones. But the way from an arrangement of single cells in a brain to the complex function of the whole system is extremely difficult. Engineers, the authors say, can be inspired by nature's approach, but should not copy biology for every detail.

Another very important point concerns the ability of humans to behave socially, namely using cooperation and information exchange, not excluding competition, between humans. The German anthropologist Arnold Gehlen made this clear back in 1940. Gehlen's insights show that man, in order to survive, is compelled to act, to work and to think, however, in a socially oriented way. The fact that he was abandoned by protective natural instincts distinguishes him from the animals that fit perfectly into nature. It is of importance that according to Gehlen, man had to create for himself a social system which he called "institutions", such as marriage, state, religion and also economy, which ultimately represent what we call culture. Gehlen says that man is a "robot of the art of survival". For Gehlen, "institutions" are the epitomes of culture, which give man the necessary security, and he defends them against a number subjectively motivated currents of his time. Language and writing are important tools for building a rapidly developing world beyond genetic evolution. Language invented 100 000 years ago is an "intelligence amplifier," says Bremen-based brain researcher Gerhard Roth, and is also an important tool for social behavior.

In this context it may be interesting that cooperation and emergent behavior with astonishing complex performance is already a main property of insect colonies and, at a lower level of complexity, even colonies of bacteria. It can be assumed that human natural intelligence thus may be based on the same fundamental principles typical for all kinds of life. Artificial Intelligence is perhaps not so artificial as it seems because it is a creation of a natural brain and obeys the same rules as all sorts of living systems. A full understanding of those rules and the consequences for Artificial intelligence may still need some more research. If it integrated in a responsible way into our society, Artificial Intelligence may become an important tool for the survival of mankind.

As a conclusion I should say that both authors represent a perfect couple to write such a book. It is very attractive to see how Chinese wisdom plus expertise in neurophysiology meets Western social science combined with information technology. When I asked myself "Do we need such a book?" my answer is definitely "Yes".

Gert Hauske

Retired Professor at Technische Universität München

Former Editor-in-chief of *Biological cybernetics*

It has been seven years since I met Professor Gu, one of the authors of this book. At that time, I was a senior student at Fudan University, interested in human mind and brain. When I searched for the scientific literature and books, I found one book Professor Gu translated: *The Quest for Consciousness a Neurobiological Approach*, and got to know him in person. Thanks to his encouragement and guidance, I continued my exploration toward human mind and started my PhD at University of California, San Diego. Throughout these years, my interests have gradually switched from consciousness and human brain to artificial intelligence and social psychology, and my career path gradually steered toward industry. Therefore, I am able to appreciate and comprehend the mind-body and human-machine intelligence problem from a broader perspective. It is a great pleasure and my honor to be one of the first group of people who can read Professor Gu's latest book on this topic. The problems of consciousness and mind are highly-interdisciplinary and they have attracted researchers from neuroscience, psychology, linguistic, social science, philosopher and computer science. The two authors of this book are from different backgrounds, yet they are bounded by the shared interests and love for this fascinating topic and they have attempted to tackle this holy-grail question from various angles.

The dialogues between them will guide you into a fantastic journey. You will learn about the latest progresses in neuroscience and artificial intelligence, you will be informed of the methodologies adopted by a variety of disciplines, and you will get a glimpse of the landscape and the important questions in different areas related to mind-and-intelligence. Most uniquely and interestingly, this book is also a previous record of the life and friendship of two researchers. This book is just as approachable and friendly to readers from all background as Professor Gu's previous books. However, there is one new flavor, which is the 1+1>2 effect created by the two authors thought sparkles.

In this journey, you will observe in first-hand how one thought evokes another, how ideas from two sources echo and trigger another layer of analysis. You will see how they reflect on the research methodology and remain critical about the authority, such about the blue brain project leader's claims. You will see how they navigate in the new waves of artificial intelligence, how they come up with new thoughts, predictions and perspectives on the future trends. Along the way, you will find that they gradually learn from each other and gain new perspectives from the other person's area of expertise. You will peek into their daily life full of humor and openness. Before you get bored, they will refresh you with anecdotes in the fields. As the book progresses, so does their friendship. You will be the witness of this process.

Perhaps you are just interested in artificial intelligence and human brain, or you have delved into a specific related area and would like to see a bigger picture, or you just want to learn something fun in your spare time. No matter what is the case, I believe that this book will bring you unique and fun experiences. I hope that you will enjoy this exploration of human mind and machine intelligence, this dialogue between from science and humanity, this big bang between a Chinese scholar and a German engineer.

Amanda Song

PhD Candidate, University of California, San Diego

This book is special, if not unique. We, the two authors, were raised in different cultural environments and have never met face to face. Karl an engineer and entrepreneur, is German, and Fanji a scientist and popular science writer on brain science, is Chinese. A mutual friend, Prof. Hans Braun, a neuro-physiologist, introduced us nearly 6 years ago and we have been corresponding since on the subjects of brain research and artificial intelligence (AI).

We have become good friends and correspond frequently, in some cases even more frequently than with some of our old friends. Our common interest in the puzzles of the brain, the mind, consciousness and AI binds our friendship, and with great enthusiasm we pay close attention to the rapid progress in these fields.

We prefer the method of rational thinking, and are always eager to scrutinize the whys and wherefores of everything, as opposed to following mainstream or scholastic thinking. Our points of view are significantly different due to our experience — Karl worked in industry and Fanji in academia.

Fanji's focus was on creating knowledge and how to teach it while Karl was engaged in how to make use of knowledge via technological applications.

After several decades of working in interdisciplinary fields with various technologies we both have reached the official age of retirement. However, scientists and entrepreneurs never retire and now we can use our time and experience to indulge our passion, and enjoy the freedom that we don't have to nurture a career anymore or impress our peers. It is a wonderful privilege to follow your interests without such restrictions and we both enjoy this extraordinarily.

As Karl has often said "we are like two free and happy birds that can sit on any tree they like and discuss whatever they are interested in".

But that doesn't mean that we have no goals or high level objectives.

First of all, it is our own need and ambition to understand as well as possible what is happening in our complicated fields of interest and to assess how they are likely to develop.

Second, Fanji has a large audience of interested and discerning readers and Karl has a number of people (young scientists, engineers and entrepreneurs and also managers in industry and government) asking his advice on these topics. Both groups can legitimately expect that what we say is neither careless nor superficial.

Our discussion started with a question Fanji had asked Hans Braun about the relevance of a specific type of neuron, after studying the technical concept of the later famous European HBP (Human Brain Project). Hans had redirected this question to Karl and so our little journey began in January 2013. It resulted in a series of e-mails which led us from neurons to the latest developments in AI, and what some people have called the technology and trade-war between China and the US.

This book is an organized collection of our correspondence. It consists of letters written in the tradition of the old debate culture in which scientists exchanged their arguments in elaborate letters and controversial discussions. Of course we didn't invent this method. In fact, it was the normal way scientists communicated and refined their ideas a hundred years ago in the golden age of science. It just looks somewhat outdated in the time of Twitter, and short message services, where everything has to be expressed in headlines, available in seconds and is quickly said and forgotten.

The adversarial tone in our letters may sound a little strange to young scientists more accustomed to a culture of consensus. It should be mentioned however that adversarial methods are very promising techniques now being introduced in the most advanced applications of artificial neural networks. To communicate in the old-fashioned adversarial way can be very time consuming and demanding, but it can also be very productive and rewarding for those who like to look under the hood in order to thoroughly understand what is going on.

Nowadays people are less used to writing long letters, but letters have a big advantage over regular publications. They are less formal, give more room for creativity and speculation, and they allow the correspondents to change their positions more easily and thereby learn more from their counterparts. You can also ask more critical questions in an easier tone and can present an idea in a non-protected form. We found this method very helpful in our highly dynamic fields of interest where nothing is engraved in stone yet and lots of myths and buzz-words are swirling around.

Especially since China has declared its ambition to become a main world center of AI innovation by 2030, many people have become interested in the questions we have been dealing with for quite some time.

To achieve clarity, a sober analysis of both the state of the art and the prospects and limits of what can be achieved in the related sciences and technologies is needed.

This is what we were trying to achieve before the hype started and we want to share our insights with those that are trying to understand what is going on in these fields and what the difference between facts, popular belief, realistic hopes, dreams and marketing buzz is.

What you have in your hands now is not a science textbook or a typical popular science book either. It is also not a systematic or complete introduction to the two disciplines.

What we did was more like a random walk; wandering from field to field, stopping to study something in depth whenever we wished. We were guided only by our curiosity, and simply put in more effort when we had the desire to understand things a little more precisely or when we felt the need to fill gaps in our knowledge. And we often enjoyed following the streams of knowledge back to their sources, including some excursions into the history of our very different cultures. But as chaotic as our journey was, we feel that through our continuous, and sometimes controversial debate, we gained insights we would have not acquired had we chosen a more systematic approach.

We both like to take the perspective of a child who asks simple questions in order to grasp what is happening. Sometimes the child can see that the emperor's new clothes are not as brilliant as they are presented. But we don't want to overdo it with this perspective because it would be presumptuous to say that we are the child in the famous fairy tale "The Emperor's New Clothes" who can see or can't see what others see.

However in the case of HBP, Karl insists that very early on, while others were still praising it, Fanji was able to recognize that there were flaws in this impressively presented project.

We spent a great deal of effort in our early letters demonstrating and assuring ourselves that more than one thing was wrong in the concept of the HBP, and that we shouldn't expect too much from it. It initiated our correspondence, and it became a good model to explore many of the basic elements of the brain and the mind and a possible link with artificial intelligence and computer-technology.

Today, after the facade of this project has been seriously damaged in public such criticism is common and our verve in the old days may seem to some as an attempt to kick a dead dog. Maybe the meanwhile common critique is even too much because in our eyes there are interesting parts in the HBP-concept that have deserved a second try.

In addition to discussing a variety of great myths about the brain and artificial intelligence, we deliberated over rational thinking and consciousness. During this period, several important events occurred both in brain science and artificial intelligence, such as the launching of the US BRAIN (Brain Research through Advancing Innovative Neuro technologies) Initiative, AlphaGo defeating the world Go game champion and a self-driving car running on the highway among many others. We followed these impressive events and discussed how to evaluate their significance. Some of the new advances supported our speculations, and this encouraged our further discussions. The progress in some areas was even beyond our best expectations, so we had to reconsider our views and learn lessons from our mistakes. All this inspired our discussion, redirected our focus, ignited new debates and sometimes made us change our mind. On some points we reached consensus, on some we still differ, and on some matters we just never reached any conclusions.

We don't expect all readers to read all letters. Some may be experts in neuro-physiology who just want to know what can be expected from "neuromorphic" chips while others may be skilled programmers of artificial neural networks and want to better understand what the problems of a link between biological and artificial neurons are. Others may be less interested in the physiological details of the brain or computer architectures but in aspects that have also occupied us in detail, like the emergence of consciousness, free will, relevance of memes, self-organization, circular causality or the organization of research in the life sciences and in engineering disciplines. Some of our friends who are concerned with science organization and that we have shown the draft of the book to, found our discussion about the misery of bureaucratic big science the most relevant to them. Others may ignore the technical details and prefer, as in a travelogue, to follow the story, as the two authors try to orient themselves in the difficult borderlands of biology and computer technology.

For those readers who are only interested in specific questions we have provided two different tables of content where they can identify letters containing such subjects.

The idea to publish our conversation occurred only after a while and the reader in the focus was the ambitious lay person with some interest in science and technology who wants to better understand what the facts and the myths around these two popular fields are. And while our discussions goes into the great detail here and there, which is hard to understand without the right background knowledge, the Chinese content of the book comes with explanations in textboxes, footnotes and drawings to give readers convenient reference material to better understand what the basics of the matters discussed are.

Adding this material takes a lot of time and also significantly expands the size of the book which makes a publication in three volumes necessary.

For those who are interested in a summary of our reasoning, and also in the consequences we have drawn, we have added a longer than usual epilogue where we summarize our most relevant findings. The curious readers who, as with a crime novel, can't wait and want to know in advance who the murderer is may read this epilogue first.

Of course, we do not claim to have answered all questions. In fact, our survey isn't complete and many questions have remained unanswered and new ones have been raised. We also don't claim that we understand the matter better than others. The self-confident and determined tone in some parts should not be misinterpreted as absolute certainty about one's own position. It is only due to the method of this old-fashioned debate culture, where a position is worked out as clearly as possible, not to protect it, but to invite others to refute it.

Many of our assumptions, conclusions and comments are probably erroneous or incomplete and will need to be corrected soon. The trouble is we don't know which of them are wrong.

In any event after we have often criticized others so generously we are prepared for this to happen to us.

We may not be happy to be proven wrong, but this is unavoidable if we want to make progress. We both believe that rational reasoning, as a continuous process of challenging theories by empirical research, is the best way to increase our knowledge. Rational reasoning alone can't replace intense wrestling with the matter in the real world and what also can't be replaced in this process are the curiosity, boldness and ambition of researchers and engineers, especially in the young generation.

For some it may be disappointing to see how little we know about the riddles of our own minds and how slowly our knowledge of them is increasing. It catches the eye that progress on the technical side of the field is developing much faster.

Others may see this as an opportunity to enter a very promising field of work in which many unbelievably rewarding discoveries await those who are ready to leave the beaten track.

We're both sure we haven't made any relevant discoveries. But we have the desire and hope that we can inspire some talents to try their luck and find new entrances to the magic castle.

Not all of them may succeed, but we hope that many readers will find our insights helpful and will enjoy this walk around the magic castle as much as we did in the last 6 years.

Karl Schlagenhauf and Fanji Gu

This series of books contains 31 pairs of letters, discussing a series of open problems about brain and artificial intelligence. These discussions and debates run through the series, but there are differences in the emphasis of each volume. Among them, the first volume *Brain Research, a New Continent to be Explored* contains 11 pairs of letters (No. Ⅰ 001 − 011), focusing on the open problems of brain research; the second volume *Mystery of Consciousness and Myth of Mind Uploading* contains 10 pairs of letters (No. Ⅱ 001 − 010), focusing on discussing different views on consciousness research and the possibility of mind uploading; and the third volume *The Third Spring of AI* contains 10 pairs of letters (No. Ⅲ 001 − 010), focusing on the potential and prospect of AI. Related scientific methodology and scientific organization are also discussed in all three volumes. The letters in the three books are sorted by time. In order to facilitate readers to understand the contents of the whole series, we provide the following list:

1. Preface(Ⅰ − 001 Fanji)

2. Scientific Methodology

"System-" and "Interest-performer" (Ⅰ − 002 Karl, Ⅰ − 003 Fanji, Ⅰ − 003 Karl, Ⅰ − 008 Karl, Ⅱ − 003 Fanji)

Nature and engineering adopt different methods (Ⅰ − 002 Fanji, Ⅰ − 003 Fanji, Ⅰ − 005 Fanji, Ⅱ − 004 Karl, Ⅱ − 009 Karl)

Different Thinking Habits in Different Disciplines (Ⅰ − 003 Fanji)

Competition and Cooperation between Scientists (Ⅰ − 003 Karl)

Academic Controversy (Ⅰ − 004 Fanji, Ⅰ − 004 Karl, Ⅰ − 005 Fanji, Ⅰ − 006 Fanji, Ⅱ − 002 Karl)

Contents

II - 001 Fanji ... 1

Consciousness; *difficult problems*; *easy problems*; *subjectivity*; *privacy.*

II - 001 Karl ... 7

Computation and information processing; *neurons and neural networks*; *single layer perceptron and multilayer perceptron*; *theory and practice.*

II - 002 Fanji ... 15

The privacy of consciousness; *the winter of artificial intelligence.*

II - 002 Karl ... 21

Singularity; *fund competition*; *NBIC* (*"Converging Technologies* for Improving Human Performance — Nanotechnology, Biotechnology, Information Technology and Cognitive Science"); nanoassemblers and nanomaterials; academic controversies.

Do not listen to every word of your professor; the singularity is near.

Speculation and religious prophecy; calculation; simulation; "when we have a strong computer, we can calculate everything" is just misleading; chaos; reverse engineering of the brain is not the best way to create artificial intelligence.

Singularity is near; The law of accelerated return; Moore's law; it is wrong to extrapolate the exponential growth trend for indefinite periods.

Moore's law; the idea of "creating artificial machines by copying biological prototypes, such as the brain" is not feasible; technology inventions and natural evolution; "to reverse engineer a machine that you do not know like a brain is the stupidest thing you can do"; the computer will be superior to humans in information processing in almost all applications someday, but it is difficult to give a timetable. Man-machine fusion.

Brain science and technology; can AI machines do everything

our brain can do? Mind uploading; subjectivity and privacy; intelligence cannot be judged by behavior only; the inner world is the real bottleneck of both cognitive science and artificial intelligence; consciousness is an irreducible basic emergent property of some complex systems; the real problem of consciousness research is under what conditions can the consciousness emerge? "Turing test" and "Chinese room" thought experiment.

Moore's law; the new trend of technological development; mind uploading; consciousness.

Mind uploading; subjectivity is an irreducible emergent property of some brain activity; the appropriate question is under what kind of brain activitycomes out of consciousness and the sufficient and necessary conditions for the emergence of consciousness? Scientists publicly criticized the HBP in an open letter to the European Union; the NIH Advisory Committee Working Group proposed a proposal for BRAIN initiative.

Mind uploading; emergence; The history of natural science is a history of metaphysical disappointments.

It is very difficult to find the necessary and sufficient conditions for the emergence of the consciousness of any subject, and the essence of subjectivity and privacy is the essence of the difficulty of the study of consciousness.

<u>II - 010 Karl</u>

Only an open mind can benefit from discussions. Computers and brains, Turing tests, the European Union Human Brain Project, the US BRAIN initiative.

Dear Karl,

Thank you very much for your kind words. And I should have written an acknowledgement attached in that article to express my sincere gratitude to you, Hans Braun, and other colleagues for discussion, I am really sorry for missing this part only owing to the limitation of the space given by the editor.

I am just wondering if I would still have been so "brave" to publish my article if I had read Markram's open letter before my submitting.☹

You must be right to analyze the whole matter so deeply, which I could not.

Your words remind me of a topic about consciousness or awareness related to the HBP. In the document "*The Human Brain Project — A Report to the European Commission*" (April 2012) there is a section titled "Consciousness and awareness", in which it declared:

"*The HBP platforms would provide an opportunity to take models that already exist, or to develop new ones, and to perform in silico experiments testing the models. Such experiments could potentially lead to fundamental breakthroughs.*" And in the section of FAQ in HBP webpage, it also declared that "*simulating the human brain will provide us with deep insights into the way the brain works: the origins of our perceptions, thoughts, emotions — our very consciousness.*"

However, in an earlier "Answers to questions about the Blue Brain Project" in the Blue Brain Project (a precursor of the HBP) webpage, the answer to the question "*Will consciousness emerge?*" was as follows:

"*We really do not know. If consciousness arises because of some critical mass of interactions, then it may be possible. But we really do not understand what consciousness actually is, so it is difficult to say."*

Again, the earlier answer is much more believable, and the later statement may give fake hope. Although there is some period between the two statements, however, there is no breakthrough in consciousness studies which can explain such a change.

Consciousness is one of the topics I am much interested in; it has puzzled generation after generation of philosophers and scientists and has not been solved. There are so many different opinions even about the basic problems of consciousness. Fortunately, I ran across a book "*Conversations on Consciousness*"[1] by British science writer Susan Jane Blackmore, to explore the state of the art of consciousness studies. She interviewed 21 leading scientists and asked them a lot of questions, basic and key to the problem. Although it was published in 2005, it at least let me know about the major different points of view about consciousness up until 2005.

At the very beginning of the interview she always asked the same question — why the study of consciousness is so special? What is the key difficulty for such studies? Most of the scientists think the problem lies in how to explain the subjective consciousness emerging from a physical brain. Australian philosopher David Chalmers called the problem the "hard problem" of consciousness studies, while the problems on how to explain the behaviors or observable functions emerging from the brain as "easy problems". Although, solving such "easy problems" is not easy at all. ☹ After reading the 21 interviews, I found that maybe the interviewed scientists could be divided into two camps. The first accepted the idea of "hard problems", they thought the special difficulty of consciousness studies lies in the subjectivity and privacy of the consciousness. In the history of science before consciousness studies, scientists always studied objective phenomena, using a third person perspective, any subjectivity was to be avoided; however, in consciousness studies, subjectivity itself becomes the topic to be studied! Thus, some of them thought the available theories could not solve the mystery of consciousness, and that new theories should be developed. Some of them thought consciousness is a basic property of some highly organized systems, just like

[1] Blackmore S J. Conversations on Consciousness[M]. Oxford: Oxford University Press, 2005.

space, time, mass etc. for all entities, which could not be explained with more basic concepts. It is irreducible. On the contrary, some scientists that came from the second camp even denied that there was any such "hard problem". They thought that to study consciousness was just the same as other scientific studies. They thought consciousness was just the same as brain activities, they are just two sides of the same coin. Therefore, there was only Chalmers' "easy problem" for them, although they never used this term. However, I doubt the latter argument and will explain in the following paragraph.

It seems to me that the subjectivity and privacy is the core of the difficulty of consciousness studies, denying the hard problem is only to avoid or neglect the subjectivity and privacy of the consciousness. However, they are still there! I could not see any possibility of solving this hard problem in the foreseeable future. Ironically, although the above two camps held seemingly opposite viewpoints, in fact, both suggested not to study how consciousness emerges from the brain at present.

The famous US neurologist Oliver Sacks told a story[1] about a US neuroscientist Sue Barry, which definitely shows that the subjective experience consists of something more than knowing the neural mechanism underlying such experience. In 2004, Sacks met Barry for the first time, and they had an interesting talk. Barry told him that she had grown up cross-eyed, and so she viewed the world with one eye at a time, her eyes rapidly and unconsciously alternating. Therefore, as a matter of the fact she had no binocular vision, and might not be able to see depth directly, but she could still judge it using monocular cues. So, she could drive, play softball etc. as everyone else. Sacks, as a neurologist, was anxious to know if she had any idea of stereovision. Did she have the subjective experience of depth perception? Barry's answer was "Yes", as she was a professor of neurobiology, she had read Hubel and Wiesel's papers, she knew almost everything about visual information processing, including the brain mechanism of binocular vision and stereopsis. Sacks described her idea as follows: "*She felt that this knowledge had given her a special insight into what she was missing — she knew what stereopsis must be like, even if she had never experienced it.*" It sounds that if you solved the easy problem — knowing the brain mechanism of binocular vision, the hard problem would have disappeared — you can also know the experience of stereopsis!

---

[1] Sacks O. The Mind's Eye[M]. New York: Alfred A. Knopf, 2010.

However, after reading the following paragraphs, you will know the above is just an illusion!

After several years, Sacks received a letter from her. She said in the letter: *"You asked me if I could imagine what the world would look like when viewed with two eyes. I told you that I thought I could... But I was wrong."* She said this because she had an operation and a variety of training after the operation, it was a success at last and now she had stereopsis! She described her experience as follows:

"I went back to my car and happened to glance at the steering wheel. It had "popped out" from the dashboard. I closed one eye, then the other, then looked with both eyes again, and the steering wheel looked different. I decided that the light from the setting sun was playing tricks on me and drove home. But the next day I got up, did the eye exercises, and got into the car to drive to work. When I looked at the rear-view mirror, it had popped out from the windshield.

...

I had no idea what I had been missing. Ordinary things looked extraordinary. Light fixtures floated and water faucets stuck way out into space... [It is] a bit like I am in a fun house or high on drugs. I keep staring at things... The world really does look different.

...

I noticed the edge of the open door to my office seemed to stick out toward me. Now, I always knew that the door was sticking out toward me when it was open because of the shape of the door, perspective and other monocular cues, but I had never seen it in depth. It made me do a double take and look at it with one eye and then the other in order to convince myself that it looked different. It was definitely out there. When I was eating lunch, I looked down at my fork over the bowl of rice and the fork was poised in the air in front of the bowl. There was space between the fork and the bowl. I had never seen that before... I kept looking at a grape poised at the edge of my fork. I could see it in depth."

From Barry's story, we can know even if the condition under which the stereovision emerges is known, you may still not experience depth perception! How such perception emerges from binocular neural activities is still unknown! Computer simulation may help us to shed light on the brain mechanism of behaviors or some observable functions, but it can't explain how subjective experience emerges from objective processes! You have no way to confirm any program run on your computer is having subjective experience, even if it could declare it had! Turing-like tests can't tell you this. IBM's Watson might tell you that it is conscious, however, it definitely isn't! From Barry's story, it seems to me, the so called hard problem can't be simply denied, there is subjective experience! Although we can study a variety of "easy problems", which are valuable for us to understand consciousness, however, without solving the "hard problem", it seems that we could not declare that consciousness is thoroughly understood!

Owing to the limitation of the space, I can't cite more about Barry's story, if you are interested, you can read Sacks' book "The Mind's Eye"[1] yourself.

Besides the HBP, now we can also hear the so called "artificial consciousness" from time to time. Even an international journal published such a title. However, owing to the above reason, I think it is still too early to study "artificial consciousness". I don't mean that artificial consciousness is absolutely impossible. Anyway, no matter how complicated the human brain is, it is still a physical system, as we know the human brain can emerge consciousness, there is no reason to conclude no other physical system can have consciousness. However, we have no idea what kind of physical systems can have consciousness. Complicated hierarchical systems? But what does "complicated" mean? Even for brains, we know the human brain can emerge consciousness. As for other species, primates or even vertebrate animals may be conscious, parrots and crows are likely, but insects are not. Where is the boundary? And why? If we can't judge which brain can be conscious, how can we judge an artifact having consciousness? Just as Jeff Hawkins pointed out that it is unreasonable to judge if a machine has intelligence just according to its behavior. It is also impossible to judge if an artifact has consciousness just according to its behavior. Therefore, it seems to me that it is still too early to discuss problems about artificial consciousness. Modeling and simulation is

---

[1] Sacks O. The Mind's Eye[M]. New York: Alfred A. Knopf, 2010.

unlikely to "*lead to fundamental breakthroughs*" of consciousness studies.

Although it is possible that consciousness is an emergent property of some kind of highly organized system, it emerges from the complicated interaction within the system, and as many emergent properties, you can't explain how such properties emerge. However, the problem is as what I argued above. We still don't know from what kind of "highly organized system" consciousness can emerge. Further, we don't know under what conditions consciousness can emerge in such a system. We may test a model to see if it behaves similarly to conscious beings, however, owing to the fact that consciousness is an inner state, it is impossible to judge if the subject is conscious or not just based on its behavior! For solving the mystery of consciousness thoroughly, the core difficulty — its subjectivity and privacy could not be avoided! Am I right?

Until now I mainly discuss the difficulty coming from subjectivity, and as the mail is already too long, I will discuss the privacy problem in my next letter.

Best wishes as always.

Fanji

Dear Fanji,

It is very kind what you said about a possible acknowledgement in your HBP article in *Brain-Mind Magazine*, but there was no reason at all. Your sharp eyes had discovered that something is wrong in the fundamentals of the HBP and that the promises made are, by far, too high. The only contribution I could provide was to say, yes, you are right.

I'm proud of you that you were brave enough to publish the article even if you didn't know how dangerous it was. But mankind owes many heroic deeds to the ignorance of danger. ☺ If people had known how risky an undertaking was then they wouldn't have started it.

We have a famous piece of literature featuring this phenomenon called "*Ritt über den Bodensee*" (*Ride across Lake Constance*). The story goes that in deep winter a horseman rushed to reach Lake Constance to cross it with a ferry boat. However, the lake was frozen over, which happens very rarely, but the horseman didn't realize and simply galloped over the thin ice. When he arrived at the other side of the lake and the people told him what he did he was so shocked that he lost consciousness and fell from his horse.

Lack of information or ignorance is often the source of heroic deeds; the other ones are threats, misery, and desperation. The same is true when it comes to great ideas and inventions.

In any event, consciousness and rational awareness, as useful as they are, in many situations aren't helpful in others. Having too much information about a problem seems to reduce creativity in some cases. This is maybe one of the reasons why major breakthroughs in science are often performed by very young people, who simply don't know or don't care about the many dos and don'ts established in a scientific discipline.

It is very interesting what you have detected about the change in hope or promises that occurred between the Blue Brain era and the HBP times in the field of consciousness and awareness. It is striking that the perspective went from realistic to overly optimistic in a short time although there was no breakthrough in consciousness studies in between.

When consciousness is a field you are much interested in you must know what is going on in this research field and therefor I trust your expertise. I'm also interested in consciousness but have only fragmentary and coincidental knowledge of the academic discussion about it. To me, this problem always seemed to be so high up in the many layers of the brain functions which we don't understand, as your friend Nelson Kiang has pointed out, that I was always a little shy to deal with such complicated matter before we understood the lower levels.

But the books you have mentioned sound promising and what you have said about the problem of subjectiveness is an indication that there seems to be a real puzzle behind it and therefore I'm happy that you've given me this impulse!

It is good that you gave me a reason to overcome my laziness, but it will take me some time to get the books, read them, think about it and come back to you with a hopefully qualified opinion about your arguments.

And while I can't do this today let me take the opportunity to come back to a few open issues and questions which you have raised in your former emails.

In your mail from August 15th (I − 010 Fanji) you asked a few very interesting and challenging questions:

a) *"As the expression of "brain-like" or "brain-style" is rather ambiguous, who can tell me how the brain computes in a clear way?"*

b) *"..., I doubt if the brain always computes like a computer to execute its function. I am not quite sure what computation means for the brain, does the computation mean here as it means for a von Neumann computer or a Turing machine? Or is it just a synonym of information processing? For me, the former seems unlikely, otherwise all the brain functions should be able to be carried out by a digital computer and I am suspicious about such a possibility."*

c) *I always take "computation" as a synonym of "information processing", although the term "information processing" itself is also not a well-defined concept. However, anyway, using "information processing" instead of "computation" may avoid the confusion considering "computation" as "calculation" or that the operations run as in a digital computer. Of course, a question can be raised: Is there any kind of computation beyond the operations carried out in a modern digital computer? If there are, then what they are, especially those performed in the brain?*

d) *It is correct to say that McCulloch-Pitts model neurons can perform logical computations, so in principle, McCulloch-Pitts neural networks can perform any computation a von Neumann computer can do?*

Your questions a) and d) are easy to answer in my opinion. The answer to question a) is "no-one can tell how the brain computes" and the answer to question d) is "yes".

The answer to questions/statements b) and c) is more complicated. But before I come back to them let me first start with the relevant remark that the answer "Yes" to your question d) is only true because of the amendment "networks" in the second part of the sentence.

As it is nicely pointed out in Appendix E (P.1585) of Kandel's "Bible"[1] a single MP-neuron can perform the basic logical functions AND and OR and by synaptic inhibition also the negation (logical NOT).

A single MP-Neuron however cannot perform the logical XOR (exclusive OR) function. At a logical XOR Gate the output is true when (in the case of two input-lines) either one of the inputs is true but not both. In the case of the OR function the output is true when either one of the inputs is true but also when both inputs are true.

For people not familiar with Boolean logic it is not comprehensible at first sight why this XOR function seems to be so tricky. And actually, it took a while until the experts found out how networks of MP-neurons could perform the XOR trick. This, by the way, reveals another scientific dog-fight story between two opponents. It was not as

---

[1] Kandel E, et al. Principle of Neural Science[M]5th ed. New York: The McGraw-Hiu Education, 2013.

rude as the one between Markram and Modha but with very relevant consequences and a tragic end.

In 1957 Frank Rosenblatt came out with his idea of the "Perceptron". This was a network of neurons to mimic visual perception. The concept is also described in the mentioned Appendix E of Kandel's "Bible" to give an example of how parallel data processing in the brain with many neurons makes our visual perception so quick. However, the presentation given there somehow hides the problem because it doesn't differentiate between Rosenblatt's "single layer" perceptron and the later "multilayer" perceptron. The difference is that a single layer Perceptron can't provide the logical XOR function while a multilayer Perceptron can. Rosenblatt's concept was one of the first feed forward neural networks.

The research was funded by the US Office of Naval Research and was promoted by its inventor with bold statements. After a press conference the *New York Times* reported the perceptron to be "*the embryo of an electronic computer that* [*the Navy*] *expects will be able to walk, talk, see, write, reproduce itself and be conscious of its existence.*"[1]

As you can see, modesty was not the strategy in the race for grants and governmental money also in the early days of AI.

And while this tonality and exaggeration sound familiar so does the reaction from competitors in the race for grants and fame.

In this case it was Marvin Minsky who later became famous as one of the great men of AI. Minsky, who happened to be a schoolmate of Rosenblatt, who had worked on similar problems seemed to have a hard time with the success and popularity his former comrade gained after he had published "Principles of Neurodynamics: Perceptrons and the Theory of Brain Mechanisms" in 1962. In any event, he felt to be obliged to tell the world how bad a concept Rosenblatt's perceptron is. Together with his colleague at MIT, Seymour Papert, Minsky published a book under the title "*Perceptrons*" in 1969 in which crucial flaws of Rosenblatt's perceptron were disclosed. One of them was that a single layer perceptron can't perform the logical XOR function. The attack of two MIT heavy weight professors was very successful in so far that not only perceptrons, almost

---

[1] https://en.wikipedia.org/wiki/Perceptron

overnight, were out of fashion but also neural networks and investments in AI. The book made some contribution to the outbreak of the first "AI winter" which brought research in neural networks to a halt for about a decade. Rosenblatt didn't have great opportunity to respond to Minsky's attack because he died tragically in 1971 on his 43rd birthday, in a boating accident.

It is not without irony that Minsky himself later turned his interest again to neural networks and that the XOR-flaw of the early single layer perceptron could be cured by multilayer networks and the introduction of the "back propagation" concept.[1]

McCulloch and Pitts had already dealt with the completeness problem of their neuron regarding Boolean logic. Pitts, who had studied with Bertrand Russell, was very well aware of the problems of formal logic and the rules of Boolean logic. And he had already found that the XOR problem could be cured when you combine more than one neuron. Nowadays it is well known that when you have a device that can perform the basic Boolean functions AND, OR and NOT you can build all the higher logical functions by combining two or more of such basic elements. It is astonishing and impressive, however, how far-sighted those pioneers of the discipline were, already 70 years ago, in this respect.

The same method is used in technical computing when the basic elements of digital electronics, Flip-Flops, and latches, are combined to build more sophisticated gates that can perform more complex functions. A further aspect comes into play when we consider whether the logic is applied time-independent (combinational logic) or time-dependent (sequential logic). The point is that sequential logic has memory while combinational logic does not. With sequential logic (which is in other words Boolean logic with memory) we can build "finite state machines" which are the basic building blocks in practically all modern electronic devices. In principle it doesn't matter whether such "finite state machines" are realized with relays, diodes, tubes, transistors or with fluidic, optical or even biological elements as long as they allow applying Boolean logic and have memory.

What should not be overlooked, however, is the fact that artificial electronic neurons or

---

[1] Minsky had already dealt with neural networks in his Ph.D. Theses: Minsky, Marvin L. Princeton University, 1954, Ph.D., Mathematics.

gates, very much like their biological relatives, are also abstract ideal models and not 1 : 1 identical with the physical world in which they are implemented.

Programmers often tend to overlook this fact, especially those who are used to producing code in high-level languages for abstract problems far above the physical world. They tend to be surprised when a logical device doesn't behave as it should. First and foremost, they suspect a programming-error when actually it is the physical realization of a device that conflicts with the logic of the program.

Programmers who write code that controls machines such as robots or IoT- (Internet of Things) applications need to have a less idealized perspective. The first thing a programmer of a microprocessor, that reads signals, learns is to think about the physical reality of the signal. The classic problem is the "bouncing" of a mechanical switch. Very much as with the action potential in a neuron, the voltage doesn't go from zero to one or from one to zero in no time. There is always a time component involved and a logical on/off signal doesn't look as clean and pure in reality as in theory. When you connect an oscilloscope to the output-side of a switch you may be surprised about the noise you can see when it is switched on and off.

The noise comes from the mechanical "bouncing" of the switch and depends on the inductivities and capacities in the system. It can be compensated by either hardware means (e.g. you may put a capacitor parallel to the switch) or by software when you put a few milliseconds of delay in the program before you read the status. Engineers not familiar with this circumstance are often puzzled that they get erratic results about the status of the switch.

I mention this trivial problem because it displays a relevant problem. We seem to have perfect theoretical knowledge about the principles of electrodynamics, and the Maxwell equations represent about the most solid foundation we have in physics. However, it can be very tricky to apply this theoretical knowledge in interaction with the real world. But machines and, even more so, biological organisms are not axiomatically defined, closed systems as pure as in mathematics. Actually, it was a similar collision between theory and practice that caused big problems in the transatlantic telegraph-cable communication in the late 19th century. It caused Oliver Heaviside to think deeply about what's happening when electric pulses travel through cables and when not just resistance but also capacitance and inductivity come into play. It turned out that a new kind of mathematics was needed which caused him to make the effort to develop it. While doing so, he also reformulated the Maxwell equations and gave them their modern form. He also invented a mathematical tool, called the Heaviside step function, which turned out to be very helpful for describing what signals do in telegraph cables and also in neural action potentials. Heaviside has achieved one of the most impressive intellectual accomplishments a single mind has ever created. It is amazing anyway how far 19th century mathematicians and logicians have come and that they were able to provide almost all the foundations that were needed to build modern information processing machines. All of them far away from the biological archetype but, nevertheless, extremely successful in some engineering domains.

By the way, the first computers in the world have been humans not machines. Our ancestors have invented methods to do difficult calculations in a smarter, time saving way. Very much as they have invented other artefacts before, like knifes, hammers, wheels, clothes, language or income tax. Such calculation tools were called algorithms and they ran on the human mind long before we had machines to run them.

Therefore, I agree when you say in your statement under c) that it is a good idea to talk about "information processing" instead of "computing". The same should apply to the expression "computer science" by the way. Philip Wadler, a professor at the University of Edinburg and former member of Bell Labs who is one of the leading experts in the world in the matters I have discussed above, also prefers "informatics" and has put it this way:

"There is nothing wrong with the expression" Computer Science" besides with the words Computer and Science. First it has nothing to do with computers and second a real science never puts the word science in its title". [1]

As funny as this statement about computers sounds, it represents a deep insight into a

---

[1] Karl remembers that Philip Wadler made this statement in one of his lectures, but cannot find written evidence for it. Wadler's "Propositions as Types" is a great introduction to the logical problems of computation and is highly recommended: Wadler P. Propositions as Types[J]. Communications of the ACM, 2015, 58(12): 75 – 84.

problem which often causes confusion in the discussion about the similarity between computers and the human brain.

Wadler doesn't only have a good sense of humor but is one of the sharpest thinkers in the discipline. I have to tell you more about him, if you like, because he will be very helpful when it comes to answering the remaining parts of your questions. But now when I realize how long this email got and that I haven't even dealt with half of your questions I rather stop and promise to be more efficient next time!

All the best as always.

Karl

23. 12. 2013

Dear Karl,

Thank you very much, as always, for your kind words and comments. Thanks also for your story about the horseman, who makes me feel easy, anyway, I did not lose consciousness, even after you told me about Markram's open letter.☺

You are right, consciousness is so complicated that, until the end of the 1980s, no serious research on consciousness from a perspective of natural science has been studied. In 1989, Stuart Sutherland said:

"Consciousness is a fascinating but elusive phenomenon: it is impossible to specify what it is, what it does, or why it has evolved. Nothing worth reading has been written on it" [1]

Although the situation has dramatically changed after that, as Blackmore pointed out in 2005 after her interviews with 21 leading scientists in this field:

"But do I now understand consciousness? I certainly understand the many theories about it a lot better than I did before, but as for consciousness itself — if there is such a thing — I am afraid not." [2]

Although almost another decade has passed, I am afraid that the basic situation remains the same. [3]

This is not a surprise for such a mystery, if considering that one decade is just a minute

---

[1] Sutherland S. Consciousness[M]. Macmillan Dictionary of Psychology. London: The Macmillan Press, 1989.

[2] Blackmore S J. Conversations on Consciousness[M]. Oxford: Oxford University Press, 2005.

[3] https://en.wikipedia.org/wiki/Consciousness

in the history of science. As I emphasized in my last e-mail, the key difficulty faced by consciousness studies is the core property of consciousness — subjectivity and privacy, which other branches of science have never faced.

In my last letter, I discussed the problem of subjectivity in great detail and said something briefly about privacy. The two concepts are closely correlated, but the points emphasized are a little bit different. The point about privacy is that one's consciousness can only be experienced by the subject and can't be shared by others. However, recently, when I read V. S. Ramachandran's classical book *"Phantoms in the Brain"* I found a very interesting argument of his about the origin of privacy. He thought that privacy came from a barrier of translation between the two languages: the verbal language communicated between the different persons, and the neural language used by the nervous system within the subjects. Verbal language could not carry all the information in neural language. Thus, he proposed a thought experiment to bypass the barrier: using a cable or a nerve tract to link the corresponding brain areas of the two subjects. He thought then the two subjects could share their experiences. He even suggested that a human being could experience the feeling of electrical sense from an electrical fish, if someone could connect the corresponding brain areas between a man and an electrical fish. However, I deeply doubt his argument, even in principle.

My main arguments are: First, no two brains are identical, and brains are always changing with time. Second, it is the brain of the receiver, especially the brain area linked, that determines the contents of the experience, not the spike train from the sender. Just think about what Ramachandran himself argued in his book many times. He explained why an amputee felt that his lost hand was touched when his face was touched — owing to the fact that the areas representing the face and hand are neighbors in the somatosensory cortex. The afferent nerve from the face now invades into the idle hand area now so that the amputee mistook the stimulus from the face as from the hand! Third, the brain is an electrical-chemical machine. Neurotransmitters, neuromodulators, and other hormones, which could not be transmitted with a cable, also play an essential role in experience. Therefore, from my point of view, it is impossible to share other's experience accurately.

Human beings do not have any electric sense area in the brain, they will never have an electric feeling, even if any area of the brain is stimulated by a train of spikes from the electrical organ of an electric fish! His feeling must still be the old one!

This problem reminds me of a story from an ancient Chinese classic "*ZhuangZi: Autumn Floods*":

Once ZhuangZi and HuiZi were sightseeing on a bridge over the Hao river.

ZhuangZi said: "*How easy the fish swims, it must be very happy.*"

HuiZi remarked: "*You are not the fish; how could you know if it is happy?*"

ZhuangZi replied: "*You are not me, how could you know that I don't know that the fish is happy?*"

HuiZi answered: "*I am not you, so I don't know your feeling; and you are not the fish, so you don't know if the fish is happy. That's all.*"

Let me come back to your arguments in your last letter. It seems that we both agree that, in the brain, there is some "computation" or "information processing" beyond any calculation or operation with sequential logic performed in a digital computer or Turing machine. We also agree that information processing is a better expression than the term computation for the brain. It seems difficult to clearly demonstrate what kind of "computation" is really in the brain. Maybe we can use another synonym "transformation" to emphasize the essence of computation; transformation simply means to lead a state or a set of elements to another corresponding state or set of elements. The problems are, what kind such computation is, whether we can describe such computation in a clear way, and how to carry out such computation in artificial systems.

From the stories about the first winter of the artificial neural network you described, so vividly, in your last letter, and also the two former winters of AI[1], in general, the lesson is that the most efficient way to spoil the reputation of a scientific field and lead it to a cold winter is to wildly exaggerate the power of this field or give unrealistic magnificent prospects. In 1965, H. A. Simon, one of the founders of AI, declared:

---

[1] https://en.wikipedia.org/wiki/AI_winter

"machines will be capable, within twenty years, of doing any work a man can do." [1]

Minsky (in 1970) also declared:

"In from three to eight years we will have a machine with the general intelligence of an average human being." [2]

However, almost half a century has passed and their predictions haven't become reality today or even in the foreseeable future. John Markoff said in 2005:

"At its low point, some computer scientists and software engineers avoided the term artificial intelligence for fear of being viewed as wild-eyed dreamers." [3]

The bolder the advertisement, the stronger its disruptive effect. Another well-known example is the Japanese Fifth generation computer project (1981—1991). Their objectives were to write programs and build machines that could carry on conversations, translations, interpret pictures, and reason like human beings. However, by 1991, the project did not reach its main goals, some of them even remain unsolved today. Its expectations had run much higher than what was actually possible.

To sum up the lesson, AI researcher Hans Moravec said:

"Many researchers were caught up in a web of increasing exaggeration. Their initial promises to DARPA had been much too optimistic. Of course, what they delivered stopped considerably short of that. But they felt they couldn't in their next proposal promise less than in the first one, so they promised more." [4]

The result was that the public and the related authority lost their confidence in AI and it

[1] Simon H A. The Shape of Automation for Men and Management[M]. New York: Harper & Row, 1965.

[2] Darrach B. Meet Shaky, the First Electronic Person[J]. Life Magazine, 69(21): 58 – 68.

[3] Markoff J. Behind Artificial Intelligence, a Squadron of Bright Real People[M]. New York: The New York Times, 2005 – 10 – 14(A10).

[4] Crevier D. AI: The Tumultuous History of the Search for Artificial Intelligence[M]. New York: BasicBooks, 1993.

became difficult for AI research to find financial support during a period — the AI winter.

Unfortunately, people are apt to forget lessons, especially when it seems that there is a rich reward ahead! Your remark *"modesty was not the strategy in the race for grants and governmental money"* is absolutely correct! The authority should remember this to squeeze the water out from the application!

Your story told us how ruthless a scientific controversy might become if it involves giant grants and other resources. Although Minsky's and Papert's criticism of the perceptron with only one layer was correct, their following statement was misleading:

"We consider it to be an important research problem to elucidate (or reject) our intuitive judgement that the extension (to multilayer perceptron — the citer) is sterile." [1]

This hinted that further studies on an artificial neural network would be just a waste of time. As their analysis of the single-layer perceptron was rigorous, it was likely for people to believe the above judgement, although it did not have any careful analysis. Their blow to the study of the artificial neural network was so heavy that it fell into a cold winter for about a decade! Almost all the researchers having leapt into neural network study leapt back into symbolic AI, the same for the financial support! Although Papert wrote years later

"Did Minsky and I try to kill connectionism?... Yes, there was some hostility in the energy behind the research reports in Perceptrons ... we did not think of our work as killing; we saw it as a way to understand." [2]

This statement reminds me of an old Chinese joke: A San Zhang buried his 300 ounces of silver underground and set up a board with the note "There is not any silver here!"

[1] Freedman D H. Brain Makers: How Scientists Are Moving beyond Computers to Create a Rival to the Human Brain[M]. London: Touchstone, 1995.
[2] Freedman D H. Brain Makers: How Scientists Are Moving beyond Computers to Create a Rival to the Human Brain[M]. London: Touchstone, 1995.

there. Now his neighbor Si Li dug it up and took it away, and also put a note "Your neighbor Si Li has not stolen the silver" on the board.

As for the story about Wadler, yes, please, do tell me! As you can imagine how anxious I am to hear your answers about the remaining parts of my questions.

Merry Christmas and a Happy New Year!

Fanji

15. 01. 2014

Dear Fanji,

Thank you for your kind letter and Happy New Year to you and all of your family as well!

The holiday season was busy, as usual, and it took me a while to come back to you. Following the religious tradition, this time of the year is supposed to be dedicated to reflectiveness and contemplativeness, but, in reality, it is just the opposite. Maybe this is because only a minority of people are really religious, or because being contemplative has become, somehow, outdated. It is a tradition to have lunches and dinners in the last days of December with business friends and employees and then there are the big family meetings with three consecutive holidays and the New Year festivities a week later!

And everybody, including us, is travelling across the country. And so it comes that although almost all businesses are closed, the holiday season is the most stressful time of the year. ☺

I have heard that you have a similar tradition on Chinese New Year at least when it comes to travelling.

Well, you understand that all this description of holiday rituals is just an excuse for my laziness which kept me away from doing my homework on the questions of consciousness and mind. I read the books which you recommended and found them very inspiring, especially Susan Blackmore's interviews with many of the leading thinkers in the field. Some of them I knew, like Christof Koch and Francis Crick while I had never heard of others. The field is very close to, or overlapping with philosophy and, as always, in philosophy you find very smart people who brilliantly present and defend the position of their school. And their arguments sound convincing even when they contradict the arguments of an equally convincing opponent from a different school. It sometimes leaves me with the unpleasant feeling that these brilliant people could represent exactly the opposite point of view with the same eloquence.

I'm always shy when it comes to taking side with a school and committing myself to a fundamental position, such as that of a behaviorist or a constructivist. It is tempting to have such a solid starting point which provides orientation and lets you fly safely as on an aeronautical radio beam. The downside of such frameworks, which are often axiomatic systems, is that you can easily be trapped in them. And debates among representatives of specific frameworks or schools are often fruitless because they are held not with the intention of advancing knowledge, but rather to demonstrate the correctness of one's own framework.

I therefore prefer a pragmatic way of asking critical questions. The most important question to all those fundamentalists is:

"What empirical evidence should happen to make you give up your position?"

And when the answer is "I can't think of such evidence" you better stay away from them. ☺

I share your skeptical view on Ramachandran's idea of a cable that links two brains to share the experiences of the individuals or to allow a human to experience the feeling of electrical sense from an electric fish. Actually, I think that this idea is totally foolish, but I'm prepared to eat all my skeptical words if he can demonstrate that such a link is possible in an experiment.

I understand your arguments but I have not yet managed to get my arms around the problem and I need to make a second run and do some more reading and thinking before I come back to you with a more qualified opinion on the question of consciousness and mind and also your very special subjectiveness and privacy arguments.

But please don't expect too much and don't forget that you have been dealing with the matter for many years!

The statements you found about the background of the AI winters and the fatal exaggerations of promises that couldn't be kept and only led to a vicious circle of ever higher promises are very illuminating! In hindsight, everybody who was bold enough to put a date on such AI-promises over the last 30 years actually made a fool of himself. But some people were more hurt than others. One of the boldest of all evangelists of AI, Ray Kurzweil, made such super optimistic statements in a book "The Age of Spiritual Machines" back in 1999. He promised, very much as you have cited H. A. Simon and Minsky, that in 25 years AI will have reached the level of human intelligence. Then we will have an intellectual supernova called "singularity" when machines and human intelligence will amalgamate and, in a hundred years, the then ruling androids will display the last remaining examples of Homo sapiens in zoos as we do with orangutans today. An impressive prophecy indeed which caused a huge debate but not about whether this is feasible at all, but about what scientists and politicians should do to prevent this from happening.[1]

Well, those 25 years will be done in 2024 and as much as we can be impressed with the performance of our modern computer equipment it becomes clear that the promised singularity has to hurry to make the prophecy come true. Kurzweil's publication also has to be seen in the context of a race for grants that started at the time and in which Hans Moravec also was involved. He published a similar, but much less popular, book labeled "Robot: Evolution from Mere Machine to Transcendent Mind" about the amazing future of AI already in late 1998. At the time, Moravec and Kurzweil were both members of President Clinton's PITAC (President's Information Technology Advisory Committee) and helped to prepare a scientific and industrial offensive program which was later called NBIC ("Converging Technologies for Improving Human Performance — Nanotechnology, Biotechnology, Information Technology and Cognitive Science").

The idea behind it was to start a kind of scientific and industrial "New Deal" fired up by a program similar to the Manhattan project to build the atomic bomb in the 1940s or Kennedy's "Man on the Moon" project. So, these books can be seen as illustrations of this offensive but also as letters of application for the desired grants. The project never took off as spectacularly as Clinton and his PITAC advisors had hoped. First, because it was launched in the days when the media was more interested in the details of the Lewinsky scandal. Second, because the new President George W. Bush showed little

---

[1] https://www.wired.com/2000/04/joy-2/

desire to push forward an idea of his predecessor which carried the handwriting of the Democrats.

One of the intellectual driving forces behind the NBIC had been US vice-president Al Gore who was a strong supporter of Eric Drexler and his idea to build a new industrial world from the Nano-scale up. Drexler, almost forgotten today, was the guru of a whole generation of technology-nerds after he published *"Engines of Creation: The Coming Era of Nanotechnology"* (1986) and *"Nanosystems: Molecular Machinery Manufacturing and Computation"* (1992). Very much like Richard Feynman, who had already back in 1959 speculated about a technology at the level of atoms and molecules in his famous lecture *"There's Plenty of Room at the Bottom"*, Drexler drafted a new kind of industry based on molecular assemblers at the Nano-scale.

When you read Drexler's ideas, which is a real intellectual pleasure even today, his visions sound very convincing. And, at the time, few people doubted that all this would be feasible one day and therefore it seemed to be quite logical to include this Nano-aspect into the concept for the new "New Deal". This changed when, under the Bush-government, other people came to the table and revised all those great ideas, especially those Nano-Assemblers promoted by Al Gore's protégé Drexler.

What followed was another scientific dogfight which in the tone wasn't much nicer than the one between Markram and Modha but with even greater consequences.

In this case it was Richard Smalley, the prestigious winner of a Noble Prize in Chemistry (1996 for his discovery of Buckminsterfullerenes), who stood up and declared that most of Drexler's ideas were nonsense and could never be realized, for fundamental reasons. In an article in the *Scientific American* (2001) he aggressively doubted the feasibility of Drexler's molecular assemblers. There followed a series of rebuttals and exchange of arguments in open letters between the two over the next three years. While most people found it difficult to follow the arguments it was the Nobelist who won the fight, at least, when it came to the question of where the government should put its money. Smalley became an advisor to the government and Nano-assemblers were out almost overnight (very much like neural networks were out after Minsky's and Papert's attack on Rosenblatt's Perceptron). The new focus of Nano-technology was now on Nano-materials. Especially carbon-nano-tubes, something

Smalley was specializing in not only in research but also in a company which he had founded (Carbon Nanotechnologies Inc.). The carbon-nano-material research field is very promising and fruitful indeed but Smalley hadn't much time to enjoy his victory. Tragically, he died shortly after winning the fight in 2005. The Nano-Assembler field never recovered from this deep-hit and there is still kind of a Nano-winter, although Nano-metrologists have been very successful in gripping and moving single atoms. I was never convinced that Smalley was right with his verdict. First, because I was involved in Nano-technology at the time and knew what we could do with atomic-force robots and, second, because I was convinced that Smalley's arguments could be overcome if engineers tried hard enough.

What he had pointed out reminded me of the arguments that were used against two physicists who tried to make single atoms visible in the 1980s at IBM's Research Labs in Rüschlikon Switzerland. Everybody, especially the best theoretical physicists, told them that what they were trying to do was in vain, because of the rules of quantum physics, Heisenberg's principle of uncertainty, and so on and so forth. They didn't give up and, to their own surprise, one day their scanning tunneling microscope worked indeed. It earned the two physicists, Heinrich Rohrer and Gerhard Binning, a Nobel Prize in physics (1986) and their invention paved the way for making the next generation of computer chips on a Nano-scale. It also allowed access to the Nano-universe of biology and not many 20th century research-results had such a dramatic and industry-relevant impact.

I had the great pleasure of seeing Heinrich Rohrer giving a speech to an audience of the leading metrology people in Germany and to talk to him afterwards. He knew that quite a few important people (at least in their own eyes) were in the audience, some of whom had declared that what he was trying to do was impossible. Now with his fresh Nobel Prize he took the opportunity to make, kind of, fun of them, but in a very kind, relaxed, and friendly way. He was about the most modest and low-key person you can think of anyway. With a good sense of humor, he disclosed that he and Binning had a problem when it came to publishing the breakthrough they had achieved because they had no new theory and especially no "formula". So, they finally found a formula which made the invention look much more scientific. He displayed it on a slide and it looked really impressive with all the typical mathematical ingredients in it. And while everybody was impressed and tried to understand what it meant, Rohrer said with a smile and his lovely Swiss accent: "Isn't it a nice formula? But believe me; it works without a formula".

So, the message was that we should not follow scholastic orthodoxy and not be blinded by our theoretical prejudices or intimidated by the boundaries postulated in grand theories. And, I would like to add, that young scientists should not be shy of opening doors with the label "forbidden" on it. I loved Rohrer's refreshing and pragmatic perspective and was very happy that I was able to tell him that, in a subsequent conversation, whereupon he told me more about the scholastic resistances he had to overcome. Not everybody in the audience seemed to enjoy the message as much as I did as I heard from the organizer afterwards. Some even seemed offended because it was he (who didn't understand their wonderful theories as well as they did) who won the prize instead of them. Holy cow, I thought, how narrow-minded can intelligent people be? ☺

There is one chapter in Drexler's "*Engines of Creation*" which is a worthwhile read even today and it deals with our problem of computing machines. He obviously thought a lot about the concept of the Turing engine and he drafts a mechanical information processing machine where the logical gates are kind of valves like in a music instrument. It's a very nice illustration that information processing machines can be built of any kind of fabric. I wonder how long it will take until a new generation will rediscover Drexler and start a new Nano-spring. In any event, this Drexler-Smalley-thing is another story worth a TV-soap.[1]

And there is another lesson we can learn from this story. Drexler's career came to an end although he didn't promise even half as bold stuff as Ray Kurzweil did. To Kurzweil, however, his science-fiction-like fantasizing was a real career-booster not only as a writer but also as a relevant player in the industry. He's a world famous author now with a huge following and about a year ago he was hired by Google as "Director of Engineering". While I'm not a great fan of Kurzweil's singularity-prophecies and take

---

[1] If you are interested, here are the details:
https://www.wired.com/2004/10/drexler/
https://en.wikipedia.org/wiki/Drexler%E2%80%93Smalley_debate_on_molecular_nanotechnology

everything he toots with a ton of salt I think Google couldn't find a better AI-evangelist.

10 years after the NBIC we saw a kind of revival of this "New Deal" approach of combining various sciences and technologies. Now the focus was on industrial manufacturing, smart robots, 3D-Printers, and computers replicating themselves and something that is now called Industry 4.0. And of course, AI again is in the middle of all of it. This time it was a journalist who got the ball rolling. Chris Anderson, the former editor-in-chief of *Wired Magazine* published a book "*Makers: The New Industrial Revolution*" (December 2012) which had a tremendous impact in the US and in Europe and it has a lot to do with China. Actually, it describes a global trend-shift in how we produce goods in industry and the role which science, technology, robots, and AI play in it. I don't know what the reception in China was, but you may tell me.

But again, I'm running out of space and time and therefor rather stop here.

Best as always.

Karl

22.02.2014

Dear Karl,

Thank you very much for your season's greeting! Although we Chinese don't have a religious tradition of Christmas, now more and more people in big cities, especially young people, also celebrate this festival with no religious meaning. A similar holiday for all Chinese people is the Spring Festival, the Lunar New Year. Several hundred million people move throughout the whole country to have a family reunion! The date of the Lunar New Year is not fixed, but generally speaking, it is between the end of January and the middle of February. As a matter of fact, the whole event may last almost one month! So you see, this also gives me an excuse for my delayed reply. ☺

Besides my laziness and the holiday season, another excuse is that, owing to lack of the knowledge about AI, I have to hurry to read some material so that I can understand your arguments a little better. I could not say that now I can fully understand what you say, as there is so much to be learned, one can't expect that a few days' reading can solve the problem. Although it is difficult for me, I think the effort is worthy, now there is so much news in the media about AI, the rapid progress in this field has given people a very deep impression. People are talking with great gusto about the victory of the chess playing computer system Deep Blue over the international chess champion Garry Kimovich Kasparov in 1997, and the question answering computer system Watson beating the former 'Jeopardy!' winners Brad Rutter and Ken Jennings in 2011. At the same time, some people are talking about an extremely urgent danger of intelligent machines overtaking human beings and being the next ruling species on the earth. Someone estimated that such risk may happen before the middle of this century. I am very skeptical about such a prophecy. However, reasonable doubt can't just be based on intuition and belief.

Your story about the controversy between Drexler and Smalley, which I did not know at all, is as interesting as the other stories you told me. Your expertise and experience in Nano-technology make your remarks more convincing. Your following comment is very inspiring: *"we should not be blinded by our theoretical prejudices … and that young scientists should not be shy of opening doors with the label 'forbidden' on it."* It reminds me of a Ramachandran's story in his classical book *"Phantoms in the Brain"*:

In the 1960s, it was well known to medical students that an asthmatic attack could be provoked not only by inhaling pollen from a rose but sometimes by merely seeing a rose, or even a plastic rose, i.e., exposure to a real rose and pollen sets up a "learned" association in the brain between the mere visual appearance of a rose and bronchial constriction.

Ramachandran was a medical student then and wondered *"If it's possible to provoke an asthmatic attack through conditioning merely by showing a plastic rose to a patient, then if it's also possible to abort or neutralize the attack through conditioning as well."* Say, give the patient a bronchodilator and show her/him a sunflower at the same time when the patient is suffering from asthma. After a period of training, is it possible that only showing a plastic sunflower can relieve the patient from asthma? Coincidently, there was a Physiology professor from Oxford University visiting, so he discussed his idea with him, the professor thought his idea ingenious but silly, so Ramachandran gave up, at least for quite a long period. However, at the end of the last century, an American physiologist Robert Ader studied food aversion in mice. To induce nausea, he gave the animals a nausea-inducing drug, cyclophosphamide, along with saccharin, wondering whether they would display signs of nausea the next time he gave them the saccharin alone. It worked. However, to his surprise, the mice also fell seriously ill, developing all sorts of infections. It is known that the drug cyclophosphamide, in addition to producing nausea, profoundly suppresses the immune system. This hints that the mere pairing of the innocuous saccharin with the immunosuppressive drug causes the mouse's immune system to "learn" the association. Once this associations is established, every time the mouse encounters saccharin, its immune system will nose-drive, making it vulnerable to infections. Ramachandran's question is not silly at all. He regretted easily giving up his idea and took it as a lesson that "don't listen to your professors — even if they are from Oxford (or as my colleague Semir Zeki would say, especially if they are from Oxford)". ☺ However, his conclusion may be a paradox. Prof. Ramachandran is a professor of course. According to his above statement we should not listen to him, including the above words. ☺ Therefore, we should modify it a little: "don't listen to every word of your professor — even if he or she is from Oxford", i.e., you must have your own judgement with critical thinking, based on your own experience and expertise. Extending a truth too far makes it absurd!

I am sorry that I hadn't read any of Kurzweil's books until I received your last letter, although I have heard of him so much. This is owing to my laziness, and the fact that I have read some books giving similar prophecies which I don't believe so that I was not too motivated to read his books. A few days ago, I borrowed a Chinese version of Kurzweil's 2005 book "The Singularity Is Near: When Humans Transcend Biology" from the library of my university. As I told you I am lazy, so I thought that reading a Chinese translation of a book beyond my expertise would be easier for me. Reading the book, I found a lot of technological jargon from Nano-technology, robotics, genetic engineering, and information technologies, which I am not familiar with. I rejoiced that I had made a correct decision, otherwise, it might be much more difficult for me to read the original version, but I was wrong! Only yesterday, I found an original copy and found that there are so many mistakes in the translation, much of what I had read is not what the author meant! I just wasted my time. ☹ The problem is that the translators did not understand many sentences themselves, and just made a literal translation. Unfortunately, as the famous Chinese linguist Shuxiang Lv said many many years ago "*English is not Chinese*", there is no one to one correspondence between the words in these two languages! I am wondering what a Chinese version would be, if it was translated by a machine translation system? Sorry, I have talked too far away from our main topic. Maybe I should not complain about the translation too much. Maybe just as Walter said in his book "*How Brains Make Up Their Minds*" (1999): "*Looking back, if we succeed, we can praise ourselves, and if we do not, we can blame others.*" ☹

My main problem is lack of knowledge about technological issues, so that I have to say that I am not in a position to give convincing judgements on topics related to these technologies. Fortunately, you are an expert in these fields, so I am anxious to listen to your analysis. At the same time, I think that maybe I should read some chapters or paragraphs of his books focusing on human brains as samples, so that I can judge his remarks for myself. Thus, I jumped to the section *" The Vexing Question of Consciousness"*, and found the following statement:

" By the late 2020s we will have completed the reverse engineering of the human brain, which will enable us to create nonbiological systems that match and exceed the complexity and subtlety of humans, including our emotional intelligence."

A prophecy very similar to Markram's, or even much bolder! Anyway, what Markram has promised is *only* to make an artificial whole human brain by 2023, which *only* matches but not "exceeds the complexity and subtlety of humans"! As I am just starting to read his book, which is difficult for me, and I have postponed my reply too long, I have to stop here and may come back to this topic again someday.

As for Anderson's book *"Makers: The New Industrial Revolution"* is concerned, there is a quite good Chinese translation. As you may know, electronic commerce is very popular in China now, but I haven't heard much about manufacturing described in this book in China. I don't know.

Best wishes as always.

Fanji

20. 03. 2014

Dear Fanji,

Thank you for your great letter!

I didn't know Ramachandran's experience with the Oxford professor but it's very instructive and very true.

Also, I, very much, liked Semir Zeki's warning not to listen to professors "especially if they are from Oxford". ☺

I told it to my daughter who has graduated from Oxford University and she immediately understood the message, laughed, and agreed.

I didn't know either, about Robert Ader and his sharp-witted work on conditioning the immune system of mice, although I should have because this is relevant stuff indeed.

I wouldn't be surprised if this effect also works on humans. The endemic occurrence of allergies as a modern disease, you could almost call it a fashion, might be caused by similar effects.

In any event, I thank you for this extremely interesting hint about the link between the brain and the limbic system. There is one body, and the brain is a part of it. You can't separate the two!

I understand that you didn't exactly enjoy reading Kurzweil.

In his case you always have to be prepared to run into rather crazy or shaky ideas and wild speculation. And he represents, for sure, not the precise, modest, and down to earth kind of scientific approach which you prefer.

I should have warned you and feel somewhat guilty that I have pushed you into reading Kurzweil and caused you so much effort and headache.

But again, your fine nose has detected that something is wrong with this popular pundit and guru! What you have detected are not short-comings in your technical competence or ability to comprehend. What you disgusted are real problems in Kurzweil's world view.

In short:

No goldmaker can fool Fanji, neither Markram nor Kurzweil.

I'm proud of you again, my dear friend!

When you find it hard to enjoy Kurzweil's writing you are not the only one and you are in good company.

If you look at the Wikipedia-Kurzweil article you will find the following comment of Douglas Hofstadter:

"It's an intimate mixture of rubbish and good ideas, and it's very hard to disentangle the two, because these are smart people; they're not stupid."

and one by Biologist, P. Z. Myers, who criticized Kurzweil's predictions as:

"being based on 'New Age spiritualism' rather than science"

and says

"that Kurzweil does not understand basic biology."[1]

I think both critics are right on the spot and I can very well understand that you, as a sober, self-critical, and serious scientist, may be frustrated when you run into arguments or predictions which are mere speculation, quasi-religious prophecies, or just plain rubbish.

Maybe you would have been less disappointed with Kurzweil's later book (from 2012) *"How to create a mind — The secret of human thought revealed"* which includes a summary of the discussion around consciousness and mind. But maybe you shouldn't

---

[1] https://en.wikipedia.org/wiki/Ray_Kurzweil

waste your time by reading the rest of the book or more books of his. The Wikipedia article on Kurzweil is enough to understand what the message of our technology- and futurist-guru is.

We have a saying that "if you want to find out whether a bottle contains wine or vinegar you don't have to drink the whole bottle". ☺

The point in Kurzweils case however is, as Hofstadter rightly says, that it's many layers of wine and vinegar. ☺

Using lots of technology and science jargon and creating clouds of fog with the intention of impressing people is the good old goldmaker tradition, and Ray is a modern master of this art.

I have to admit that I like him as a science fiction writer because he seems to have read the same SF-stories in his youth as I did, and similar speculations about the future of technology have inspired my own interest in science and technology when I was a boy.

And could there be more thrilling ideas for a 16 year old than eternal life via mind-uploading and the amalgamation of biological and technical machinery? A world controlled and shaped by scientists and engineers evolving mankind into a super-human future where engineers can do everything and we nerd-savants are the masters of the universe. This is what my favorite SF-author Van Vogt with his idea of "Nexialism", which I have already mentioned, made me dream of at the time. Kurzweil seems to have dreamed the same dream — but he never woke up. ☺

Speculation per se is not as bad a thing as it is often characterized in earnest scientific discussions. Actually, all major scientific discoveries have started as assumptions, conjectures, or speculations. A cautious scientist may call them hypotheses as long as there is no empirical evidence to support them. But speculations are helpful when they inspire the inventor and other researchers to test or modify them and find ways to show empirical evidence. A speculation is nothing else than the previously mentioned "Einfall" (an idea) as Max Weber called a starting point of scientific theory building. So, there is nothing wrong with speculations as long as they are not declared as true insights without any supportive evidence in a fundamentalist way or as religious-like prophecies in rational disguise.

It is this prophetic and apodictic tone which makes Kurzweil so hard to digest for sober, rational scientists. It is especially hard for those who would rather comprehend and check what he's predicting than become evangelized followers of his church of singularity.

And as I said, I'm taking his statements always with a ton and not just with a grain of salt. I hadn't read any of his stuff for long and when this "How to create a mind" came out I was interested to see what is new and whether he had changed his position or method.

But although I tried to approach the new Kurzweil-book in a relaxed and unbiased mode it lasted only to page 5 of the introduction until he managed to put me in anger with a whole series of rubbish.

He displays the same optimistic point of view about the help of computers and simulations as Markram and his arguments have a lot to do with the questions we discuss since a while. I'm optimistic myself about the useful role of computers and simulations, but he aroused my annoyance by the false representation of what we had already achieved with it.

It starts with the statement that

"*The complexity of proteins for which we can simulate protein folding is steadily increasing as computational resources continue to grow exponentially. We can also simulate how proteins interact with one another in an intricate three-dimensional dance of atomic forces. Or growing understanding of biology is one important facet of discovering the intelligent secrets that evolution has bestowed on us and then using these biologically inspired paradigms to create ever more intelligent technology.*"[1]

And then he goes on to praise Markram's Blue Brain Project (the predecessor of HBP) as

"*argueably the most important effort in the history of the human-machine civilization.*"

---

[1] Kurzweil R. How to Create A Mind: The Secret of Human Thought Revealed [M]. New York: Viking Penguin, 2012: 4.

And after stating that humans are likely to be the only intelligent species in the universe he goes on with:

"From this perspective, reverse-engineering the human brain may be regarded as the most important project in the universe."

Well, one goldmaker is praising another and Markram would probably have liked this praise and maybe some politicians who had to decide about the money to support the HBP were also impressed.

A closer look, however, reveals that almost all arguments are misleading or wrong.

The number of proteins the folding of which we can simulate is pretty small and the number of interactions of proteins we can simulate is even smaller. Actually, the numbers are ridiculously small. But the major problem is that we talk about simulations and not the real folding or the real interaction. These simulations are all based on mathematical models which are about as appropriate as our models of the neural networks in the brain.

Patricia Churchland made the point in her interview with Susan Blackmore, in the book which you recommended to me, when she said

"The problem of how proteins fold was thought to be an easy problem; ...but we still don't know how proteins fold".[1]

That was true in 2005 when Churchland said it and it is still true today.

The situation is even worse when it comes to understanding how proteins interact.

We have only vague ideas about how this works with some small proteins or subsections of larger proteins. Many people, and I believe Kurzweil belongs to them, have no idea how big and complex building proteins can be, how many of them interact, and what the signals are they transmit. What you typically see in textbooks is some proteins

---

[1] Blackmore S. Conversations on Consciousness[M]. Oxford: Oxford University Press, 2005.

floating in a cell or sitting on a cell membrane. This simplification is made for pedagogical reasons in order not to confuse the students. In real life, a cell is stuffed with an incredible number of proteins and the cell membrane is populated by proteins as a spring meadow is covered with grass and flowers. And the communication pathways and signal cascades, in which they are involved, besides other small molecules, are from an astronomical complexity. We have understood some of them at the lowest level of the metabolism and also deciphered some signal cascades.

The understanding of DNA was a major achievement, but even 60 years later the major part of the inter-cell-communication, especially inside the brain, between the brain, and the limbs is still as mysterious as ever.

As you know, I have been involved in a project of building instruments to unfold proteins with the help of atomic force spectroscopy where the best bio-physicists and mathematicians tried to understand the folding and binding phenomena Kurzweil is displaying as easy and almost solved if we only throw more computer power on it.

Yes, there is some progress and especially the imaging and metrology techniques at the atomic and molecular level are getting better. But we are far from understanding this "dance of atomic forces". And we also don't understand the signals and the code on the higher levels in our bodies which keep the metabolism going, control the immune system, or perform this mysterious homeostasis. Not to mention higher level functions like memory, language, or consciousness.

The fact that we still know so little even about some basics is the reason that despite all those wars on cancer, decades of the brain, Alzheimer's projects, and the many billions spent on them, there is so little progress in medicine.

When it comes to understanding the functioning of our bodies, we are facing the very same problems which you have pointed out in the case of brain simulation. There is no theory about this "dance" and how can you simulate something on a computer which you don't understand in the real world?

Kurzweil wants to make his readers believe that all this can be overcome by more computer-power and applying mathematics.

As an example, he talks about

"Bernoulli's principle which states that there is slightly less air pressure over a moving curved surface than over a moving flat one. The mathematics of how Bernoulli's principles produces wing lift is still not yet fully settled among scientists, yet engineering has taken this delicate insight, focused its power, and created the entire world of aviation".[1]

This made me shake my head because it is about the most inappropriate and misleading example you can think of.

First, because a wing doesn't have to be curved to make an airplane fly.

Second, because it wasn't Bernoulli's equation or any other mathematical algorithm which guided engineers to create the world of aviation.

The only true part in the statement is that *"the mathematics of how Bernoulli's principle produces wing lift is still not fully settled among scientists"*.

This curved wing (as the main reason that airplanes can fly) nonsense you find even in physics textbooks and it is a perfect example of the scholastic way of teaching what's in the books without thinking. One might easily add it to the examples Feynman is complaining about when he talks about his depressing experiences with the scholastic university system in Brazil.

In *"Understanding Flight"* David Anderson and Scott Eberhardt give a good insight into how difficult and delicate this seemingly simple problem of wing lift is.[2]

It clearly shows that mathematicians have to make unrealistic idealizations of the physical reality (like gases being uncompressible or the absence of vortices in a stream of gas) to be able to calculate such a relatively trivial phenomenon as a gas streaming around an object.

The engineering of aircrafts wasn't inspired by mathematics but by the dream to fly and was realized in a long series of trial and error. These were always accompanied by

[1] Kurzweil R. How to Create A Mind: The Secret of Human Thought Revealed [M]. New York: Viking Penguin, 2012: 5.
[2] Anderson D, Eberhardt S. Understanding Flight [M]. 2nd ed. New York: Mc Graw-Hill, 2010. See also: https://en.wikipedia.org/wiki/Lift_(force)

physical assumptions, which turned out to be incomplete, and attempts to do mathematical calculations were often only made after the thing worked in the real world.

Kurzweil wants his readers to believe that everything in the world can be calculated if you only have the right algorithm. Actually, many people do believe this and I have the impression that the less they understand about natural sciences and math the more they seem to believe in this myth.

The fact is, as you know, there is a relevant class of phenomena which are resistant to algorithms and also to numerical calculations. In the typical textbooks about brain research or the mind you can often find this hint. However, it is mostly justified with the problem of indeterminism resulting from quantum mechanics and the principle of uncertainty.

We don't know how relevant this latter argument really is, but physicists and mathematicians had already discovered that another problem exists which makes fully deterministic systems incalculable and unpredictable even in good old Newtonian mechanical physics.

19th century physicists and mathematicians like Henry Poincaré, had to realize that it is very hard to calculate the longtime movement and the positions of the planets in our solar system. Even small systems with just three bodies like the sun, earth, and moon are very difficult to calculate as the famous three-body-problem demonstrates.

Only under unrealistic, idealized restrictions, can we calculate primitive 3-body systems with point-masses and there are no solutions for larger systems as the n-body problem demonstrates. The clean world of Kepler and Newton, with planets orbiting around the sun on perfect eclipses, only exists on paper and in our mind. In the real world, each celestial body is disturbing each other via gravity. This makes all objects wobble a little and gives them slightly imperfect trajectories which are hard to describe but may have huge effects over time.

And this already happens with the idealized assumption of point masses orbiting in the model. But idealized point masses, very much like idealized point neurons, do not exist in the real world. Each celestial body has a certain size and is neither perfectly round as a sphere nor is the mass distributed evenly. If you consider the uneven distribution of mass and the complex elasticity of a planet consisting of unevenly distributed firm matter or liquid stuff like melted iron even a two-body system like the earth and the moon cannot be calculated for a long time.

In practice, this means that we cannot calculate when or whether a constellation will appear or when one of the planets in our solar system will be ejected from it or crash into another object.

Poincaré's dealing with those problems led to the development of chaos theory.

And chaotic systems can be very hard to calculate. Global analytical solutions exist to the (restricted) three-body problem in the form of a convergent power series as Karl Sundman and your fellow countryman Qiudong Wang have shown. But those solutions are of little practical value because such series converge very slowly. If you want to use a Sundman series for astronomical observation "*computations would involve at least $10^{8,000,000}$ terms*".[1]

So you need a big computer and lots of time. ☺

This problem of deterministic systems that can't be calculated has been known for more than 100 years but many people just prefer to ignore it.

And it is not so, that these are exotic and rare systems. The world of chaos and turbulent vortices, as described by the Navier-Stokes equations, is everywhere from streaming fluids around wings, to weather, and oscillating stock-markets. And our body is also full of chaotic systems!

Trying to solve such problems by means of numerical computation only helps for short time frames. Weather forecasts have improved considerably in the 1 to 3 day timeframe since we have more computer power to do the numerical calculations. But in the 2 to 3 weeks forecast you still can play dice as well. And this will remain so because the event space in the model explodes over time. We will see improvements, but if you want to predict the weather in Shanghai for Monday in a year even a computer as large as planet Earth wouldn't help much. Simply because (as every mathematician knows) there are various kinds of infinite — small ones, big ones, and infinitely big ones. ☺

---

[1] https://en.wikipedia.org/wiki/Three-body_problem

And remember, we are talking about good old Newtonian physics and primitive systems with just a couple of elements. It is remarkable, by the way, that quantum mechanics allows a much better calculation on the molecular level by means of the most wonderful and useful tool mankind has developed so far — the Schrödinger equation. One of the reasons why some physicists like Dieter Zeh recommend giving up Newtonian physics and switching completely to quantum-mechanics. But this is a different cup of tea.

To make a long story short: when Mr. Kurzweil is telling us that "computational resources ... continue to grow exponentially" this means nothing in regards to the problems he wants to solve.

Unfortunately, this "we can calculate everything when we have enough computer power" isn't the only misconception in Kurzweil's perspective.

The other is the belief that we need to understand the biological original, e.g. the brain in order to reengineer it (especially by finding the algorithms on which it is built) and then create artefacts that come close to the original.

If anything is rubbish in Kurzweils world-view this is the very core of it!

But I'm appalled now to see that this email is already so impolitely long and I'm only half way done with my critique of Kurzweil.

I'm not sure whether you are interested at all in those questions. So please forgive me if I've bored you with all this stuff. I've been dealing with my discomfort with this astoundingly popular Kurzweil philosophy for many years and I just used the opportunity to put it into writing. So, in case I'm run over by a bus, now you will know what my arguments are — or at least 50% of them. ☺

In case you are interested in the remaining 50% just let me know.

In any event, I've done more of my homework around consciousness, mind, and subjectiveness. Although I'm still not fully settled and, for sure, not on par with your expertise I'm very much looking forward to discussing it with you, maybe right after Kurzweil.

Good night for now and best to Shanghai.

Karl

20. 04. 2014

Dear Karl,

I enjoyed your great letter very much and am sincerely grateful for your kind words. Your arguments about Kurzweil's problem, about which, I only had a vague feeling before reading your letter, are as clear as a crystal.

Yes, I must admit that I did not enjoy reading "*The Singularity Is Near*", although I had expected that I should, as it was the "best book of 2005" as reported by the media, and was praised by many celebrities, including Bill Gates. This reminds me of Dr. Ramachandran's words again.

Although I did feel there must be something wrong in his book, at the very beginning, I did not realize what the main problem was. In my first impression, he always talked in a way as if what he said must be the truth, he used so much scientific jargon in a variety of fields which many readers, including me, are not familiar with, just as you correctly pointed out:

"*Using lots of technology and science jargon and creating clouds of fog with the intention of impressing people is the good old goldmaker tradition and Ray is a modern master of it.*"

People are apt to regard something which seems to be too profound to be understood with reverence. He jumps to conclusions without thorough proof as if it were self-evident, so that if a reader can't follow him, he or she might think that must be owing to his or her own ignorance. How such a smart genius being wealthy in knowledge can be wrong! You are right again that "*the less they understand about natural sciences and math the more they seem to believe in this myth.*" The other point is just as you cited from Douglas Hofstadter: "*It's an intimate mixture of rubbish and good ideas, and it's very hard to disentangle the two, …*" which reminds me of a story from a Chinese Kunfu novel "Luding Ji". Wei Xiaobao, the leading character of the novel, was very good at telling absurd stories to his opponents to make them believe what he said. The trick which he played was to talk nonsense at the key point, while giving a lot of true details on secondary matters. It seems that all the gurus, no matter western or eastern, ancient or modern, adopt similar strategy!

I liked your following quote very much: "*if you want to find out whether there is wine or vinegar in a bottle you don't have to drink the whole bottle*". Your words and the Wiki-encyclopedia about him you recommended to me saved me from further suffering from struggling with his books, although maybe I shall read some part of them someday if I have time. Anyway, I have read a few chapters of "*The Singularity is Near*", so I have already drunk a few gulps from his bottle, and can understand what you criticized. You are right that the key problem making his work so hard for me to follow is not just owing to my technical competence or ability to comprehend, but mainly owing to his world view and approaches. As you said that I tried to "*comprehend and check what he's predicting*", and only found "*speculations without any supportive evidence in a fundamentalist way or as religious-like prophecies in rational disguise*". This is why I dislike his book. Of course, I don't want to be an *evangelized follower of his church of singularity*.

You correctly pointed out that his core argument, "*we can calculate everything when we have enough computer power*", is wrong! Your letter has already explained this in detail, I won't repeat it again. You destroyed the cornerstone of his and other similar guru's prophecies. Really good!

As for the other cornerstone of Kurzweil's super-intelligence rhapsody, finding the algorithms on which brains are built to reengineer it. We have already pointed out that it is very unlikely to reach this goal in the foreseeable future when we discussed the feasibility of the EU's HBP. Now I see that Kurzweil and Markram are really birds of the same feather.

I thought that Kurzweil's so-called "The Law of Accelerating Returns" is the third cornerstone of his prophecy, according to which technology will develop exponentially. [1]

---

[1] https://en.wikipedia.org/wiki/Accelerating_change#The_Law_of_Accelerating_Returns

This is an extension of Moore's law, which declares that the power of integrated circuits increases exponentially. Although Moore's law has held for about half of a century since 1971, it is not a natural law, but an empirical rule. It can't hold forever. Now the size of a semiconductor chip is already close to 10 nm, further decreasing it will make its size comparable to tens of atoms, thus heat dissipation and quantum mechanical effects would become a serious problem. Even Moore himself said in 2005:

"It can't continue forever. The nature of exponentials is that you push them out and eventually disaster happens". [1]

Kurzweil argued that whenever a technology, such as the traditional semiconductor industry, approaches some kind of a barrier, a new technology will be invented to allow us to cross that barrier. He selected historical data supporting his ideas to support his speculation, very similar to Franz Gall picking data to support his phrenology. We know that the human population although increased approximately exponentially before 20th century, now the speed has dramatically decreased. In many developed countries there is even a trend to be zero or negative. Olympic records also tend to approximate some saturation level. These examples show that exponential trends can't last forever. Kurzweil can't deny the above facts. He asked himself in *" The Singularity Is Near"* [2] : *" So isn't there a comparable limit to the exponential trends that we are witnessing for information technologies?"* He answered himself: *" The answer is yes, but not before the profound transformations described throughout this book take place."* Then he just said that using technologies he imagined could save energy to a very small amount and the problem would vanish. Although he knew the criticism to his " law" :

"It's a mistake to extrapolate exponential trends indefinitely, since they inevitably run out of resources to maintain the exponential growth. Moreover, we won't have enough energy to power the extraordinarily dense computational platforms forecast, and even if we did they would be as hot as the sun." He just used the vague promise as his answer:

---

[1] https://en.wikipedia.org/wiki/Moore%27s_law
[2] Kurzweil R. The Singularity Is Near: When Humans Transcend Biology [M]. New York: Penguin Books, 2006.

"Exponential trends do reach an asymptote, but the matter and energy resources needed for computation and communication are so small per compute and per bit that these trends can continue to the point where nonbiological intelligence is trillions of trillions of times more powerful than biological intelligence. Reversible computing can reduce energy requirements, as well as heat dissipation, by many orders of magnitude. Even restricting computation to 'cold' computers will achieve nonbiological computing platforms that vastly outperform biological intelligence."

Even so, his answer is only for the second part of the criticism, but not for the first part — the essential one! And he pretended that as if he had solved the whole problem!

As a matter of fact, the fact that population growth obeys an S-curve instead of an exponential curve is no coincidence, it means that the resources for our survival are limited, the larger the number of our population, the slower the growth speed, when the whole population reaches a given scale. Therefore, for the survival of our species, the population number has to reach a saturation level sooner or later. This is an essential constraint to the speed of technological progress: more people means more potential inventors, and more inventors means faster technological growth. This may explain the speed of technological progress in the past centuries. Now as the size of the population approximates to a saturation level, could the speed be maintained forever?

In summary, to extrapolate from an empirical summation without considering the condition under which the data were collected to give a prophecy is dangerous.

Oh! You are really a modern Schehera-zade, who always reached the most interesting point when daylight came and asked the Sultan if he wished to know the end of the story. Of course, he wished! Although I am not the Sultan, and you are not Schehera-zade, please do tell me the other half of your criticism!

As for the problem about the subjectivity of the consciousness, as this e-mail is already long enough, after your example, I would like to say something more in my next e-mail.☺

Best wishes as always.

Fanji

18. 05. 2014

Dear Fanji,

You won't believe how much I enjoyed your last email! I especially liked what you described as the method of Wei Xiaobao. *"The trick which he played was to talk nonsense at the key point, while gave a lot of true details on secondary matters."* And even more so the resume you took: *"It seems that all the gurus, no matter western or eastern, ancient or modern, follow a similar strategy!"*

The examples you gave from Kurzweil's argumentations are a perfect fit.

I also liked your discussions about exponential growth and Moore's Law. This is a delicate field, however, and belongs to what Hofstadter described as *"It's an intimate mixture of rubbish and good ideas, and it's very hard to disentangle the two, ..."*

The rubbish part of it is the naive inductive belief that processes in the future will behave as they did in the past, which you have rightfully criticized. There are technical issues, however, that make it possible that Moore's law, not on a chip-level but on a systems-level may go on for a long time and may even accelerate. This is all but trivial and not easy to decide. Actually, it is as difficult as deciding whether a mathematical series will converge or diverge. Often, you don't see this at first sight. Sometimes, it takes a long time to find out and sometimes a decision is not possible at all. I would like to discuss this in more detail because it is crucial in Kurzweil's argumentation. However, I don't think it's the major flaw in his theory.

So as not to provoke another "when daylight comes" break, please allow me to discuss this later and let me start with what I offered to do, namely state the remaining 50% of my criticism of Kurzweil's position. It's very kind of you to accept my offer, but maybe there was no chance of rejecting it. ☺ But I'll try to do my best to demonstrate that it is also relevant to the questions about consciousness and mind in which you are interested.

At the end of my last letter I said that if there is one thing that is rubbish in Kurzweil's research program, it is the idea that reengineering the biological original, e. g. the brain, is the way to build artefacts.

The history of mankind's inventions, tools, and artefacts is impressive indeed. And Kurzweil is right when he emphasizes that this happened in a surprisingly short time frame. Engineers, have indeed, managed to improve almost all the weak natural senses and capabilities of humans within a few hundred years after Mother Nature made this possible after a ramp-up phase of some billion years. But, the point is that this was never done via understanding the biological archetype first. On the contrary, it was only possible because engineers didn't waste their time in fruitless attempts to understand, in detail, how birds fly, the eye sees, or the ear hears. Scientists and researchers tried to do this but not engineers. Engineers start with dreams and idealized thoughts and materialize them in machines. Sometimes they were inspired by biology, mostly totally wrong as in the case of the brain, and then went on based on principles that had nothing to do with the world of biology.

When machines apply forces which exceed those a biological muscle can provide, by orders of magnitude, by use of levers, pistons and hydraulics, explosives or electromagnets, the principles have nothing to do with the biology of the muscle. When we improve eyesight down to the level of making atoms visible, or to see distant stars, and look into frequency ranges inaccessible to human eyes it's all done based on processes totally different from the methods biology uses. The same is true with hearing and smelling. And the most impressive functions engineers made available don't even have a biological model. Telecommunication and the internet are the most prevailing ones. And when engineers invented fast locomotion, and finally, even flying, the old dream of mankind, they didn't follow the examples of horses or birds. Leonardo da Vinci tried to understand the principles of bird flight and built flight-machines following those principles. He failed miserably, like all the others who tried to do it this way. The problem was the power to weight ratio. At that time there was no material that was stable and light enough, and above all, no suitable energy source. The era of aviation only started with the availability of combustion engines, for which, there is no biological model at all.

Realizing their imaginations and their machines in completely new ways, engineers have outdone human performances in almost all disciplines, including many brain functions.

We are used to focusing on what computers and AI are still not able to do when we compare them with the functions of our brain. And there is still a remarkable gap. However, we shouldn't overlook that there are many aspects of memorizing and information processing which machines can do much better than biological beings. The fact that I'm sitting here, typing my ideas in order to send it to you via email after I have searched on my laptop and on the internet to find Feynman's Brazil article is the best example. Our grandparents wouldn't have believed that this would ever be possible. And here, as well, machines aren't only able to do what the biological model can do, they are typically much better at doing it.

The human brain is very bad in doing numerical calculations and remembering correctly. Actually, human memory is an awkward thing. It isn't only poorly understood, it also works in a very unreliable way. Searching and finding is a dreadful job for a brain and our memory is often incomplete, misleading, or outright wrong. We not only forget things; our brain even pretends to memorize things that never happened! No reasonable engineer would design or want to copy such an unreliable and faulty system. Computers can memorize much better and with incredible reliability. They are also better in many kinds of information processing. Actually, I agree with Kurzweil that there isn't that much left to demonstrate the superiority of the machines over their inventors.

But the point is, that all those machines are based on axiomatic systems which, unlike biological systems, are fully understood because they are rationally defined by the inventor and designed to follow very specific rules.

You can find all those clean and ideal things that exist in math but not in the real world in the toolbox of engineers full of Lego bricks like logical building blocks. And the rules how they fit and how they work together are implemented in their design following a blue-print. There is no perfectly round circle in the whole universe. But it exists in mathematics and in the mind and blue-prints of engineers. Of course, those building blocks of machines aren't fully virtual and axiomatic because they consist of real matter and interact with the real world as the example of wing lift demonstrates. And mechanical shafts may break because of material fatigue, a transistor may burn because of overvoltage, but, in general, those building blocks do what they are designed for and follow the rules quite reliably. They represent rationally and intentionally planned systems, which are, in any case, much more comprehensible and predictable than the world of biological systems that have resulted from meandering and random paths of evolution.

Engineers are so successful because they don't care about the muddy world of biology and rather build a new world out of clean and calculable building blocks. Inductivism, in the sense of believing that tomorrow's world will behave exactly as it does today, is a very poor and unreliable principle when you want to distil the laws of nature in the real world of open systems. But, it is extremely successful in the world of machines intentionally built by engineers as axiomatically, defined, closed systems. This is especially true for information processing machines, from the very first computers that used vacuum tubes as logical elements to the latest "neuromorphic" chips realized in CMOS.

Computer-engineers and scientists from McCulloch and Pitts to Modha were often fascinated and, kind of, inspired by the biological brain. But, their technical creations have always been lightyears away from a biological neuron. A neuron, first of all, is a cell as any other cell in our body. Which means that it contains a nucleus with the DNA and all the other cell compartments like mitochondria, ribosome, and so on. Even a single neuron is a complete factory much more complex than a nuclear power plant or an oil refinery. It doesn't only fire and send action potentials along the axon. It can also replicate, modify its shape, link and communicate with other cells (not only via thousands of synapses and dendrites) and probably do many more tricks we still have no idea of.

So, calling an electronic switching device in a network an artificial neuron is about as appropriate as calling a candle an artificial sun. Actually, it is plain ridiculous. The only thing artificial neural networks have in common with the neural nets in our brain is the naming.

The amazing fact, however, is how it can be possible that a technical device so primitive and so totally different from the biological model can provide similar and, in many respects, better information processing than the original. (By the way I do believe that your friend Walter Freeman is absolutely right when he questions whether the brain is an information processing machine at all! But this is a different cup of tea.)

However, when we look at other examples of machines outperforming human capabilities the fact is less amazing. Reducing a machine to a few logical principles and leaving out the ballast, that has accumulated in biology over the course of evolution, and its innumerable wrong turns seems to be a good design strategy.

When information processing machines still can't do important things our brain and mind can do, the question occurs whether this is caused, and generally limited, by the inappropriate simplicity of artificial "neurons" and networks and whether it can be overcome by more intense engineering on the chosen technical platforms. Some people, like Kurzweil, are convinced that the latter is the case while others believe (or seem to hope) that there will always remain a secret part which will be reserved to us humans and make our mind and consciousness superior to our machines.

This is an open question and the engineers are in the race for reducing the secret part. They are doing quite well so far, and the question is; why should they apply a different strategy to improve human abilities just in the case of the brain when they never applied this method of copying the biological model and principles in other fields?

The design and improvement of machines is often done in an empirical and heuristic process as a kind of continuous wrestling with the matter. But, when these machines are finally working and can be replicated, following a blue print, they represent applied logic and mathematics. They are models based on axiomatic elements and rules. It may be that some modules contain capsuled elements, about which, little more is known besides the fact that they somehow work. But you don't have to understand the details of the phenomenon that a specific amount of voltage applied to the electrode of a field effect transistor opens or closes a gate and thereby can represent a bit of information.

Of course, such machines don't escape the limitations of physics or logic like the ones which have been discovered by Gödel, Turing, and others. And their elements aren't fully ideal in a platonic sense because they import some of the "dirtiness" of the real world as soon as they leave the drawing board and are manufactured out of real fabric. But, they are, by orders of magnitude, more predictable, calculable, understandable, repairable, and improvable than living biological objects.

And this is the very reason why our competence in technology and engineering, kind of, explodes while progress in the understanding of biological systems and in medicine is so disappointingly slow!

So, it's just the other way round as Kurzweil describes it!

And trying to reverse engineer a machine like the brain, which you don't understand, is about the most foolish thing you could do, as you my dear friend Fanji have realized and expressed in your article on the HBP project long ago!

This is why people like Modha, meanwhile, prefer to say that their computers are not "brain-like" but just inspired by the brain. Most engineers in the field are quite aware of what I tried to point out here, but they are also clever enough to call their computer-chips "neuromorphic". They simply sell better with a fashionable and mystic label, especially when emperors, politicians, and grant givers seem to be obsessed with the idea that we should copy the brain to get the best computers and general AI.

However, engineers should be fascinated by a characteristic of biological information processing. Namely, low energy consumption. It's a clear indicator that there are much more efficient ways to do information processing than by running von Neumann machines on our established CMOS-chips. But this they already know from the days of Leo Szilard and Rolf Landauer. Actually, we know quite well what minimum energy is required to write or delete a bit of information. The human brain is pretty good at doing it in an efficient way which is, orders of magnitude, better than in computers based on the actual CMOS platforms. But it can be done better on all kinds of other platforms and, very likely, also better than the brain. It's just a matter of effort and time to find out how to do it.

In so far, I'm generally with Kurzweil when he predicts that, someday, computers will outdo humans in terms of information processing in almost all kinds of applications. He's far too optimistic, in my eyes, when it comes to the timeframe. Instead of 20 to 50 years, I would rather plan on a 100 to 500 year timescale, which would still be pretty fast. But I understand that you can't promise an emperor or a silicon-valley tycoon to achieve eternal life in 500 years. ☺

I doubt however whether we will see those technical-biological hybrid systems because of the problems mentioned above. It is much easier to build a better information processing machine than the brain based on other principles than to interface with such an awkward system that we don't understand. Therefore, I also don't see the possibility of downloading a mind to a machine.

I wouldn't say that this neat idea which fascinated me as a boy is totally impossible, but Ramachandran's cable problem is much more difficult than Kurzweil believes.

The question is whether it is needed at all. A full-blown AI system of linked computers in the future would contain all the knowledge and available information in the world anyway. At least the non-subjective one. And almost all kinds of human feelings and emotions have been described in millions of books and movies. And people will have documented their histories of all that they have seen, said, heard, written and "liked" from email to Twitter and from Facebook to WeChat or whatever the systems will be that document human activities in the future.

So, what is left is the very subjective knowledge of an individual and what is called "qualia" whatever that may be. Everything that can't be observed or communicated by language is difficult, especially when it has to do with physical experience. For example, it is very difficult to describe, in words, why it is a pleasant feeling when we touch a silk scarf or a polished metal surface, especially if the person to whom we want to explain it has not yet had the corresponding experience.

This may be very relevant for an individual of course but from a general knowledge perspective this may not be very interesting for humankind. And beyond what Kurzweil calls singularity, it may be as irrelevant for mankind as the former caterpillar-phase is to a butterfly.

I don't want to downplay the relevance of the subjective feelings and memory and qualia for an individual. But I'm asking myself what these cryonic people, who want to be deep-frozen when they die and un-frozen 100 years later, will do anyway? Build a club with other cryonics and talk about how great it was in the 21st century? ☺

Well, there would be a reason to hibernate for a century as David Hilbert has pointed out when he was asked what his first question would be if he could come back in a hundred years. "Has Riemann's assumption been proved" he said. I like this question and would add "Have the secrets, hidden in the Navier-Stokes equations been unlocked?" ☺

There is a religious aspect, of course, in Kurzweils singularity prophecy. He and his followers typically start from a strict materialistic position, rely on Darwin's Theory of Evolution, and are convinced that the concept of "intelligent design" (the idea that the world is designed and controlled by a deity[1]) is only for religious backwoodsmen.

But when mankind can design machines that amalgamate with humans and this new species steers the process of its own evolution then, voila — you have a new kind of "intelligent design".

We are our own gods and Ray is our prophet! ☺

The idea is not new, of course, and everybody who thought about self-organization, self-replication, evolution of consciousness, and the problem of teleology came to think about this aspect.

And it comes as no surprise that the resonance to such secular prophecies is especially high among the technical elites in the USA who are busy giving the world a new shape by means of scientific knowledge and technology. They believe that almost all dreams can be realized with rationality and technology and that they are the ones who can accomplish this, ideally during their lifetime. Well, you know how optimistic and enthusiastic the feeling of life in the Silicon Valley can be and nobody can express it better than Kurzweil. Reading Kurzweil isn't always fun but even when he's wrong, or especially when he's wrong he can be inspiring. And every time I'm in California I'm tempted to think how fascinating it would be if he's right. And therefore, despite all his big-mouth tooting I like him to a certain extent because he's all but boring, and in case that a new Noah's arch had to be built I would recommend taking him on board as kind of intellectual jester to entertain the crew. ☺

This dazzling opalescence combined with his prophetic certainty seems to secure him his enormous flock of followers. It isn't difficult to see the similarity between Kurzweil's singularity message of technical salvation and the biblical promise of Moses leading his people to the promised land of milk and honey.

---

[1] https://en.wikipedia.org/wiki/Intelligent_design

Actually, I never found out whether Kurzweil has a real problem in differentiating fiction and dreams from reality or whether he's just a good salesman.

But as I said, in any event, Google couldn't find a better evangelist and I'm curious to see what he can do in the real world.

Your example of Robert Ader brings us back to your questions about consciousness and the aspects of privacy and subjectiveness. In one of my last emails I said that I haven't gotten my arms around this problem. I still can't claim that this is the case, but I do better understand now what your points are. What I found very helpful are the interviews Susan Blackmore made with some of the leading people in the discipline. She managed to make them express their positions in a short and explicit way. I understand that you are very skeptical about the chances of discovering the mystery of consciousness by means of studying the material substrate of the brain.

When you first mentioned privacy and subjectiveness I interpreted this in a different, more technical, way than after I read those various positions. But I'm still not sure whether I have really grasped the problem.

To better understand your point, it might help if you could tell me which of the people interviewed by Blackmore come closest to your position and which ones you don't like or are skeptical about.

Thank you for your patience and best for today.

Karl

PS Google has lately bought DeepMind, a company involved in deep learning which claims to have the most advanced technology of neural networks and represents the spearhead of AI. I don't know whether Kurzweil has been involved in this move and what the plan is, but I will try to find out.

Dear Karl,

I am glad to hear that you liked my last e-mail. As for my comments about exponential growth and Moore's Law, I must acknowledge that I am not so sure. Anyway, I cannot deny the fact that technological progress has accelerated. I am just wondering if everything can develop exponentially without any limitation. It seems very likely that technology may develop this way in the foreseeable future, but can it develop like this forever? Can an empirical rule be extrapolated infinitely? I am glad that you will discuss this matter further. I appreciate your attitude towards Kurzweil's ideas very much, you don't deny every idea proposed by him just owing to the fact that it is proposed by Kurzweil, but distinguish the rubbish from good ideas! I should follow your example and have an open mind! People are apt to have prejudice.

The story you told in your last e-mail is just wonderful, which clearly demonstrate why Kurzweil's following idea is wrong: "*reengineering the biological original, e.g. the brain, is the way to build artefacts*". As a matter of fact, I think that Kurzweil is not the first one to propose this idea, it has been popular for a long time. I must acknowledge that I was a believer of a similar idea for quite a long time, and there are many colleagues of mine who believe it. Most of them are from physics, mathematics, and informatics, and only know a little about biology. They admire the sophistication of the living creature, and believe, that owing to the struggle for existence, the mechanisms of the surviving creature must be optimized or at least be good enough. This even almost becomes a belief. My colleague Prof. Fuchuan Sun, who is one of the smartest in my domestic friend circle, by the way, criticized us as "believers of a bionic religion"! Your letter explained why the main dogma of the "religion" is wrong. I could not find better arguments than yours. Of course, this does not mean that we should not seek any inspiration from living creatures at all, if a species can survive during the struggle for existence, there must be some merit in its mechanism, at least Mother Nature gives us "existence theorems" to tell us what kind of function can be realized. In addition, just as Francis Crick said that to bind one's one arm behind his own back is not a good strategy for fighting, to seek inspiration from Mother Nature is beneficial, but to take it as the main or even the only way for technical development, especially to copy every detail is absurd.

Your following argument is very forceful:

"The history of mankind's inventions, tools and artefacts is impressive indeed. And Kurzweil is right when he emphasizes that this happened in a surprisingly short time frame.

Engineers, have indeed, managed to improve almost all the weak natural senses and capabilities of humans within a few hundred years after Mother Nature made this possible after a ramp-up phase of some billion years.

But, the point is that this was never done via understanding the biological archetype first. On the contrary, it was only possible because engineers didn't waste their time in fruitless attempts to understand, in detail, how birds fly, the eye sees, or the ear hears. Scientists and researchers tried to do this but not engineers.

Engineers start with dreams and idealized thoughts and materialize them in machines.

Sometimes they were inspired by biology, mostly totally wrong as in the case of the brain, and then went on based on principles that had nothing to do with the world of biology."

Your words remind me of Walter's following comment:

"In the 1940s, neurobiologists and logicians reconceptualized the functions of neural activity. They conceived of neurons as binary switches performing Boolean algebra and forms of Aristotle's logic. Action potentials of neurons were no longer interpreted as electrical pulses, but as bits of information, binary digits that represented on or off, yes or no, one or zero. This led rapidly to the development of programmable digital computers, as starting from a mistaken view of how neurons work." [1]

---

[1] Freeman W (2002) Preface for the Chinese version. In: Freeman W. Neurodynamics: An Exploration of Mesoscopic Brain Dynamics[M]. Hangzhou: Zhejiang University Press, 2002.

To tell you the truth, when I first heard his argument that

"cellular neurobiology has flourished with the discovery of the role that neural action potentials play in communication among neurons ... the conception was replaced by the idea of action potentials as units of information, that is, as binary digits. That inference was incorrect, ... However, even though this new idea was wrong in brain science, it led to the birth of neural networks and digital computers. These devices have provided the technological foundation of international society in the Information Age, showing that a theory does not have to be correct to be fruitful." [1]

in his preface for the Chinese translation of his book *"Neurodynamics: An Exploration of Mesoscopic Brain Dynamics"* of which I am one of the translators, I could not realize the deep meaning at the first sight. Only after our long correspondence, did its meaning become clear to me, and not only limited to the special case of computers!

In your letter, you raised a very interesting and important open problem: Could an artificially intelligent machine do everything our brain can do? For me, in principle, I can't exclude such a possibility, as I have said above, our Mother Nature has already given "existence theorems" about all functions of our brain! Since brains are also physical systems, there is no reason to deny the possibility of building other systems which can carry out the same functions. Thus, it seems to me it is not a problem to predict that *"computers will outdo humans in terms of information processing in almost all kinds of applications someday"*. The problem is when? I agree with what you said e.g. that 20 to 50 years may not be the case. 100 to 500 years seems more practical, but I am not sure. It is too difficult for me to estimate!

As for the second question, I have to acknowledge that once I thought simulating our brain was the only way to do so. However, your arguments convincingly show this is not the case. Just as you said that *"And trying to reverse engineer a machine like the brain which you don't understand is about the most foolish thing you could do"*! And we have discussed the very different approaches adopted by engineers and our Mother

---

[1] Freeman W (2002) Preface for the Chinese version. In: Freeman W. Neurodynamics: An Exploration of Mesoscopic Brain Dynamics[M]. Hangzhou: Zhejiang University Press , 2002.

Nature.

By the way, I always don't understand what "downloading a mind to a machine" really means. Mind is not a state, but a variable dynamic process varying with the internal and external environment from time to time. Just as I argued in my previous letters on 2013/11/20(II - 001 Fanji) and 2013/12/20(II - 002 Fanji), mind, just as consciousness, is subjective and private. In my letter 2013/12/20, I argued that even using a cable or nerve tract to connect two brains, what the receiver brain feels or experiences is still its feeling or experience, not the sender's! Only if the two brains are exactly identical in every aspect and with all details, no matter how tiny it might be, and only the two brains can be connected point to point, not only for electrical activity, but also for chemicals, then the receiver brain may feel or experience what the sender feels or experiences exactly. However, in such case, the former is just the latter itself! If a wetware machine such as a brain cannot download another brain's mind, how can an artefact do it?

Let me come back to the problem about subjectivity and privacy. I thought they are at the core of the problem about consciousness, or even intelligence, mind, feeling and the like. No one of them can be defined only by the subject's behavior; there is something in the subject's inner world.

In science history, behaviorists tried to deny consciousness, they thought only the behavior could be studied, all brain functions are just a series of reflexes, but they failed, and a new discipline of cognitive science has been developed as one of the frontiers of modern science. In engineering and technology, it has also been tried to develop artificial counterparts based on the similarities of behavior. As a matter of fact, the famous Turing test is a test based on behavior. Although no machine has passed Turing test yet, could we say that a machine passing the Turing test really has intelligence? John Searle used his famous "Chinese room" thought experiment to give a negative answer. Jeff Hawkins said in his book "*On Intelligence*" :

"In my opinion, the most fundamental problem with most neural networks is a trait they share with AI programs. Both are fatally burdened by their focus on behavior. Whether they are calling these behaviors ' answers ' , ' patterns ' , or ' outputs ' , both AI and neural networks assume intelligence lies in the behavior that a program or a neural

network produces after processing a given input. The most important attribute of a computer program or a neural network is whether it gives the correct or desired output. As inspired by Alan Turing, intelligence equals behavior.

But intelligence is not just a matter of acting or behaving intelligently. Behavior is a manifestation of intelligence, but not the central characteristic or primary definition of being intelligent. A moment's reaction proves this: You can be intelligent just lying in the dark, thinking and understanding …" [1]

The above remarks denote that human intelligence has something which is lacking in artificial intelligence, artificial neural network, or artificial consciousness: the thing happened within the subject's head, which is subjective and private! Of course, from a point of view of functional application, maybe such a subjective and private aspect is not so important, but it would be a big problem from the viewpoint of ethics. The real difficulty is that we don't know how to explain how such subjective experience emerges from physical brain activities, or even how to define an inner world in our head, I think it is the real bottle neck both for cognitive science and artificial intelligence.

As for the problem *"which of the people interviewed by Blackmore come closest to your position and which ones you don't like or are skeptical about"*, it is difficult to give you a definite answer, as Blackmore asked many questions concerning many aspects of consciousness. I may agree with one point from A, and agree with another point from B. If I must select one viewpoint with which I agree most, then I would like to choose David Chalmers'. I agree with his following remarks:

"Subjective experience can't be reduced to a brain process … If it turns out that the facts about consciousness can't be derived from the fundamental physical properties we already have, like space and time and mass and charge, then the consistent thing to say is, 'OK, then consciousness isn't to be reduced. It's irreducible. It's fundamental. It's a basic feature of the world.'

So what we have to do when it comes to consciousness is admit it as a fundamental feature of the world — as irreducible as space and time. Then we need to look at the

---

[1] Hawkins J, Blakeslee S. On Intelligence[M]. New York: Levine Greenberg Literary Agency, Inc, 2004.

laws that govern it, at the connection between the first person data of subjective experience and the third person objective physical properties. Eventually we may come up with a set of fundamental laws governing that connection, which are akin to the simple fundamental laws that we find in physics." [1]

However, while I agree that consciousness is an irreducible and fundamental property of some complicated system, but it is not a property of every object. The open problem is; which system can have such a property? And under what conditions can such a property emerge from such systems?

Christof Koch, although having many views different from Chalmers, also declared in his classical book *"The Quest for Consciousness"* :

"I assume that the physical basis of consciousness is an emergent property of specific interactions among neurons and their elements. Although consciousness is fully compatible with the laws of physics, it is not feasible to predict or understand consciousness from these." [2]

Here, I want to emphasize that consciousness is a macroscopic emergent property of global brain activities, very likely chaotic. It is a property, instead of a thing! As an emergent property of some complex system, it cannot be reduced to the properties of its elements. As for the relationship between consciousness and brain activities, I prefer the word "emerge" to "generate", "give rise to" or "produce". For the former, it implies that consciousness is an emergent property of brain activities, it accompanies or goes along with proper brain activities simultaneously; while the latter may imply the two are different entities. For emergent properties, we rarely ask how it emerges from the activities of the elements at the next level, but ask under which conditions, the property emerges. Therefore, it seems to me, although the "hard problem" is hard, it is not a proper problem for scientific studies. A little similar to mathematical axioms, mathematicians never ask why the axiom is true or how the axiom comes. So, for me,

[1] Blackmore S. Conversations on consciousness[M]. Oxford: Oxford University Press, 2005.
[2] Koch C. The Quest for Consciousness: A Neurobiological Approach[M]. Greenwood Village, CO: Roberts & Company Publishers, 2004.

the question is under which conditions consciousness emerges? In addition, I would like to emphasize that the core of consciousness is subjectivity and privacy, which I assumed is an irreducible emergent property, which may be the core of our inner world. Up to now, all the available artefacts are lacking subjectivity, so they are not conscious or have real intelligence. However, there is also a big problem here: how we know they are lacking subjectivity, owing to the subjectivity and privacy themselves!

In the above, I have just said something about the subjectivity and privacy of consciousness. There are many other arguments I haven't discussed, such as, is consciousness global or localized? Can consciousness be measured? And so on and so forth. However, it would be too long to discuss all these topics in one letter. Following your example, I would like to discuss these topics in my later e-mails. Here I just want to say that I am very skeptical about the quantum theory of consciousness proposed by Stuart Hameroff, Roger Penrose, and others, as it seems to me that it is only a speculation lacking empirical evidence. But, maybe I am not in a position to criticize this theory, as I know little about quantum mechanics, which I only had a course in when I was a student more than half a century ago.

Thank you very much for telling me the news that Google has bought DeepMind. I did not know that. Very interesting, let's see what will happen.

Fanji

07. 07. 2014

Dear Fanji,

Thank for your kind e-mail and for liking some of my thoughts.

Your observation that people are apt to prejudice is very wise and should be obeyed more often. I have to confess, however, that as much as I admire this virtue I often fail to stay with it in everyday life. And while I have observed that I'm not the only one with this problem I have the suspicion that this has something to do with the fundamental principles in the architecture of our minds.

A similar and probably related phenomenon is the asymmetry in our ability to detect failures in our own and in other people's ideas and behaviors.

Why is it so much easier to detect such failures in others than in ourselves?

I have a hypothesis about this but would rather come back to this question later.

Please let me first make a comment on the point of Moore's Law which you have raised.

Of course, you are right with your critical question whether an empirically derived rule can be extrapolated infinitely. Of course, it cannot, just for mathematical and logical reasons! And, therefore, the original version of the " Law" will come to an end. However, this doesn't mean that the performance-growth of computer systems will slow down soon. On the contrary, it may even increase and accelerate dramatically. Please allow me to go into some technical details to explain these paradoxically sounding statements.

When Gordon Moore published his observation in 1965 that the number of objects on an integrated electronic circuit doubles every 2 years he also predicted that this pace may last for another decade. This prediction has come true and after the first microprocessors on silicon chips were introduced, it has become a tradition to use the number of transistors ("transistor count") per chip, as a measure of the advancement in chip technology. Since the mid-1970s, the number of transistors on integrated circuits went up from a few thousand to a few billion. The feature size on the semiconductor wafers produced in the fabs (fabrication plants) went down from 10,000 nanometers (Nm) to about 20 Nm with further reductions maybe down to 3 Nm until 2020.

This is pretty small even in the world of biology, where viruses are in the range between 15 and 400 Nm. How small this is may be better illustrated with the fact that only 30 hydrogen atoms fit into 3 Nm. So, this could be about the limit which is technically feasible in size because, at this level, the isolation between conducting parts may get critical because of quantum tunneling effects.

I wouldn't bet on this limit, however, because such barriers have often been declared in theory and engineers, nevertheless, found ways to overcome them.

But even if 3 Nm is the limit in feature size, it doesn't mean the end of Moore's Law.

Nor is it the clock-rate of the processor, the other physical parameter that determines the performance of a computer. It tells us how many cycles per second a CPU can do to work on a stack of tasks. This clock-rate went up from less than one Megahertz in the early Microprocessors to a couple of Gigahertz today. When I mention this physical unit "hertz" for cycle per second I'm not modest enough to hide the fact that it is named after Heinrich Hertz, who confirmed the existence of electromagnetic waves here in Karlsruhe in 1886. ☺

Raising the clock-rate is the easiest and cheapest way to increase the speed of a computer. Very much like raising the revs per minute of a combustion engine is the cheapest way to squeeze some extra-horsepower out of a motor. This is why it's the common way (called "overclocking") in which computer-gamers are tweaking their machines to ultimate performance.

It comes with a price however (in cars and in computers) and this is heat dissipation. Actually, heat is the biggest problem in modern chip technology and the major advantage biological data processing (at relatively cool 37 Celsius) has over computers.

While all conductors inside an IC have a resistance, all currents through them create thermal heat. This waste energy has to be transported away in order to prevent the Integrated Circuit (IC) from damage or from being melted down. It's all too easy, by the way, to damage an IC by overheating it and, therefore, the cooling system in a PC or in the racks of an IT-Center is a crucial device. And cooling is expensive, especially in those huge datacenters which e.g. handle internet-traffic and transport this email. The energy to cool down the computers can be more than what is needed to run them. And cooling is difficult when the surface of an IC is small and the CPU is a hot spot almost gleaming red. When feature size goes down on an IC, thermal efficiency gets better and it may allow the engineers to squeeze out some more Gigahertz but the possibility is quite limited.

Over the years chip-engineers have been unbelievably creative in finding ways to tweak extra performance out of the good old von Neumann machine which dates back to the late 1940s. Just as automotive engineers have repeatedly found ways to improve the performance and efficiency of internal combustion engines whose concept is more than 100 years old.

And so we still have von Neumann machines on most modern computer-boards and chips are still made of silicon wafers (or maybe of Gallium-Arsenide based on similar processes). This isn't necessarily so because alternative concepts exist for both the architecture and the materials used. The reason why the industry is sticking with these old technologies is that it is cheaper to improve a well-known technology than to move to new platforms.

There is still a lot of room for improvement on a small scale, as predicted by Moore's law without leaving the silicon— and von Neumann-turf.

Putting more than one CPU on a chip as we see it in the smartphone ARM processors with many cores, is the next step. Integrating more devices into one chip and building complex data processing centers on one chip as we see it in the SoC (System on a Chip) concept is a further step.

You gain a lot of speed (and safe power) when you transport data from one IC to another on one chip instead of transporting the data between separate ICs by wires.

And in a 3 Nm world you can pack an incredible amount of transistors and functional units on an SoC. And you are not limited to transistors because you can also build mechanical, fluidic, or pneumatic structures on silicon.

And when you run out of space even in a 3 Nm flat-world, you can go 3D and build cubical SoCs. Just using the third dimension which means stacking multiple layers of ICs on top of each other will allow many more Moore's cycles.

Using light instead of electricity to communicate on the chip, between the stacks and to outside devices is another possibility to increase speed and safe power.

And again, this is all possible without leaving the well-known silicon world. Much quicker and less energy consuming materials are known and we are still not talking about real revolutions like light computers or quantum computing.

Of course, you are right with your skeptical question about the limited timeframe within which Moore's law can hold true. It can't be in place for ever because the effect of exponential growth by factor two will eat up all those means some day. Very much like in the old story of Sissa ibn Dahir who played chess against Indian sovereign Shihram and who suggested a seemingly tiny prize. In case he won he asked for one grain of rice on the first field of the chess board, two on the second field, four on the third and so on for all 64 fields.

Well, since then, not only mathematicians, but also sovereigns know that 2 to the power of 64 (minus 1 — of course ☺) is a pretty big number.

In the case of Moore's law, the limit might be roughly guessed just by calculating how many times the number of transistors in a modern computer can be doubled until we reach 10 to the power of 80, the estimated number of atoms in the known universe. The result is the number of Moore's cycles needed until all atoms in the universe have to be transistors of a super computer.

As crazy as this idea sounds it isn't crazy enough that a creative inventor couldn't have already had it. Actually, it was Konrad Zuse, the computer pioneer, who in 1945 had the idea that the whole universe is nothing but a huge computer.[1]

A dyed in the wool fanatic of Moore's Law probably wouldn't give up when confronted

---

[1] https://en.wikipedia.org/wiki/Calculating_Space

with the argument that one day each atom in the universe has to be a transistor if the Law is right and then finally has to come to an end. He or she may refer to Hugh Everett's many-worlds interpretation of quantum physics and insist that all these worlds could be used for computing. ☺

But when we come back from such cosmological dimensions to our near future of, let's say, the next 25 years there is no reason why computer power couldn't be improved at the same speed as we have seen in the last 50 years.

The actual pace, however, is less determined by what is technically feasible but rather by what increase we can afford, i.e. what we are willing or able to pay for it.

Moore's law, in the past, was no prediction of the possible speed in which computer technologies could be improved. It was the engineer's roadmap where they promised their colleagues, in the marketing department, the pace at which they could provide new products with enough power (and subsequently more features) to make a customer buy them.

In the old days, when IBM controlled the market for data centers almost like a monopolist, it was a few thousand data centers in the world that had to pay the cost for the next generation of computers. That base broadened when Microsoft and Intel, in the PC revolution, started to take the lead from IBM and soon it was hundreds of millions of users who paid the cost for the next generation of fabs. Then it was the internet and Google, Facebook, Baidu etc. and all their huge data centers which made the demand for computer power explode; very often making these data centers the largest power consumers in their areas. And finally, it is now the owners of smartphones, soon to be counted by the billion, which make their contribution.

And they do this, often, every other year when the manufactures show up with their new models. This happens very much in sync with Moore's law and it's not a coincidence, but because it is the plan set out in the roadmap of the industry.

In the days of the cold war and the prestigious space projects, it was the government that paid the bill from their military and research budgets.

Today, it is the consumer who pays for better computer power, AI, and all kinds of supportive technologies around it.

The reason why the rate of improvement isn't higher and beyond Moore's prediction is based on the circumstance that the required investments are huge.

The costs for a new semiconductor factory are in the billions of Dollars. TSMC (in Taiwan, China) has already announced in 2010 that it will invest close to $ 10 billion in the new 300 mm fab called FAB 15.

Now we are heading towards 450 mm wafer sizes and 3 Nm structures and this requires even more money.

By the way, to me, it looks like TSMC is going to outperform former leader, Intel in fab-technology and I wouldn't be surprised if they will be the first to enter the 3 Nm territory.

Asia, in general, and especially China became a heavyweight in this domain after the US had chosen to make it their factory of modern computer technology. For a long time, the American big brands like Apple believed that they were able to control the market by the sheer power of marketing and brand recognition very much as IBM was able to control the market of data centers for 40 years.

However, this is not a given anymore. It is amazing how quickly Korean and Chinese manufacturers have learned not only to produce components for American brands, but also to create their own products and brands. And this also applies to the related software technologies. Baidu e.g. is on par with Google in searching technologies since long and so is Alibaba compared to Amazon. And I'm sure the same is true when it comes to the use of AI.

When my guess is correct that the new generations of IT will be paid for by the consumer, then those players with a large consumer base have a substantial advantage.

Take this all together and you can understand the wakeup call Chris Andersons is giving to the American industry in his book "Makers: The New Industrial Revolution" which I mentioned the other day. The message is quite simply that it was problematic for the US industry to make Asia and, especially, China the factory of the world for the latest generation of computer technologies.

AI is a driving force of the technology process in the next round from autonomous cars to speech recognition, robots and factory automation; and AI eats computer-power like no other application.

So, there may be a need to accelerate the process beyond the pace of Moore's law.

Oh my dear Fanji, again you have tempted me to deviate and to make a far too long excursion with just one simple question. ☺☹

But it's not your fault but only mine and I can only hope that you didn't find those technicalities too boring.

But although this email is already long enough let me please tell you how much I appreciated your comments on Walter Freeman's words about digital computers "*starting from a mistaken view of how neurons work*".

I didn't think that I was the first one to find this truth but I'm especially happy that it is Walter Freeman who agrees with it.

I generally find Freeman's view very refreshing. Also, his perspective that information processing in computers is different than what happens in our mind which you also cited.

I was as perplexed as you were when I read this — really impressive!

I found a more intense explanation of the same thought in his book "How Brains make up their Minds" in chapter "Meaning and Representation" (from P. 13).

And here he comes to the conclusion that the whole idea of the brain as an information processing machine is wrong and misleading. He says (again) that this idea was imported by computer people and logicians and that the discipline of mind-research is confused since this wrong perspective became the paradigm.

I had to read the related pages a few times before I grasped what he meant.

But I think he's right, at least with the finding that this perspective was imported from outside.

I'm not sure whether his alternative to "information" as the mind's currency which he postulates as "meaning" is the ultimate solution to the problem. But in any event, I can say that your friend Walter has changed my point of view in a field where I believed to be on safe ground.

Such insight doesn't happen very often and, therefore, I'm grateful one more time for this great hint.

It's a pity that Walter was not among the researchers interviewed by Susan Blackmore.

His writing isn't easy to read because it is full of real content and he's unusually independent and original with his very own world view. But to discover his world is a sheer delight and he's, for sure, not one of these loud tooters and pretenders!

When you say that you don't understand what *"downloading a mind to machine"* really means, I believe that Kurzweil simply means what you can see in many sci-fi-movies. There it is no big deal to copy a human mind into the memory of another human or a machine. So, he probably believes that just everything can be copied, including subjective feelings.

I don't think that he cares about the, so called, "zombie problem" as long as the individual behaves as the original. I have to admit that I also believed in this possibility until I learned more about the structure of the brain and its embeddedness in the rest of the body.

This question of course is highly related with your interest in subjectiveness and qualia.

When I had asked you who among the researchers, interviewed by Blackmore, comes closest to your own perspective I had hoped that you would name Chalmers and I'm happy that you did.

Not because I liked his position best among the many alternatives offered, but because I found his position hard to understand. And while I want to learn from you who is dealing with these questions longer than I have I hope that you can make it more understandable to me.

Maybe you can start with arguments Chalmers would probably use against the heavy critique Patricia and Paul Churchland expressed against him. Patricia is especially very harsh and even denies that Chalmers' famous differentiation between the hard and the easy problem of consciousness makes any sense. The same might be interesting for Daniel Dennett's arguments which deny that there are any qualia at all and who seems to take himself as the know-all in the business. In principle, I like his position because it comes close to my own, maybe naive, starting point when I look at the problem.

I read one of his books and although I like his general approach, what disturbs me is the high amount of self-confidence (or should I say Kurzweilness?) in his writing and that he needs so many repetitions and so many words to express what he means. He's smart, for sure, but I prefer Freeman's style.

But what I say here is more emotional, irrational, and subjective in the true sense and maybe I'm totally wrong and haven't grasped the real thing and what qualia are.

But you will tell me!

In any event, I will try to do my best to stay with the principle of not to becoming a victim of my prejudices. ☺

Best for today.

Karl

Dear Karl,

Thank you very much for your deep analysis of the problem related to Moore's law. Especially, thank you for giving me a detailed technical background of the problem, of which, I know just as little as most of the general public. You showed me many ways to speed up computation which I did not know before, and several of which are just in their ramp-up phase! Therefore, although the problem is simple for you, it is complicated for me and maybe others. Your detailed explanation is just what we need!

After reading your arguments, I have to admit that my previous argument is somewhat book-worm like. Although my argument may be correct, in principle, on an infinite time scale, however, it may make no sense for the technical progress in the foreseeable future. More practically, just as you said in your last letter:

"But when we come back from such cosmological dimensions to our near future of, let's say, the next 25 years there is no reason why computer power couldn't be improved at the same speed as we have seen in the last 50 years."

I must learn a lesson from that.

As for the problem of "downloading a mind to machine", I think that the main problem is we don't even know what the mind is in a clear way. Just as your remark in your letter on 20/03/2014 (II-003 Karl): *"how can you simulate something on a computer which you don't understand in the real world?"*, if we don't know what the mind is exactly, how can we copy it to a machine? If the mind was just special verbal thought, then downloading it is possible, you could even consider this letter as a copy of what I am thinking now — some downloading from my mind! However, mind is not only memory or thought, it is not static, it is always changing. It can be emotional, and something that isn't able to be expressed with words. It seems to me, mind is emerged from the global brain activities interacting with its external and internal environment, how could you copy all of these factors to a machine?

As for the debate between Chalmers and Churchlands/Dennett, Chalmers' expression which I cited in my last letter is closest to my idea, although I can't say that I agree with his every point, and I suppose that his idea may also change with time for some points. Therefore, I will not defend his every point. Of course, I can't expect all my ideas about consciousness to be correct, just as you warned me in your letter on 15/01/2014(II - 002 Karl): *"But please don't expect too much and don't forget that you are dealing with the matter for many years!"* I think this is not only true for me, but also for Chalmers, the Churchland couple, Dennett, and others. In addition, I have to admit that I haven't read every book of theirs, what I learn about their ideas is mainly from Blackmore's interviews, thus, maybe I misunderstood their arguments at some points. Please don't hesitate to denote my mistakes when you find them.

When Blackmore asked Patricia and Paul Churchland why the problem of consciousness is so special (difficult), they denied the problem is special. Paul said that *"I discovered that the kind of intellectual befuddlement we feel now when we look at consciousness or qualia is by no means a new thing."* Patricia added *"lots of people in the early stages of any scientific theory are very surprised."* They thought if people knew enough about the brain, the problem would be solved naturally. They think that many "puzzles" have been solved in science history, these puzzles seemed very hard before they were solved just as consciousness today. From their point of view, consciousness and brain activities are just the two sides of the same coin. It seems to me that they haven't demolished Chalmers' main argument: consciousness is subjective, while the others to be studied in science history are objective. To say consciousness is identical to brain activities is not so convincing.

I have to admit that I know little about Daniel Dennett's view, I have even not read his *"Consciousness Explained"*, what I know about him is only from Blackmore's interview. From her interview, I agreed with Dennett's following statements:

"I think the reason that we find consciousness so hard is that we have evolved a certain capacity for self-knowledge, a certain access to ourselves which gives us subjective

experience— which gives us a way of looking out at the world from where we are. And this just turns out to be very hard to understand." "let's see if we can figure out what the sufficient conditions are in purely material terms for there to be something that it is like something to be; something that has an inside; something that has a subjective point of view." [1]

However, to find such "sufficient conditions" is really hard. Within my knowledge, all the conditions scientists, including Gerald M. Edelman and Giulio Tononi, have figured out are only necessary, but not sufficient! I think that I should read " *Consciousness Explained*" someday so that I can discuss him with you better.

Now let me come to discuss David Chalmers' point of view. As I wrote above, that his words cited in my last letter inspired me a lot. I also agree with his remark: " *The heart of the science of consciousness is trying to understand the first person perspective* (subjectivity — citer)." However, I don't completely agree with him on his following statement:

"one of the crucial questions in this field, ' How are we going to be able to explain subjective experiences in terms of the objective processes which are familiar from science? How do 100 billion neurons interacting in the brain somehow come together to produce this experience of a conscious mind with all its wonderful images and sounds?' " [2]

If the last question is his "hard problem", then I have to say that this is not a proper problem to ask. It sounds as if consciousness is one thing, and brain activity is another thing, and even the latter "produces" the former! For me, the former is an emergent property of the latter. As all other emergent properties of a hierarchical nonlinear system, you cannot ask how the emergent property is emerged from the system. The proper question is under which condition such property emerges? Therefore, I agree with his words which I cited in my last letter. Subjectivity is an unreducible basic emergent property of some kind of brain activities. Just as Chalmers said that

[1] Blackmore S. Conversations on consciousness[M]. Oxford: Oxford University Press, 2005.
[2] Blackmore S. Conversations on consciousness[M]. Oxford: Oxford University Press, 2005.

consciousness "*is a basic feature of*" some kind of brain activities, but not "*of the world*" as he said. The proper question is: What kind of brain activities consciousness is emerged from? This problem is really hard. The final answer to this question should give a necessary and sufficient condition, which no one knows yet. Therefore, although subjectivity is a fundamental feature, not every system has such property, even not every brain activity has such property! This is my view inspired by Chalmers and others (even some of his opponents). Of course, just as I cited from your words, I couldn't expect that my view would solve the dispute. I am expecting your comments and even criticism.

Maybe you already know about the open message to the European Commission concerning the Human Brain Project undersigned by 156 neuroscientists on the 7th of last month [1], in which they criticized that the HBP focused on an overly narrow approach, leading to a significant risk that it would fail to meet its goals. They especially criticized that the HBP removed an entire neuroscience subproject — cognitive neuroscience, the only one using a top-down approach. They appealed that

"*the HBP is not on course and that the European Commission must take a very careful look at both the science and the management of the HBP before it is renewed. We strongly question whether the goals and implementation of the HBP are adequate to form the nucleus of the collaborative effort in Europe that will further our understanding of the brain.*" [2]

They requested that the coming review should be transparent, independent, and could reflect the diversity of approaches within neuroscience. Based on the review, they asked that the HBP should have a radical reform both on its academic goal and management, otherwise they called for the European Commission and Member States to reallocate the funding currently allocated to the HBP core and partnering projects to broad neuroscience-directed funding to meet the original goals of the HBP — understanding brain function and its effect on society. In the event that the European Commission is

---

[1] http://www.neurofuture.eu/
[2] Open message to the European Commission concerning the Human Brain Project (http://www.neurofuture. eu/).

unable to adopt these recommendations, they would urge scientists to boycott the HBP.

A very interesting and serious event! Let's see what will happen next!

Best wishes.

Fanji

P.S. In my letter on 12/10/2013 (I −011 Fanji) , I mentioned that a working group of the Advisory Committee to the Director (ACD), led by neuroscientists Cornelia Bargmann and William Newsome, was established in NIH, which would propose a report to lay down the scientific goal of the US BRAIN initiative to the head of NIH. Now I heard that ACD's proposal has been sent to the head of NIH, and was accepted on 05/06/2014. It is said that it proposed that the project should be supported with 4.5 billion US dollars in 10 years from 2016. It is suggested that in the first 5 years, it focus on developing new technologies, and in the last 5 years the technologies will be applied to brain research. The goal sounds not so ambitious as the HBP and more practical, however, I haven't read related material with details, so I cannot give my comments at the moment. Maybe we can discuss later.

30. 08. 2014

Dear Fanji,

Huh, that's interesting news. The HBP is under fire and it looks like there is trouble ahead for our friend Markram. From what I can tell from a distance, however, it looks more like an action motivated by jealousy or injured vanity than by serious doubts about the general concept like the ones you and I share.

We know that Markram is not very sensible when it comes to dealing with other people. But orchestrating such a monster project needs a very high degree of empathy and very gentle manners indeed. Not exactly his sweet spots I believe.

Well, we'll see how he counters this attack.

But let me come to your arguments from your last and your previous email.

Now we are approaching the core of the problem of consciousness and meanwhile I begin to better understand what your major concern is with subjectivity and privacy.

I agree that understanding consciousness is very difficult, if not impossible, just because the representations of external sensations are " stored " in a different way in each brain. And I also agree that on top of subjectiveness we are facing the delicate problem that the " *mind is not only memory or thought, it is not static* " as you have emphasized.

Yes, you are right!! We are talking about a continuous and highly dynamic process and a temporal snapshot wouldn't help much. This is a very strong argument indeed which most naive "mind downloading" fans overlook!

On top comes the problem that the brain can't be seen as an isolated information processing machine because it is always embedded in the biochemistry of the rest of the body and is continuously bombarded by internal and external sensory signals.

The more I learn about this, the more it gets clear that the attempt to catch a mind this way is as illusory as trying to catch a rainbow and store it in a box.

If you really want a copy of your mind it would be necessary to copy the entire body atom by atom. And even this wouldn't be enough because in order to catch the dynamic changes of e.g. the travelling fluids, action potentials, and electromagnetic waves you need the additional information for each atom and maybe even for each electron in which direction and at what speed they are moving in the very moment. And you need all these data instantaneously!

A related scan performed by a super-metrology device, which might be feasible in principle, would only be useful if it would supply the information for all objects at the same space-time.

So, this scanning machine either needs as many sensors as objects have to be scanned, work at a speed faster than light, or if the scan is done sequentially at a slower speed the positions of the objects already scanned have to be calculated to determine their correct place, orientation, and speed when they are composed together.

Not an easy task and you don't need to refer to Heisenberg's principle of uncertainty to realize that such attempts are an illusion.

They may be good enough to serve as a thought experiment however. At least in science fiction movies the trick is no big deal because when the crew of space ship Enterprise goes to a planet via "beaming" they use exactly this method.

The body of a person is scanned, dematerialized, and transported to another place, rematerialized and off she goes — no problem. They not only do it all the time but also enjoy the comfort of storing copies of a person which is very useful when someone was killed because people can be recreated as perfect copies that way, including qualia and everything. As I said, Kurzweil must have read the same sci-fi books as I did as a boy because he still thinks that such things will be possible soon.

Beaming via dematerialization avoids the conflict of multiple copies of an individual but raises another problem much severer than the ones mentioned above. Scanning an object down to the level of electrons is difficult, but beaming an average body, means transforming 70 kg into radiation before it is rematerialized in the next place. This means the release of energy equivalent to, quite a few, atomic bombs. Well, to be fair with Ray Kurzweil, he didn't promise beaming on top of mind uploading. But I wouldn't be surprised if he would. ☺

But let me come back to the thought experiment and the key question I want to ask you. If we would have a scanning machine as described above that allows us to make a perfect 1 : 1 copy of a human body including the brain, would this identical copy contain the same conscious and qualia as the original?

Chalmers would probably say "no", while Dennett would certainly go for "yes, of course".

What do you think?

I'm on the "probably yes" side while I'm not sure whether I could agree with Dennett when he goes even a step further and says that consciousness is an illusion that somehow develops automatically when a brain gets complicated enough.

I'm with you when you make the point that, with consciousness, we are confronted with an emerging phenomenon that cannot easily be derived from the underlying physical-chemical substrate. But this is true for many emerging phenomena in the physical world such as all these puzzling phase transitions where matter shows totally different properties from one moment to the next. Liquid water that freezes to ice is probably the most common and best-known example.

But, although this freezing phenomenon was very hard to explain theoretically and needed deep understanding of quantum physics, it could ultimately be achieved by means of rational physics.

I like what Christof Koch says about emergence in the passage you have cited. He argues in the tradition of Nobel laureate Robert Laughlin who makes the point that physicists should generally take more care of emerging phenomena and phase transitions because there they may find more interesting things than in the "comfort zone" of continuously differentiable equations between these enigmatic singularities. Ilja Prigogine, also a Nobelist, highlights the importance of those amazing and thrilling zones far away from the equilibrium where even the most solid law of all, the second law of thermodynamics, seems to be out of place on the micro level. The world of self-organization of matter is full of such disturbing phenomena of emergence but Laughlin and Prigogine see no need to do what many people do when they refer to emergence, which is, to give up the method of rational physics and to retire into the cloudy world of metaphysics.

When it comes to Chalmers' position on the question of whether consciousness, including our subjective feelings and qualia, whatever that may be, is entangled with physical matter or a separate entity I guess that his interpretation of emergence is more strict than yours. And I suppose he will insist that it is not connected with the material world but that it is a separate and independent entity.

What I do not like in Chalmers' position is the metaphysical touch in his position which reminds me of the discussion of "vitalism" in the 19th century and of "élan vital" in the 20th century.

The history of the natural sciences is a history of continuous disappointment for metaphysicians.

Every time they postulated a limit which essentially separated the world of dead matter from the world of living matter not accessible by the rational means of natural sciences this limit had to be moved back sooner or later. Vitalism, the belief that the organic world is fundamentally different because it contains mysterious non-physical elements, suffered a terrible hit when Friedrich Wöhler managed to synthesize the first organic molecule, urea, from inorganic matter in 1828.

And when Miller and Urey succeeded in creating an environment in 1959 where amino acids emerged spontaneously, this seemed to be the final blow to traditional vitalism.

However, the idea occasionally comes back like a recurrent theme or "Leitmotif" also in relationship with consciousness. Dennett says that this is because many people find it just uncomfortable to live in a mere materialistic world and that there is a deep-rooted desire that there should be more than mere matter. Something like the essence of life that makes the difference between the rocks, the animals, and us humans — something that makes us special.

In Europe, this idea is highly related with the tradition of the scholastic university dominated by the belief system of the church for a long time. And the priesthood always insisted that there are other things beyond mere matter, something like the soul, totally inaccessible to rational reasoning and the methods of the natural sciences.

But I think that there is a similar idea deeply rooted in Asian, and especially, Chinese philosophy.

I don't know how far the concept of a "living force" in Qi is related to the idea of vitalism and whether those concepts have influenced each other. But I'm sure you will know much better.

I understand that you are not agreeing with Chalmers on everything he says.

And before I start to build my own man of straw or a metaphysics-dummy of your position just to be able to attack it more easily, I would rather ask you to answer my physics-metaphysics question in the Chalmers-Dennett case mentioned above.

Maybe you have a third alternative or can offer me a more helpful perspective.

In any event, I look forward to hearing your opinion!

Best for today.

Karl

28. 09. 2014

Dear Karl,

You are absolutely right; the core problem of consciousness puzzling me is its subjectivity and privacy!

I think that we share the consensus that mind-uploading is almost impossible, at least practically.

As for your thought experiment, my opinion is, that if your scanning machine can really make a perfect 1 : 1 copy of a human body including the brain, down to every atom, and if all the atoms of the same isotope are identical (I don't know, however, I don't have any reason to oppose), then my answer is "yes" to your question: "*would this identical copy contain the same conscious and qualia as the original?*" My logic is, that as consciousness is an emergent property of neural activities, if all the processes of neural activities are identical, then their emergent properties have to be identical. However, there is still a problem. As I emphasized in my last letter, mind is dynamic, it is changing all the time, depending on the interaction between the subject and its environment. Therefore, even if your scanning machine can make a perfect 1 : 1 copy of a human body down to every atom, to make it have the same consciousness and qualia as the original, you have to place the copy at the same position of the original (however, it is impossible for two different bodies to occupy the same spatial position), otherwise you have to make a perfect 1 : 1 copy of all the changing external stimuli to the copy too, i.e., you have to make a perfect 1 : 1 copy of its external environment too. Therefore, it is impossible that there are two identical minds, or to download a mind into another system practically.

I have to admit that there is something ambiguous in my expression about emergence and phase transition. You correctly pointed out that the emerging phenomenon "*cannot easily be derived from the underlying physical-chemical substrate.*" You put an important word "easily" here, which implied that some emerging phenomenon may be derived from the properties of and interactions between its elements. Therefore, to say consciousness is an emergent property of neural activities does not mean that it is impossible to show how it emerges from neural activities. Of course, it also does not mean that consciousness must be able to be explained by neural activities. I suppose that Chalmers' following argument may still be correct here with some modification: consciousness is an irreducible basic property of the world. However, it is only a property of some kind of complicated systems such as an awake human brain, but not of everything. I don't think that consciousness is an irreducible basic feature of the world! The essential problem is; what kind of systems should it be which could have consciousness? If we know some kind of agents can be conscious, such as living human beings, another essential point is to give a sufficient and necessary condition for neural activities, under which subjectivity emerges. If we knew such condition, then "the hard problem" would vanish for human agents. However, no one knows the answer yet. And I doubt if people can find such conditions universal to all kinds of agents. Subjectivity makes all such studies extremely difficult, if not impossible, as you cannot prove if a system with complicated behavior is conscious, no matter how much it appears to be conscious, because consciousness is subjective and private, you cannot share it! Maybe it is only a zombie!

Your argument to compare the debate about consciousness with the one about life is strong. Chalmers and his supporters argued that consciousness is unique, owing to its subjectivity, which scientists have never faced, so the victories of science in other fields could not mean that the puzzle of consciousness must be solved at last as the other "puzzles". Your argument is that life was also unique in the history of science, but now it is solved. That's right. "Uniqueness" could not be the very reason for a problem which could not be solved. However, subjectivity may be the reason, maybe it is irreducible. At last there must be something irreducible, just like axioms in mathematics. Except for acknowledging the fact, you could do nothing! Owing to the subjectivity, you cannot, rigorously, judge if another agent is conscious or not. For human beings, because all have nervous systems of almost the same structure, maybe we can take the following as an axiom: if other people can behave just like conscious me, then he or she must be conscious. Higher animals have brains, maybe we can also adopt a similar axiom, maybe there is not much difficulty to admitting that chimpanzees have consciousness, or even dogs, cats, and parrots, but what about the lower animals? Do you believe a worm is conscious? And what about the machine? Owing to the subjectivity and privacy of the consciousness, we cannot know the subjective experience in any other agent, what we can do is only observe its behavior. However, just as John Searle argued in his famous "Chinese room" thought experiment, you cannot judge if an agent is conscious or not just based on its behavior, except for having some axiom, say, human beings having behavior similar to me is conscious. Could you take the Turing test as an axiom? Searle's "Chinese room" thought experiment seems to say "No". Even so, such "axiom" is rather ambiguous, what does "having behavior similar to me" mean exactly? Different nations have different traditions, they may behave differently under similar circumstances, can I deny a foreigner with different behavior from me, in some cases, having consciousness? A kid has many behaviors different from me, could I deny that it is conscious? And so on and so forth. The answer, obviously, should be no.

In my opinion, at the present moment, we can study the necessary conditions, under which a human agent is conscious. Recently, there have been many interesting studies on "neuronal correlates of consciousness" (NCC) and "signature of conscious access", they have made big step towards this goal. However, these studies mostly focused on some aspects of consciousness — some special contents of consciousness, or conscious access — a special aspect of consciousness. As for the concept of conscious access, I will discuss in a future letter, otherwise the letter would be too long! Maybe someday, when the data from the above studies accumulates richly enough, people will be able to find the sufficient and necessary conditions for a human being's consciousness, as you and some other scientists expect, and the "knot of the world" would be solved. If the sufficient and necessary conditions are known and fulfilled for a human subject, consciousness emerges naturally, and the hard problem may vanish. I could not say this is impossible, but for me, this is very unlikely in the foreseeable future. The real problem lies on how you can know the agent other than human beings has subjective experience! To find the sufficient and necessary conditions for being conscious of any kind of agents is really hard!

I feel ashamed to say that I have never known what the concept "Qi" really means, it is rather ambiguous. In ancient China, people thought "Qi" was one of the basic elements of everything, which, similar to gas, flows everywhere. It also flows in the body to guard the body and gives energy support. Of course, now we know there is no such Qi in the body.

Talk to you later.

Fanji

Dear Fanji,

Your consciousness puzzle and the subjectivity and privacy problem are quite inviting and inspiring, and I have to admit that your argumentation is very smart. And again, I owe you quite a few hours of wonderful reading and exploring new and old ideas in order to gain an overview of the various positions regarding the mystery of the mind. Of course, I'm not done with it, but here is the preliminary result.

I like your pragmatic interpretation of the problem but not Chalmers' position. To me, he's trying to protect his theory by immunizing it in a well-known way. When he defines subjectivity as something to which we have no access, any attempt is in vain to test the theory. When you do this the argumentation gets circular. You can't prove subjectivity because subjectivity is unprovable. The same works with soul, Qi, Psi, or any kind of phantasy-invention.

I'll come back to this but let me first come to your arguments about my thought experiment on the perfect copy of a body including the brain.

I'm happy when you say that a 1 : 1 copy of a body would also contain consciousness and related qualia. However, it could be difficult, to impossible, to create such a copy. Technically, you are, absolutely, right when you insist that the copy of the body has to be put exactly at the same place in time and space as the original, because of the dynamic of the processes and the continuous flow of information from the sensors to the brain. I was thinking about the problem of dynamic and, therefore, asked that not only all the atoms are in the right place but also have to have the correct speed and orientation in space. If you want to be very accurate, you also have to take care of the spin of the electrons and all the subatomic particles to guarantee that all flowing fluids and travelling action potentials and the related electromagnetic fields are represented in their dynamic and everything is in the correct phase.

So, these requirements make a copy-machine already practicably unfeasible in the real world, but your correct argument that we would have to place the copy at exactly the same point in space-time as the original is a much harder requirement. To save the thought experiment, just for the sake of continuing the discussion, I might only resort to Everett's idea of multi-universes in quantum mechanics. So, we would just create a new universe in which the copy exists. This is not as crazy as it might sound at first sight, at least not for Dieter Zeh, famous for his concept of "Decoherence" in quantum mechanics. He has good arguments why such parallel universes are created anyway all the time. This is an unusual perspective but if you also have ever felt a little uncomfortable with the Copenhagen interpretation of quantum theory and its mysterious "complementarity," Zeh might offer you an alternative. I'm not sure whether Einstein, Schrödinger, and de Broglie would be happy with it.

At least I was, because I truly shared the discomfort of these three founders of modern physic about this cloudy "complementarity". They always refused to become believers of the "Bohr-Church", very much to the displeasure of Niels Bohr and his flock. However, Zeh's liberation from complementarity comes at the cost that you have to accept Hugh Everett's many-worlds concept. And yes, it also contains the idea of many-minds.[1]

But, in Zeh's interpretation it's not more absurd than complementarity, and after I have dealt with the subject for a while I would prefer to live in a Zeh-Everett world rather than in a Bohr-Heisenberg world. But this is a different cup of tea and I don't want to try to escape your arguments through a smokescreen.

So, I'm happy to agree that your argument is very strong and you win this round. ☺

A more pragmatic argument might be, however, to ask how precise and detailed a copy of a brain has to be to convey subjectivity and consciousness. From the material world that is surrounding us, we know the phenomenon that physical objects can be very robust on a macro-level although they are composed of those highly dynamic objects that are so difficult to measure and localize on the micro level. The moon as a macro-object orbiting the earth can be localized and measured with incredible precision although it

---

[1] https://en.wikipedia.org/wiki/H._Dieter_Zeh

consists of those micro objects, all poisoned with uncertainty, when you want to measure them.

On the other hand, we have to ask how much we lose when we are not that accurate. And how robust are subjective feelings and qualia against transformations? Do they substantially change when you stand up and when you walk, run, or do a somersault? Maybe some of them do, but others don't, and at least you will be able to detect it. You may even be able to tell me what the effect was. It may be very similar to people having been able to describe the effects of drugs that, for sure, can change perception and consciousness.

Some parts of your qualia may be very robust against most kinds of mechanical changes or chemical influences while others are less stable. What I'm out for, is the point that we may get access, at least to some part of your subjective consciousness, by means of detecting changes in it. It's not trivial but it's not impossible, in principle, and if researchers try hard enough we may see experiments able to chase the mysterious thing. Sometimes, it takes a long time and huge efforts to find empirical evidence for a theoretically predicted phenomenon. When Einstein predicted gravitational waves in 1916 he was convinced that they exist because it was a consequence of his general theory of relativity. But at the time, there was no technology available to detect the related extremely week signals and Einstein himself was skeptical whether it would ever be possible. We still have no proof of the existence of gravitational waves, but the theory tells us what we have to look for and what effects we have to expect. Instruments with the necessary resolution are in progress and there is hardly a physicist who doesn't expect us to see the evidence soon.[1]

What I would like to ask Chalmers is what should happen in an experiment so that he's willing to give up his dualistic position. If he says that this cannot happen because subjectivity is inaccessible by definition, his theory is worth nothing and is more like a religious dogma. With the same method you can defend the theory that consciousness is controlled by anything arbitrary, e.g. something as absurd as a flying spaghetti monster

---

[1] This was the case when Karl wrote the letter. However, LIGO observed gravitational waves produced by a fusion of two black holes in September 14th, 2015. Three scientists were awarded the 2017 Nobel Prize in Physics for this discovery. Since then several gravitational wave occurrences have been observed.

(FSM). FSM is the god worshiped by the church of Pastafarianism, an organization founded to demonstrate the absurdity of the belief-system in many religions, especially those who insist in intelligent design.[1]

It's a fun story, of course, but not everybody can laugh about it. I hope you can. ☺

So, I understand that you do not insist in a metaphysical position for consciousness, but I'm not sure what Chalmers would say.

I'm not sure either, however, whether Dennett is right and everything about subjectiveness is just a pseudo-problem. We may really need a new kind of science beyond physics and chemistry to get access to consciousness and subjectivity. But, before I'm willing to surrender and opt for metaphysics, I would rather try everything possible in the realm of physics and rational methods.

Your reference to the new research discipline of "neural correlates of consciousness" (NCC) is very helpful because it demonstrates that there are people on the way to attack the mysterious phenomenon. Thank you for giving me advice on how to better support my own position — you are really a good sport! Actually, I wasn't aware that this discipline was already quite established.

They seem to follow John Searle's advice who also takes a very pragmatic position, similar to my own, in his interview with Susan Blackmore. He states that it's wrong to think that consciousness cannot be objectively researched and that it is possible to do objective scientific research on an ontologically subjective sector and that we just have to be creative and try hard enough. I liked his answer to the question, similar to yours, whether he believes animals and which animals have consciousness. He says that we don't know whether termites are conscious, but "my guess is they are". And I also liked his answers to the following questions about what the experts could possibly tell us. He said that in case we find out that very specific processes xyz in the brain are responsible for consciousness — in a way that it would be possible to give back consciousness to patients with specific brain damages — we may check the brains of dogs, cats, and primates for xyz. We may find that they also have xyz and, therefore, must be conscious. Then he says we have to go deeper on the phylogenetical scale and

---

[1] https://en.wikipedia.org/wiki/Flying_Spaghetti_Monster

finally may find that termites have xyz but not snails.

Besides the problem that it is unclear what xyz might be, I find this a nice outlay for a research program of NCC.

Another empirical way to approach the subjective part of consciousness may be the way of communication. Although we can never be sure that the red you see is the red I see, we can learn to get closer to another person's impressions. I can supply a personal experience to this. I have a total addiction to the color blue. Especially to a kind of blue like the Bugatti race cars had, and later, Triumph TR sports cars (in a little brighter version) under the name of "pageant blue". All my friends know about my special addiction to blue and Adi of course knows best. And she's perfect in doing so because she can exactly tell you whether a blue is too dark, too bright, too creamy, too violet or just perfect Karl's blue.

We all learn to understand or interpret subjective feelings of others not just by language but also by empathy. Although it is difficult and not everybody can do it equally well, the more experience we have, the better we get. The better we know an individual the better we can assess how he or she feels, in what mood a person is, and whether somebody feels good or uncomfortable in a given situation. Sometimes we can even detect that a person feels just the other way as he or she is trying to tell us.

Like in a transmitter receiver situation, our competence in detecting signals in verbal and in non-verbal communication gets better the longer we communicate, simply because the index-table gets more and more complete and the likelihood that a new type of content occurs gets smaller over time. In verbal communication, once you have started and established a few basic words which you understand you can explain more complicated terms by using primitive ones. This is a bootstrapping process of guessing and better understanding in a continuous adaption circle of trial and error. And this is also the natural way how children learn a language. They don't do it by applying abstract grammar rules which have been distilled by professional philologists. The often fruitless attempt in our schools to teach languages via rules und grammar is another example of the failure of the attempt to reengineer a system which you don't understand. Of course, it's interesting what linguists found about the logic of grammar. But their formal knowledge is virtual and artificial as artificial neurons. Such formal and logical principles have nothing to do with what is going on in our brains. And this is the reason why teaching languages the normal academic way, similar to mathematics, is so inefficient and so stressful. A child learns to speak by immersion in the communication-situation and by adaptive exploring (very much like Freeman's salamander explores his odor-environment) and not from abstract rules.

This immersion in "doing the thing" also allows you to learn foreign languages or learn more about the internal state of another human, an animal, or a machine. Interaction allows you to gradually complete the knowledge about the landscape in which you operate. Of course, it is easier when you communicate with an individual from a near culture and more difficult when you communicate with a person from a more distant culture. It takes longer when you have to communicate about terms which don't exist in one culture, but you can get closer with exchanging more details. Over time and with many error-correction cycles, we can improve the process until we achieve a reliable communication. There are probably no two languages that are more different than Chinese and German and hardly two cultures that are as different as ours. But nevertheless, we have succeeded in communicating successfully even on complicated issues by using a third language, English, which we both do not speak perfectly (at least I don't). And while our choice of words and grammar may not always be perfect in the eyes of an English native-speaker, I think we have learned to exchange our thoughts sufficiently well. And in doing so, we have also acquired some knowledge about subjective feelings of the other person. I don't know how substantial this growth of knowledge is, but I think that it's more than zero.

What I'm saying here, you have probably realized it, is inspired by Walter Freeman's perspective. His major point is that the continuous interaction with the environment is the key to building awareness and consciousness. And I'm sure that he agrees with Searle's and my positive opinion about the understandability of consciousness. When I looked around for his newer work I found a very interesting lecture of his which he gave in a Berkeley seminar just a few months ago. The title was " The evolution and neurodynamics of the action perception cycle — When, where and how consciousness and quantum field theory enter"[1].

---

[1] https://archive.org/details/UC_Berkeley_FOM_2014_05_02_Walter_Freeman

In case you haven't seen it yet you may like it, especially because of the progress he has made since he wrote "How Brains make up their Minds" in 1999. Basically, he's still dealing with the same subjects and the examples with the olfactory system of the salamander. So, it will all look familiar to you. I'm not sure whether adding the quantum field aspect is really necessary, but I think that he has well-honed his general idea of how consciousness evolves from interacting with the environment.

I particularly like Freeman's idea that, in the brain, we are dealing with neural networks creating electromagnetic fields and oscillations which extend over larger areas of the brain and represent coded patterns of an analog system. This is in sharp contrast to the concept of specific regions or even specialized neurons were information is stored as on digital computer hard-disks. Freeman always followed Karl Lashley's idea of mass action, and also, his (Walter's) teacher Karl Pribram's idea of dispersed storage of information, as it happens in holograms. But now he is getting more precise on how these fields work.

You know his work much better than I do and I would be very grateful if you could have a look at it and tell me whether I'm right with my impression that Freeman has made progress with his theory. Here is the video and below it there is an introduction to his work and a summary of his publications.

What I can't asses is the relevance of the added idea of quantum fields. Something Freeman has published in a 2007 publication with the lesser known physicist Giuseppe Vitiello. It doesn't seem to have had much resonance and it is conspicuous anyway that Freeman seems to be ignored by some of the big names in the business. It is surprising to me that he isn't mentioned once in Kandel's "Bible". In my eyes, he deserved more attention. Maybe he's just too solid a worker focused on the one enigma and isn't interested in playing on the Kurzweil-keys of the science-show-business-piano. Maybe he doesn't care at all, but if there is any "Interest-performer" out there in this field then it's Walter Freeman III.

But, sometimes fame and popularity do not necessarily result from the inner quality of a work. Often, it is completely different things that decide the success or failure of an idea, theory, or invention. Sometimes, it can be completely non-essential things that have nothing to do with the matter itself, but have a huge effect. In the interview with

Susan Blackmore, John Searle makes an interesting comment in this regard when he is asked about the effect of the publication of his famous "Chinese Room" thought experiment in 1980. It attracted an enormous amount of critique and nevertheless made him world famous because it is one of the most cited publications of modern science, and the "Chinese Room" became almost a global household name. He says that the outrage it created surely came from the fact that many people saw it as an attack against their world view based on a reductionist attitude towards consciousness. But that wasn't the real problem Searle says. The real problem, he says, was that he had overlooked the fact that at the same time it was a threat to many grants, careers, and expected money. Many people, he realized, consume enormous amounts of money based on the wrong premise that they can rebuild the human spirit. He said this long before there was an HBP, by the way. ☺

I'm not sure how all these arguments for detecting consciousness in animals are applicable for machines. AI, as we have it today, is still too primitive to tell. However, it already becomes obvious that the Turing Test, alone, is too weak a test because it is restricted to language. Therefore, some people are already calling for a more complex test, called Total Turing Test (TTT) that might be needed for the next generation of machines.

All in all, my message is that Freeman, Searle, and Karl see good chances that there are ways to get access to consciousness and subjectivity by rational means and within the realm of the natural sciences.

If this doesn't impress you, here is something for you which at least may make you smile a little. It is the first of the famous three laws which sci-fi author Arthur C. Clarke once postulated[1] :

"When a distinguished but elderly scientist states that something is possible, he is almost certainly right. When he states that something is impossible, he is very probably wrong."

Good night for now and best from Karlsruhe.

Karl

---

[1] https://en.wikipedia.org/wiki/Clarke%27s_three_laws

20. 12. 2014

Dear Karl,

I am glad that you agree with some of my points on the problem of subjectivity and privacy of consciousness. Of course, as you said, I couldn't expect that all my points would be correct, which just provided something to think over and discuss.

As for your question *"how precise and detailed a copy of a brain has to be to transmit subjectivity and consciousness"*, it reminds me of Sebastian Seung's paper you recommended in your letter on 07/09/2013[1].

I like the paper very much, so I even ordered another of the author's books *"Connectome: How the brain's wiring makes us who we are"* (2012), in which, the author declared that you are your connectome. However, he did not stop at this argument, he went even further and said:

"Is there anything about the brain that is fundamentally incompatible with the framework? One difficulty is that neurons can interact outside the confines of synapses. For example, neurotransmitter molecules might escape from one synapse, and diffuse away to be sensed by a more distant neuron. This could lead to interactions between neurons not connected by a synapse, or even between neurons that do not actually contact each other. Because this interaction is extrasynaptic, it is not encompassed in the "wiring diagram:" the connectome. It might be possible to model some extrasynaptic interactions fairly simply. But it's also possible that the diffusion of neurotransmitter molecules in the cramped and tortuous spaces between neurons would require complex models.

If extrasynaptic interactions turn out to be critical for brain function, then it might be

---

[1] https://www.scientificamerican.com/article/massive-brain-simulators-seung-conntectome/#

necessary to reject the hypothesis 'You are your connectome' the weaker statement 'You are your brain' could still be defensible, but this would be much more difficult to use as a basis for uploading [your brain into a computer, as some futurists have proposed]. We might have to throw away the abstraction of the connectome and descend still further to the atomic level. One could imagine using the laws of physics to create a computer simulation of every atom in a brain. This would be extremely faithful to reality, much more than a connectome-based simulation.

The catch is that a huge number of equations would be necessary, since there are so many atoms. It seems absurd to even consider the enormous computational power required, and is completely out of the question unless your remote descendants survive for galactic time scales. At the present time, it's difficult to simulate even those modest assemblies of atoms called molecules. Simulating all the atoms of a brain is almost beyond imagining. Limited computational power is not the only barrier. There is also the difficulty of obtaining the information to initialize the simulation. It might be necessary to measure all the positions and velocities of the atoms in the brain, which is far more information than in a connectome. It's not clear how to collect that information, or how to do it in a reasonable amount of time." [1]

I think that his arguments and yours are forceful enough to reject any possibility of transmitting or downloading/uploading the mind practically.

As for the problem of whether we can know another's feeling, my guess is "yes" and "no". Here, I think that maybe we should distinguish empathy from sharing feeling exactly. All the arguments I talked about in my previous letters are about sharing consciousness exactly. Owing to the reasons discussed above, it seems that to share another's feeling or qualia exactly is very unlikely. The very underlying reason is that no two brains are identical. Therefore, no two subjects can experience just the same. Think, blue cheese is so delicious for Italians, but terrible for most Chinese; on the contrary, the preserved egg is delicious for most Chinese, but terrible for western people. People could not have the same feeling of the other exactly! However, this doesn't mean that they can't share their feelings approximately. A Chinese would think

---

[1] https://www.scientificamerican.com/article/massive-brain-simulators-seung-conntectome/#

that the feeling of an Italian who has a taste of preserved egg must be similar to his when he takes a bite of blue cheese, as both have a center for disgusting. The center must be activated in a similar way, when they have similar feelings. Thus, empathy is based on the similarity between brain structures and activities, owing to the fact that no two brains are identical, empathy can only be approximate. Therefore, I agree with your argument that *"we may get access at least to some part of your subjective consciousness"*, but never your experience accurately.

Then you may raise a strong opposition that as there are no two identical macroscopic objects in the universe, why should I emphasize the impossibility of identity of experiences of different subjects? Should we consider that there is no sharp gap between accurately sharing subjective experience and empathy as in a continuum? For the first question, I think this is just owing to the fact that people used to think that if the stimulus is the same, then their neural representation should be the same from an information processing view. However, in fact, the corresponding perception may be quite different. Therefore, Walter Freeman emphasized that the brain's function is not to represent the stimulus, but to make its meaning with its internal experiences, and thus the meaning will be unique to the subject. As for the second question, maybe it's correct, as both of us admit, that experience is emerged from brain activities, then we have to admit that if activities in two brains are identical, then their experiences must be identical, although such probability is tiny if not zero. Thus, can we say that if activities in two brains are very similar, then their experiences should also be similar too? It sounds reasonable, however, I am not so sure, as the brain is a nonlinear system, chaos may play essential role in its activities, thus the similarity between brain activities could not guarantee the same degree of similarity in their experiences.

I appreciate your attitude that *"before I'm willing to surrender and opt for metaphysics I would rather try everything possible in the realm of physics and rational methods."* My opinion is that subjectivity may be an unreducible emergent property of brain activities. Thus it is not a proper problem to ask how subjectivity emerges from brain activities, instead, we should try to find the sufficient and necessary conditions for emerging consciousness, firstly of human beings. If we find such conditions, then subjectivity will emerge naturally when these conditions are fulfilled, and Chalmers' "hard problem" will evaporate. Only after that, may we be able to extend such study to other agents including animals or even artefacts. However, to find the sufficient and necessary conditions for emerging consciousness even for human beings is hard. The key difficulty is still the old problem: how do you know the agents have subjectivity? If the condition is sufficient, then the agent must be conscious if such condition is fulfilled, however how do you know the agent is conscious? For human beings we may take it as an axiom that all the human beings behaving normally is conscious, although how to judge the behavior as "normal" is also a problem. However, what about other agents? For machines, we may take your "Total Turing Test", although no machine can even pass the original Turing test up to now. As far as other animals are concerned, obviously we cannot adopt the same strategy, as an animal's behavior must be different from a human being's someway!

As for John Searle's "xyz theory", I have a question. Let's have a thought experiment, if we had a patient with a normal thalamus and cerebral cortex but damaged brainstem, and suppose we found a way to cure the brainstem damage, then the patient should awake and have consciousness again, could we say that the brainstem is responsible for consciousness? I don't think so, otherwise, you would get a conclusion that all the vertebrate animals are conscious as all vertebrate animals have brainstems. Christof Koch excluded such brain activities from NCC, and called them "enabling factors" to be distinguished from "necessary conditions". The enabling factor is some premise. Normal blood supply is also a premise for consciousness, but obviously is not a part of NCC. How to distinguish necessary conditions from enabling factors is also a problem. By the way, it was reported this year that it was found that when the claustrum was stimulated with high-frequency electrical impulse, the subject lost consciousness, while he recovered once the stimulus stopped, as if the claustrum was a switch of consciousness. However, it is not clear what other conditions should be met at the same time, in order to be conscious, except for normal activities of the claustrum. So, it is still unclear that if normal activities of the claustrum are sufficient for consciousness, although it seems to be necessary.

Therefore, it seems to me, a practical way to study consciousness nowadays is to find the necessary conditions for human beings under which consciousness or some of its aspect emerges. At the same time, we should also pay attention to see if these conditions are sufficient, although this is difficult.

Recently, I read a new book published just this year: *"Consciousness and the Brain: Deciphering How the Brain Codes Our Thoughts"* by Stanislas Dehaene, very interesting indeed. The author and his colleagues tried to explain how consciousness emerges from the brain with experimental studies, theoretical hypothesis, and computer simulation. He tried to turn this problem into an experimental question. I am not sure if you have ever read this book. Let me try to summarize what he has done and thought and give my comments.

As there is still no clear and generally accepted definition of consciousness, different people have different views of the term under different contexts. To avoid confusion, Dehaene focused his studies on "conscious access" as he called (the phenomenon that a subject becomes aware of an attended stimulus, and can report to others), which few people would deny as an essential aspect of consciousness and is the gateway to more complex forms of conscious experience. I appreciate his strategy very much, just as Francis Crick pointed out many years ago:

"It is characteristic of a scientific approach that one does not try to construct some all-embracing theory that purports to explain all aspects of consciousness. ... In a battle, you do not usually attack on all fronts. You probe for the weakest place and then concentrate your efforts there." [1]

Contrary to behaviorist's attitude to rejecting introspection radically, they took subjects' reports, telling the experimenter if they are aware of the stimulus, as valuable raw data rather than research approach. Using masking, binocular rivalry, and other methods, they showed that although the stimulus kept constant or almost unchanged, the subject's perception may change radically, say from unconscious to conscious, or vice versa, thus, conscious access can be treated as the only variable which can be experimentally manipulated. Then, they just looked for patterns of brain activity which appear when, and only when, the subject had conscious access to the corresponding stimulus. They took these patterns as the markers of conscious access, or "signatures of consciousness" as they called them. They found such signatures as follows: (1) an amplification of

---

[1] Crick F H C. The Astonishing Hypothesis: the Scientific Search for the Soul [M]. New York: Charles Scribner's Sons, 1994.

lower level brain activity, progressively gathering strength and invading multiple regions of the prefrontal cortex and parietal cortex; (2) a late sudden rising P300 event related potential in brain waves; (3) a late amplification of high frequency oscillations; and (4) a synchronization of activities across distant brain regions. So far so good, they rigorously showed that their "signatures of consciousness" are closely related to conscious access, which tell us what happens in the brain when the subject experiences a certain conscious state. He concluded:

"During that conscious state, which starts approximately 300 milliseconds after stimulus onset, the frontal regions of the brain are being informed of sensory inputs in a bottom — up manner but these regions also send massive projections in the converse direction, top-down, and to many distributed areas. The end result is a brain web of synchronized areas whose various facets provide us with many signatures of consciousness."

"During conscious perception, groups of neurons begin to fire in a coordinated manner, first in local specialized regions, then in the vast expanses of our cortex. Ultimately, they invade much of the prefrontal and parietal lobes, while remaining tightly synchronized with earlier sensory regions. It is at this point, where a coherent brain web suddenly ignites, that conscious awareness seems to be established." [1]

Maybe we can view their "signatures" as the necessary conditions for conscious access of human beings. A big step forward in consciousness studies!

Based on the signatures of consciousness, they found the process from an unconscious state to a conscious state is like phase transition, the underlying brain activity must be over a threshold, and then become self-amplifying, the brain suddenly seems to burst into a large-scale activity pattern, thus they proposed a "global neuronal workspace hypothesis" trying to explain the neural mechanism of consciousness. The author said:

"The proposal is simple: consciousness is brain-wide information sharing. The human brain has developed efficient long-distance networks, particularly in the prefrontal

---

[1] Dehaene S. Consciousness and the Brain: Deciphering How the Brain Codes Our Thoughts [M]. New York: Penguin Books, 2014.

cortex, to select relevant information and disseminate it throughout the brain. Consciousness is an evolved device that allows us to attend to a piece of information and keep it active within this broadcasting system. Once the information is conscious, it can be flexibly routed to other areas according to our current goals. Thus we can name it, evaluate it, memorize it, or use it to plan the future. Computer simulations of neural networks show that the global neuronal workspace hypothesis generates precisely the signatures that we see in experimental brain recordings. " [1]

He said: "*When we say that we are aware of a certain piece of information, what we mean is just this: the information has entered into a specific storage area that makes it available to the rest of the brain.*" Subjectivity vanishes! The author mentioned several criticisms to his hypothesis in the last section of his book, for example that Ned Block said that the hypothesis cannot explain qualia; David Chalmers said that it will never explain the riddle of first-person subjectivity, or the hard problems of consciousness. Nevertheless, he only gave very brief answer to these criticisms:

"*My opinion is that Chalmers swapped the labels: it is the " easy" problem that is hard, while the hard problem just seems hard because it engages ill-defined intuitions. Once our intuition is educated by cognitive neuroscience and computer simulations, Chalmers's hard problem will evaporate. The hypothetical concept of qualia, pure mental experience detached from any information-processing role, will be viewed as a peculiar idea of the prescientific era, much like the vitalism — the misguided nineteenth-century thought that, however much detail we gather about the chemical mechanisms of living organisms, we will never account for the unique quantities of life. Modern molecular biology shattered this belief, by showing how the molecular machinery inside our cells forms a self-reproducing automaton. Likewise, the science of consciousness will keep eating away at the hard problem until it vanishes ... Once we clarify how any piece of sensory information can gain access to our mind and become reportable, then the insurmountable problem of our ineffable*

[1] Dehaene S. Consciousness and the Brain: Deciphering How the Brain Codes Our Thoughts [M]. New York: Penguin Books, 2014.

experiences will disappear."[1]

I appreciate his studies on signature of conscious access very much, his global neuronal workspace hypothesis for explaining the signatures seems OK, but I am skeptical about his hypothesis being able to explain consciousness or even conscious access itself. My main points are as follows:

(1) "Signature of conscious access" is not "conscious access" itself, just the same as my signature is not me myself. The noisy clicks a railway train makes is a signature of a running train, you can hear it when, and only when, the train is running. The underlying mechanism making such clicks may correlate with the train running, however, you cannot explain how the train runs with the mechanism of click making. Of course, this comparison may be too extreme, however, it does show that you cannot explain the mechanism simply with the mechanism of corresponding signatures. Explanation of signatures may give some hints to the explanation of its owner, but not give an exact explanation to the latter. Strictly speaking, his studies only showed that if the subject has conscious access, then there are such signatures in his/her brain, but not the contrary.

(2) Except for using subjects' subjective reports to judge if they are conscious of something, Dehaene's work does not touch the subjectivity problem of consciousness. All his experiments and theory are based on objective facts. He uses objective signatures to replace subjective conscious access. Therefore, even if his hypothesis illuminates the mechanism of how these signatures are originated from some special patterns of brain activities, and even if his statement can be extended to the mechanism of conscious access itself, his theory, at most, is like the binocular disparity theory about stereovision, which does explain under which condition stereovision emerges, just as 3D movies have proved, however it still cannot explain how we can have such qualia. Just as Dr. Barry told Dr. Sacks. Dr. Dehaene could not explain why it would be wrong when Dr. Barry said she was wrong! Of course, I do not mean that his work is not worthy at all, similar to disparity theory of stereovision, his work may be a

[1] Dehaene S. Consciousness and the Brain: Deciphering How the Brain Codes Our Thoughts[M]. New York: Penguin Books, 2014.

big step forward to finding the necessary conditions for conscious access emerging, but not the sufficient conditions.

(3) Although their computer simulation seems to support his hypothesis, a neural network model based on which was constructed, its behavior mimicked the four signatures, a phenomenon like phase transition happens, however, even he himself admitted that the simulation was *"far from to be conscious"*. Thus, their simulation may prove that his global workspace may only be a zombie, just opposite to his own view that these signatures cause consciousness! It seems that to have conscious access needs something more than his hypothesis!

(4) Although his phase transition notion or avalanche metaphor sounds interesting, we can also use these metaphors to describe epileptic seizure, a state absolutely without consciousness. So maybe these signatures are only necessary, but not sufficient for conscious access. Many theoretical studies on consciousness have similar flaws, they confused necessary condition with sufficient condition.

In short, his signatures and hypothesis may give necessary conditions for conscious access, but not sufficient. His simulation shows that even a network constructed on his hypothesis having signature-like features is still far from being conscious, there must be something lost! The question is what is lost? Subjectivity? I think so, but is there anything else?

If people find necessary conditions enough, is it possible that they become sufficient too? Could subjectivity emerge abruptly when these conditions are fulfilled? We don't have the answer yet, but as you emphasized, people should try and not give up. In case this happens, should we say Chalmers' "hard problem" evaporates? Let me repeat again, in my opinion, it is no use to ask how subjectivity emerges, it just emerges when some conditions are fulfilled, the proper question is what are these conditions? First, for some aspects of consciousness, such as conscious access, then for consciousness in general. First, for human beings, and then for other agents. What is your idea?

Sorry, my letter is too long and has to stop here, the other questions you raised in your last letter we may discuss a little later.

Merry Christmas and a very Happy New Year! The time flies so fast!

Fanji

19. 02. 2015

Dear Fanji,

Happy New Year to you as well!

It is way too late for the German New Year, but I'm happy to see that it is just in time for your Chinese New Year. In any event, I wish you all the best for this year under the sign of the sheep.

I have to apologize that it took me so long to answer your interesting letter, filled with so much input, mostly new to me, and full of tricky questions. I didn't want to just send you an honest "I don't know" answer and therefore had to do some reading and thinking. I'm still not sure, however, whether I can give you satisfying answers. You are dealing with those questions about consciousness and subjectivity for many years and maybe some of your questions nobody can answer. But I'm starting to understand better what you're getting at, and I'm trying to do my best.

Reading Dehaene's book "Consciousness and the Brain: Deciphering How the Brain Codes Our Thoughts" was great fun. Thank you for recommending it to me! I chose the German translation, which is pretty good, in my opinion, which you as the master of translation may like to hear.

Similar to you, I don't agree with everything Dehaene proposes, but he can be inspiring also in places where he's on the wrong track in my eyes. What really annoys me is that he too ignores Walter Freeman which is a pity not only for your friend Walter but also for Dehaene. He would have profited a lot if he would have obeyed Freeman's advice not to think of the brain as an information processing machine but as a system to detect meaning.

What I also find problematic is the claim of Dehaene, which you have cited:

"that computer simulations show that the global neural workspace hypothesis generates precisely the signatures that we see in experimental brain recordings".

This statement makes me suspicious because when you know what has to come out, it is easy to make a computer-simulation which shows precisely what you have expected. You just have to tune the model and the weights of the parameters until you see what you want to see. Very much as bible believers feed specific texts of the bible into a computer-program that reveals mysterious messages about the end of the world or the date when the Titanic sunk. Actually, it's pretty simple to do this when you start with the message you want to get. Every reasonable programmer is able to perform this Bible-code wonder by writing an algorithm with parameters that allow the right spacing and shifting of words and pages and voila you can decipher the recipe for, let's say, Peking Duck from the bible or the German national anthem from the DNA of drosophila. The trick is pretty simple. You start retroactively with the sentence you want to create and search for the words or letters you need in the selected source in the first step. Once you have found them, all you need to do is to define the order in which the words or characters are to be read in an algorithm. This results in a reading rule that determines how to jump from page to page, from line to line and from column to column. In this way, any desired result can be read out from any source.

The miracle is not the algorithm, but the fact that you can still make headlines with such hocus-pocus. ☺

Of course, I'm not saying that Dehaene has produced his findings in a fraudulent way! What I'm saying is that I'm generally not surprised or impressed when scientists find what they are looking for in large amounts of data when they use computer statistic-models and simulations. There are too many little radio-buttons on these systems which can be tuned to fit. I worked long enough in computer statistics to know the temptation to turn the many knobs on these magical instruments until you get the result you expected to see.

So, somebody else should do experiments similar to Dehaene, as it is good practice in science.

The criticism of Ned Block and Chalmers, which you cited, is correct in principle, but it would not shock me if the rest of Dehaene's theory takes us closer to explaining the phenomenon of consciousness.

Especially Chalmers fundamentalist statement that such results "*will never explain the riddle*" reinforces the suspicion, which I have expressed in my last mail. To me it seems that he's not willing to give up his position whatever evidence you will show him, because, to him, subjectivity may simply be defined by not being accessible. But maybe you will tell me that I'm wrong. I should be happy that Dehaene supports my view that we will solve the consciousness and subjectivity problem step by step. While I agree with him in this prediction, I dislike, as I said, his information processing approach and the idea that computer-simulations will be of great help. Maybe he just added this computer stuff because he was part of the HBP, in the beginning at least, and to do simulations was required to become a member of Markram's club. But I don't want to be unfair with the man, who is a respectable researcher, for sure, and you may know better anyway.

Your four critical arguments seem all valid to me. It's typical Fanji-style and hard to argue against. The first one is especially crucial in my eyes, because signature of consciousness is not conscious access as you rightfully say!!

I'm not sure, however, whether you are leaning too much on the Chalmers side because you, as well, may have fallen in love with the idea that subjectivity is (and maybe should remain) inaccessible. ☺

This is OK with me because I can't exclude that Chalmers and you are right. But I think that we can agree that progress is possible, more reasonable research can be done, and that Dehaene and his critics gave interesting hints where this should happen. Very much in the sense of the research program John Searle has outlined in his interview with Susan Blackmore.

And I may not give up hope that one day Siri, Alexa, or your personal robot will talk to you about their subjective feelings. Just in case this happens you should tell them to call me and let me know about my victory. ☺

I'm very happy that you also found Sebastian Seung's ideas about the connectome so helpful.

I didn't know his book "Connectome: How the brain's wiring makes us who we are" and knew his position only from the article I had sent you.

Thank you for telling me! In the meantime, I have also looked into the book to see what his position is and this was very, very helpful indeed!

But what I found I will only tell you in my next letter.

Good night for today and best to Shanghai.

Karl

01. 04. 2015

Dear Karl,

Thank you very much, indeed, for spending so much time and energy thinking about the questions I raised, and even ordering books to read for answering my questions! I cannot find another friend being able to do so.

Yes, I also feel surprised that Walter's theory has been ignored, only a few people have noticed his contribution, and considered the brain as a machine creating meaning, instead of an information processing system only. Considering the brain as an information processing system only is still the main stream in neuroscience, especially in the field of computational neuroscience. This may be another important difference between biological brains and technical artefact systems.

Freeman emphasized that the brain is a machine creating intentions and meanings, a nonlinear dynamic system, which is influenced, only a small amount, by the sensory stimulus and, a much larger amount, by its own ongoing activities, including its history, attention, action, and emotion. Intentions and meanings precede consciousness. Owing to his above points, there is no one-to-one correspondence between stimulus and intention, meaning, or perception. Meaning and perception are mainly created in the brain, they are private. He said:

"A sensory stimulus from an object does indeed induce the formation of a pattern in the brain, but when it is given repeatedly it does not induce precisely the same pattern in the same brain, let alone in any other brain. This is to be expected because not only does the same object mean different things to different people, its meaning for the same person is continually shifting. My conclusion is that meaning cannot be transferred directly into and between brains in the way that information and knowledge based in representation can be transferred into and between machines.

...

All that brains can know has been synthesized within themselves, in the form of hypotheses about the world and the outcomes of their own tests of the hypotheses, success or failure, and the manner of failure. This is the neurobiological basis for the solipsistic isolation that separates the qualia of each person from the experiences of everyone else ..." [1]

According to him, privacy is rooted deeply in intention and meaning creating, which even precedes consciousness. Perception is not a representation of sensory stimulus; it also depends on the subject's previous experience, and thus is private too.

His arguments may help me to answer your big question of whether another's consciousness can be accessed, although a definite answer is almost impossible. As a matter of fact, we have discussed so much about the subjectivity and privacy of consciousness in our previous letters. It seems to me, at least, up to now, no method can be used to thoroughly access another's consciousness owing to the reason Walter explained above, and I don't expect that there will be a solution to this problem in the foreseeable future. However, at the same time, I don't deny that people can have empathy, which may be a rough version of accessing another's feelings. People may even say that mirror neurons are responsible for empathy. However, even so, they still cannot explain how mirror neurons fire when the subject has a feeling similar to another's. Just as I argued in my previous letters, if some sufficient and necessary conditions for being conscious for human beings are found (this may be possible, although difficult, especially how to know the other is conscious? Maybe just based on common sense?), then maybe no one will ask how consciousness emerges from a human being's brain activities again, and the problem "evaporates". As the structure of human brains are similar (but not identical) to each other, similar brain activities may emerge similar feelings, this may be the basis for empathy. However, a question is still there, how one's brain activity triggers another's similar brain activity? As for artefact machines, maybe we just use a Total Turing Test. Although we still don't know if the machine really is conscious just by behavior. We can neither exclude nor confirm such a

---

[1] Freeman W J. How brains make up their minds[M]. New York: Columbia University Press, 1999.

possibility. Although these criterions are not satisfying enough to me, I can still accept. However, what about other animals? I have no idea. Maybe you can propose a criterion.

I cannot deny that, it seems to me, that the barrier of subjectivity might never be able to be overcome thoroughly; you can never accurately experience what others experience. However, I admit that to experience another's feeling partly or roughly may be possible. Even so, you may still not be able to solve the "how" problem. Do you think that science can solve all "how" problems? Say, how the electron has a negative charge? Maybe for those very basic properties, we have to admit the fact, take them as some axioms and not ask a "how" problem.

Just similar to what you said, I cannot exclude the possibility that I am totally wrong. And I agree with what you said that progress in consciousness studies is still possible, especially in finding necessary conditions for a human being's consciousness. I think that Dehaene's work on finding his signatures of consciousness is a big step in this direction.

I cannot exclude the possibility that someday my Siri would talk to me about her subjective feelings, however, I will still be doubtful if it is really conscious, or if it is just a zombie. According to your TTT, I may admit that they are conscious, but at the bottom of my heart, I may ... Alright, I think that both of us have explained our points of view quite clearly, I don't think that we can get an easy consensus, maybe let us watch the progress in this field, see what will happen and think over our arguments again and discuss later on this topic.

I totally agree with your comments on the limitation of simulation, especially when there are too many parameters which can be adjusted freely. In such cases, you can get any result you wish to get. Such simulations can hardly explain anything. Of course, if almost all the parameters are decided by biological experiments, that would be another story.

Now let me shift my topic to another very interesting event, which you may have already noticed. Just last month, a mediation report[1] was announced by the mediation

_____

[1] http://www.neurofuture.eu/media/official_HBP_mediation_report.pdf

committee (MC) of the HBP, which is responsible for mediating the dispute between the HBP and more than 800 scientists who set up an open message to the EU in July of last year.

The mediation report strongly supports the criticism of the open message to the HBP on its goal, approach, and management. It points out that the project is managed improperly, its core target to simulate the whole human brain is unrealistic, and the project has lost the trust of the public and the academic community. The HBP must carry out an essential reform to recover such trust. The report was approved by almost all members of the MC except for two members. (Within all the 27 members of MC, there are 10 members from the HBP)

The report proposes concrete suggestions on both academic contents and administrative affairs. I shall only list the most important ones as follows.

As far as the scientific contents are concerned, the task and the goal of the HBP must be more concentrated. To simulate the whole human brain is unlikely to be reached in one decade, thus it is suggested that the contents of the brain simulation subproject should be reconsidered. The report suggests that brain function is too rich, the available knowledge on brain function is still very limited, thus a reliable bottom-up simulation of a whole human brain is hardly to be expected to be reached. Therefore, the HBP should focus its target on some concrete targets which can be expected to be reached within a limited period and with limited resources. The report suggests that the HBP should develop a neuroinformatic approach and technology as its main task. The HBP should take advantage of available data from neuroscience experiments, instead of filling all "gaps" both on structure and function by itself. Only those experiments which have clear purpose and are necessary for developing information technological platforms should be done. These platforms should be developed and checked by an interdisciplinary cooperation, cognitive neuroscientists, and system neuroscientists should take part in such developing. Cognitive neuroscience and system neuroscience studies should also be re-integrated into the HBP again and should be a subproject with 3 or 4 aspects. The developing of such platforms should be closely related to concrete studies such as spatial navigation or decision making and so on. The report also points out that the HBP should not exclude studies with non-human primate animals, which are essential linkages between mouse brains and human brains.

The report criticizes that the news released by the HBP over-exaggerated its potential, such as understanding main brain functions, or diagnoses, and treatment of degenerative disorders of the nervous system within a decade. Such exaggeration makes the public have an unrealistic expectation, and the project loses scientists' trust. The reputation of the HBP relied on its convincing scientific results and information technological platforms, which can be widely used by neuroscientists, at last.

As far as the management is concerned, the report almost criticizes Markram directly, it said:

"the coordinating scientist, ... is not only a member of all decision-making, executive and management bodies within the HBP, but also chairs them and supervises the administrative processes supporting these bodies. Furthermore, he is a member of all the advisory boards and reports to them at the same time. In addition, he appoints the members of the management team and leads the operational project management." [1]

Of course, one cannot be both sportsman and referee at the same time.

The report requests that the fund distribution should be transparent, and the established IT platforms should be used by all the members conveniently. The report asks that the responsibility of the HBP and the role of the coordinator should be transitioned from EPFL, to which Markram belongs, to a new legal entity jointly represented by those institutions that most strongly contribute to the project.

Facing the serious criticism of the report, the HBP had to take some urgent measures. Even about one month before the release of the report, the HBP had to dissolve the three-member executive leading group. All three persons, including Markram, had to resign. The report was approved by the HBP board of directors on the 18th of last month. A leading working group was established, in which many leaders of international scientific organization took part. Several different groups dealing with academic, administrative, and financial affairs were also established. A big reform seems to be on its way.

---

[1] http://www.neurofuture.eu/media/official_HBP_mediation_report.pdf

I am curious about what Markram thinks of the mediation report, and what role he will play in the HBP in the future? Do most of the members of the HBP board of directors really agree with the report? The reform will not be easy, I think.

I agree with the main points of view of the report, although it is not ideal. As a matter of fact, it seems to me that no reform will be ideal for such a project with these congenital deficits. If the suggestions in the report are totally accepted, then the goal of the HBP will become to develop new information technology (IT) and platforms for understanding the human brain and its disorders. Although this goal is limited, compared with the original one, it is still quite magnificent. The HBP and the US BRAIN initiative will compensate each other, the former focuses on developing IT, while the latter focuses on developing novel technology for observing, recording, and imaging neural circuits.

In fact, technological invention has played an essential role in scientific development in the history of science, although, in which, the scientists using these technologies to make significant discoveries are mostly praised, instead of the inventors of these technologies. Galileo improved the early telescope invented by Hans Lipperhey, and used it to find that there were four satellites rotating around Jupiter. His discovery showed that not all celestial bodies orbit the earth, as people believed before. At least, there were four satellites orbiting another planet. Similarly, in the history of neuroscience, Cajal improved Golgi's dyeing method to observe a variety of neuronal tissue section, which led him to proposing the neuron doctrine, which became the foundation of neuroscience. Using an electronic amplifier opened a new era of electrophysiology, and functional brain imaging technologies made observing brain activities, in an awake subject, possible when he or she completed some mental activities, promoting the development of cognitive neuroscience. Therefore, developing new technologies, of course, is very important for brain science. The problem is how to combine such invention with important topics in brain studies, but not just for developing new technologies themselves. Otherwise, the HBP will become a pure technological project, but not a project for brain studies, in such a case, if it takes a big share of the financial support dedicated to brain studies, then it is hard to say if the HBP would do any good for brain studies. If developing IT technologies can be closely related to important topics in brain research, if the new technology will greatly promote such studies, and the latter will propose new demands to IT technology, that would be great. However, I don't know if there is any consensus about which topics are essential for current brain research and are expected to be solved in a foreseeable future. Perhaps, this problem should be seriously discussed in the neuroscience circle. I am not sure if this can be done within and/or beyond the HBP. Anyway, you and I can do nothing about this matter. Just let us watch what will happen in the future. The letter is already too long, I have to stop here.

Best wishes.

Fanji

14. 05. 2015

Dear Fanji,

Thank you for your beautiful letter and your kind words. Well, I have to tell you that it is sheer fun to follow your hints and questions which always contain so much wisdom and add new perspectives. Without you I would never have taken the effort to immerse myself in all those thrilling matters you've been drawing my attention to or to formulate ideas I was dealing with for quite some time. Expressing a thought in writing forces one to formulate it in a much more concrete and precise way than it may have existed in a more diffuse state of mind for some time.

This is time consuming and requires a certain amount of energy of course. But when you do so, and repeatedly develop and rework a train of thought, and when you feel that you get a better grip on the thing and start to grasp the problems of the matter, I believe it is one of the most rewarding things a human mind can experience. As an experienced researcher you must have felt this kind of reward much more often than I and you will know for sure what I'm talking about. Of course, I'm still far away from mastering the field but I can feel an increase of knowledge and a better ability to orientate myself on this playground. And if I'm a little less ignorant than I was before, I owe it to our conversation and therefore to you, my dear friend.

My own learning experience and Freeman's emphasis on meaning instead of information as the main currency of the brain made me think about how this "formation of patterns" in the brain works that allows us to filter meaning from sensory input and relate it to intentions. It is clear that this can only happen in a very subjective way because each brain has gathered input on different channels, and in a different order of succession. Each new input to such a system can only be interpreted and given meaning based on the previous state of the system. And while the original states of two systems may be quite similar in some cases (e.g. identical twins) even small differences in the input sequence can soon result in substantial differences of the meaning linked to it. And as a result, the system will soon be able to develop very different intentions. The "meaning filters" built of synaptical connections between neurons have to be dependent on each other and may either be piled up or linked like functional building blocks in arrays. And every new input has to be embedded into the existing neural network and, therefore, creates a new state of the system where meaning is filtered from perception and interpreted in relation to expectations and intentions. And this network has to be very different in each brain because the sequence of (external and internal) sensory input, experience and latent intentions is unique when a body moves through space and time. Even identical twins must have totally different synaptical connections 10 minutes after birth. Actually, their brains can never have been identical anyway just because the amount of information that can be coded in the DNA is by far not enough to determine the astronomical amount of synaptical connections that already exist in a "naked" brain.

You gave an excellent and crisp interpretation of Freeman's position about how the brain deals with perception, meaning, expectations, and intentions. Actually, it's the best and clearest I have seen anywhere!

It took me quite some time to grasp the essence of Freeman's position and I wished I would have had your help earlier. But maybe the process of comprehension needs many internal runs and, like playing the violin, continuous practicing.

Freeman published his idea about the relevance of oscillating fields in the brain in "Mass action in the nervous system" already back in 1975. He's been continuously working on the very same phenomenon for forty years and is still fascinated and obsessed by it. That's an amazing dedication and performance. Of course, this doesn't mean that his theory is correct. But he's made continuous progress, for sure, and accumulated incredible competence which you can literally feel when he talks about the brain. To me, he gets closest to the problem among all the authors I have read so far.

And I fully underwrite what you cited as his interpretation of privacy:

"According to him, privacy is rooted deeply in intention and meaning creating, which even precedes consciousness. Perception is not a representation of sensory stimulus; it

also depends on the subject's previous experience, and thus is private too."

I'm not sure whether Freeman is more on my optimistic side to believe that we may be able to trace consciousness by means of the natural sciences. But if you and I can agree that subjectivity, based on this definition, makes each of us unique I'm happy to agree. And I also agree when you say that we have made our positions in this question clear enough.

When it comes to further research about consciousness, I would take a pragmatic view for the moment as you have expressed it with your "common sense" statement. I don't believe that we will be able to solve all "how" problems, but I'm convinced that engineers will, again, outperform natural scientists and will build more intelligent machines much faster than biologists will comprehend the functioning of the brain, not to mention such complicated things as consciousness. My guess is that engineers like those who invented Siri and Alexa will steam ahead and build ever better systems of AI. Robots will move and behave more and more man-like, will talk to us, and create the illusion of intelligence, and maybe even empathy. It will probably take longer than Kurzweil predicts until they develop into real zombies, but their zombieness may be less frightening than many people fear.

And if an android behaves nice, friendly, and supportive and is full of wisdom, thoughtful, and humorous, the question is whether people wouldn't prefer such a zombie to some aggressive, hostile, dumb, and grumpy biological originals?

I agree that people like Dehaene have good ideas, but my guess is that technical progress is developing at an exponential rate, while progress in biology and understanding the brain is only linear.

I'm not sure whether multi-billion dollar projects planned and organized like industrial factories will change this for the better. I'm even concerned that this could make it worse.

But I was so busy over the last weeks with closing my knowledge gaps that I didn't find enough time to follow your references to the BRAIN initiative you gave me already some time ago and also on the palace revolution in Markram's HBP Empire. Both are very interesting and important events, indeed, which we should discuss in detail.

But please let me first come back to tell you what I promised to do at the end of my last letter which is to tell you what I have learned from Seung.

Before I read Seung's book I thought that this American Human Connectome Project, launched by the NIH in 2010, was a precursor to the large project which the Obama administration had announced under the name BRAIN Initiative (Brain Research through Advancing Innovative Neurotechnologies) in 2013. But I had wondered why this NIH connectome project is relatively small (30 million $) compared to the large numbers of the HBP.

What I've learned now, from Seung, is that here we talk about two different kinds of connectomes. Here is what he says about the NIH connectome project (page 181):

"Most people don't realize that this project is only about regional connectomes, and has nothing to do with neural connectome"

Oops, I have to admit that I also belonged to this group of "most people". ☹

What he explains is very helpful. When doctors and neurologists talk about the connectome they mean the link between functional regions in the brain, like in the Broca-Wernicke-model of languages, where one area responsible for understanding language (Wernicke) is linked to another area responsible for producing language (Broca). To call those kinds of links connectomes made quite some sense in the days when the knowledge about those functional regions was generated. At the time, tools for detecting and imaging, and their resolutions only allowed identifying relatively spacious regions.

When Seung talks about connectome as in the parts which you have cited, he talks about neurons and molecules at the lowest level, very much, as we did in our previous discussion. That makes quite a difference!

The cited complication which he adds to the picture we have discussed so far is also relevant. It's true when he says that neurons are not only linked via synapses. They, indeed, can also be influenced by all kinds of molecules, such as neurotransmitters, floating around anywhere in the spaces between neurons. And I find it remarkable that he's willing to question the paradigm "You are your connectome" in case extrasynaptic interactions turn out to be critical for brain functions.

Too me, it's pretty likely that they do because, as I've already mentioned, a neuron, first of all, is a regular cell, populated by lots of transmembrane proteins, all representing switches just waiting to start or stop all kinds of processes. Nicotine, caffeine, alcohol, and many other drugs can influence a fully connected biological neural network, and its well-known how such chemical substances can influence our perception, awareness, and consciousness. It is possible that this happens via interference at the synaptical gap and hereby disturbing the connectome, but it is also possible that the interference happens on the level of other cell mechanisms of the neuron.

But, in any event, it's clear that the neuron, as it appears in the neural-network models as a transistor-like device, only exists in the fantasy of logicians and computer scientists but not in the real brain. Unlike many people who play with abstract logical neuron models and simulate them on computers, Seung has grasped the essential difference, and also the negative consequences for mind uploading and other nice brain-tricks people are promising us. Of course, this has nothing to do with the fact that computers running neural network applications based on idealized neurons can do wonderful things, many of them even much better than the biological brain.

So, the consequence is, that although the complexity of the brain-system exploded by adding the aspect of connectome, it is still not enough to catch the content of a brain and to identify the personality of an individual. And again, we have to say "no, you are not your connectome" very much in the same way as we had to say, "no, you are not your DNA" and later "no, you are not your proteome".

In any event, I'm glad you have discovered an expert, skeptical of mind uploading who came to the same conclusion as the two of us.

It's not a very popular insight and bad news for the Kurzweil-flock, but probably a true one.

So, trying to copy a brain on a 1 : 1 level is not a valid option when you want to get access to consciousness and qualia. But maybe this isn't necessary at all. And maybe we should use even more coarse building blocks instead of ever refined Nano-tools to understand the brain functions on a molecular level.

There seems to be a race going on in neuro-biology to have ever finer instruments with ever better temporal and spatial resolution. But the problem, even with the best and finest microscope, is to find the right focus-level. When you look at a cell with the naked eye you don't see much, but when you use an electron-microscope you may see details that are too fine and therefore irrelevant. Very much as when you make a snapshot of a person's face and zoom in down to the level of the skin pores. When you look at the snapshot you may get very interesting information about the skin of this person, but you will not be able to detect who this person is.

Now, when we have instruments that extend to the very atom, the researcher's choice of the level of granularity they want to apply to detect effects is huge. If you don't have a theory that tells you where you have to look, you can just poke around and hope to be lucky enough to find the needle in the haystack.

Some time ago, a biologist, Yuri Lazebnik, who did his PhD-thesis on apoptosis ("cell death", a possibly crucial phenomenon to understanding cancer) wrote a provocative article with the title "Can a Biologist Fix a Radio? — or, What I Learned while Studying Apoptosis". (It's available on the Internet and easy to find).[1]

The question he asked is whether biologists (he could as well have said neuro-biologists) with all their wonderful instruments would be able to understand and repair a transistor radio, applying the same methods they use on human cells. Well, as you may have guessed, the answer is no and he gives good examples why biologists wouldn't even grasp an information processing device which is, many orders of magnitude, more primitive then a cell (which they also don't understand). I'm not sure whether his advice (establishing a more formal approach in the life sciences, similar to mathematics and the use of a more precise research language) would change the situation for the better, but his analysis is full of wisdom. It reads like a Feynman-satire on the misery of the modern life sciences bemoaning that it produces little output besides papers but needs ever more money.

---

[1] Lazebnik Y, Can a Biologist Fix a Radio?: or, What I Learned while Studying Apoptosis[J]. Cancer Cell, 2002, 2: 179 – 182. Inspired by Lazebnik's question Jonas and Kording analyzed a much more complicated artifact, a digital microprocessor, with neuroscientific methods and got quite sobering results. Jonas E, Kording K P. Could a Neuroscientist Understand a Microprocessor? [J]. PLoS Comput Biol, 2017, 13 (1): e1005268. doi:10.1371/ journal.pcbi.1005268

Lazebnik also raises the question about the use of the enormous amount of data produced by ever more refined experiments and says that the stage we have reached "*can be summarized by the paradox that the more facts we learn the less we understand the process we study*". And related to my above question where we should look for the "*anticipated gold deposits*" he cites a Chinese saying: "*It is difficult to find a black cat in a dark room, especially if there is no cat*".

I'm not sure whether this is really a Chinese saying, because people in the West tend to label everything that bears such kind of deep and funny wisdom with *Chinese*, but you may tell me. In any event, it describes, very well, the situation we face in neurobiology and the not so successful attempts to find the origin of mind and consciousness in the brain.

Maybe we are really looking in the wrong place or on the wrong level. If Freeman is right and we are primarily confronted with oscillating fields it won't help to create ever finer maps of the connectome or to simulate networks of logical gates on digital computers when the original is an analog filtering system.

This all sounds very desperate and almost hopeless. But on the other hand, when no two brains in the universe are the same, and when, in fact, everything on the Nanoscale is unbelievably complicated, how is it possible that two brains are able to exchange thoughts, and not only rational information, via speech and also via empathy just by modulating the voice or by touching a person. And this works astoundingly well, even when a Chinese brain and a German brain speculate about the architecture of their brains in a third language via the restricted means of email. And while the speculation may be mostly erroneous, the two conscious parts in the two brains are pretty confident that doing so is similar kind of fun for both brains which are obviously so different!

Don't you agree? ☺

I'm dealing with this communication-puzzle maybe as long as you are dealing with the puzzle of consciousness and subjectivity and would like to discuss this in more detail if you like.

This also has to do with the question you asked about criterions for consciousness tests and with the puzzling phenomenon of language. But this is a totally new playground, at least one or maybe several levels higher in the functional hierarchy of the brain and therefore let me rather close here.

Good night for today and best to Shanghai.

Karl

Dear Karl,

Thank you so much for your kind words again, although I feel ashamed that I am far from being as smart as you said. On the contrary, your words are just the very ones for describing my feeling about your letters. Especially, *"Without you I would never have taken the effort to dive into all those thrilling matters or to formulate ideas I was dealing with for quite some time. Bringing a thought into writing forces you to formulate it in a much more concrete and precise way as it may have existed in your mind already for some time in a more diffuse state."*

Discussion, especially with a smart friend, is the best whetstone to sharpen one's mind. Many puzzles become suddenly clear. Some ideas, which one has never thought of, would come to one's head during discussion. And the result *"is one of the most rewarding things a human mind can experience."*

I agree with your evaluation of Freeman. I should, ashamedly, admit that I began to realize his great idea about meaning and intention only recently, although I knew him personally for more than 25 years. I think that his idea is ahead of the time, and not many people realize its significance even today. People are still talking about the brain as an information processing system, nothing more. This may be the most significant difference between the brain and the available artefact machine. I also agree that, although we cannot conclude that his theory must be correct, he inspired us to think about brain function in a new way, which is deeply rooted in his studies on olfactory systems for more than half a century.

I should also, ashamedly, admit that I used to underestimate technical progress. After our long correspondence, I have to admit your following conviction must be correct: *"engineers will again outperform natural scientists and will be much quicker in building more intelligent machines than biologists will learn more about the functioning of the brain not to mention such complicated things like consciousness."* I used to overestimate the role of copying biological mechanisms in developing new technologies. Of course, inspiration from living creatures is important for novel technology, however, mimicking only works sometime, but not always. Your guess that *"while technical progress goes at an exponential rate, progress in biology and understanding the brain goes linear"* may be correct, at least qualitatively. In case that novel technology must follow understanding biological mechanisms, the significant discrepancy between the development speed of technology and neurology could not be explained.

I cannot exclude the possibility of building an android passing the TTT someday. According to an axiom we agreed on, I may acknowledge that he may be conscious, at least, I cannot find a strong enough reason to exclude such a possibility. However, at the bottom of my heart, I may still doubt if he is really conscious, as I cannot share his experience, and I know that he is radically different from me with his material and structure. I am not sure if there is anything inner in him, something similar to our feeling, experience, thinking, meaning, intention, or consciousness, even if he declares he has. Or, if he just looks like, sounds like, or even feels like a human being only in view of behavior, without such inner mental activities at all. A zombie! Although it seems impossible that there is a biological zombie, a technological zombie may be possible.

In addition, I have a question about the TTT, or even the Turing Test itself. What does it mean that you cannot judge if the agent is a human being or not. Generally speaking, I think that means that if the agent does not make any mistake a normal person would not make, then he must be a human being. However, what if the agent is too smart, if his or her response is too quick, too perfect, so much that no normal person can do the same. In such case, I suppose most people will doubt that the agent is a human being. We often describe a genius as a person who is different from others. Therefore, it seems to me, that the Turing Test may be a good criterion to judge if the agent is intelligent or not, but not a criterion to judge if the agent is conscious or not. Even for judging intelligence, it is only a sufficient condition, but not necessary. Why must an intelligent agent behave just like a normal person? What do you think?

I am sorry to acknowledge that I have never heard of Lazebnik. ☺ Based on what you described about him, he must be very smart. However, I have never heard of the proverb *"It is difficult to find a black cat in a dark room, especially if there is no cat"* in old Chinese sayings, maybe your explanation is correct.

Your argument about the communication-puzzle must be one of the most difficult problems. While the privacy of the mind seems undeniable, different brains can still effectively communicate each other as you strongly argued. How can the brain do it? How can a brain have the empathy that another brain has? Giacomo Rizzolatti, V. S. Ramachandran, and M. Iacoboni suggested that mirror neurons do it, however, Greg Hickok and others doubted. There seems to be no generally accepted theory about empathy yet, although the empathy phenomenon seems undeniable. There must be some neural circuits responsible for it, is it mirror neurons or the mirror neuronal system? I don't know.

Your letter raised so many important questions, which are not easy to answer, and some of them I have never thought of. They are really inspiring, and I have to read more and think more.

Now I have to shift my topic to the progress of the US BRAIN initiative, although I am awfully sorry to give my comments so late, almost one year later. On the other hand, the delay gives me more time to think about the project.

As you know, although the BRAIN initiative came from BAM (Brain Activity Map project), the Obama administration seemed to be expanding its goal a lot. Many scientists found that the goal of the BRAIN initiative was far beyond the one from BAM, which was thrilling and exciting, but not practical to realize. On May 6th, 2013, the US National Science Foundation (NSF) organized a symposium on the BRAIN Initiative. No consensus was made. As Van Wedeen, the organizer of this symposium, said, all the participants believe there would be a big leap forward, however, as for what kind of leap it would be, and where the leap should go, people debated each other and could not reach a conclusion. At last, all the participants were asked to write a one page proposal, in which, the key topic faced by neuroscientists should be described. In such a way, although no consensus could be made, it was still an important step to investigate which topics are important for brain research, which would be important for laying down a reasonable plan. Now the goal and topics will not be decided by the authority or individual experts only but by many leading neuroscientists. A working team (so called "dream team") with 15 members was organized by Collins' Advisory Committee to the Director (ACD) of the NIH (National Institute of Health) to lay down the goal, contents, timelines, milestones, and cost estimates, which was co-chaired by William Newsome and Cornelia Bargmann, the latter was a critic of the BRAIN Initiative. The task of the team was, firstly, to propose a plan for the first year (2014) and hand it to NIH by September 2013, then lay down a long-term plan and hand it in by June 2014.

Four symposiums were organized by the BRAIN working group to hear different opinions from a broad neuroscience circle. The topics of these symposiums focused on the following fields: molecular technology, large-scale recording technology, computational neuroscience and theoretical neuroscience, and the human brain. There were more than ten invited speakers for every symposium. The speaker discussed with the "dream team" after their talk. Bargmann emphasized that privacy was necessary to allow scientists to speak freely — and sometimes critically — about different experimental approaches. Even so, owing to the broadness of neuroscience, many neuroscientists, including molecular neuroscientists, cellular and developmental neuroscientists, and clinical neuroscientists still felt that their fields hadn't got enough emphasis.

The working group handed in a 2014 fiscal year report to Collins on Sept. 16, 2013, and it was accepted. The long-term report was handed in on June 5th, 2014 and accepted. In both reports, the proposed tasks included: generating a census of brain cell types; creating structural maps of the brain; recording activities of all the neurons at the level of neural circuits; developing a suite of tools for neural circuit manipulation; linking neuronal activity to behavior; integrating theory, modeling, statistics and computation with neuroscience experiments; developing platform sharing with normal and pathological data about human brain; and improving and developing novel technologies for the above tasks. It was proposed that in the first five years, the tasks focus on developing technologies, and in the next five years, transfer to applying these technologies to brain research. Of course, there is no sharp gap between the two stages, only the focal points are different.

From BAM proposed by a few scientists to the BRAIN initiative proposed by the Obama administration, and then to organizing "the dream team" to collect different opinions from leading scientists over broad areas of neuroscience, it seems to me that US scientists persisted in a practical and realistic principle without obeying the will of those in authority, they didn't propose unrealistic goals just for getting giant financial support from the government. Both the goal and the approach became more practical and realistic after repeated discussions.

As a matter of fact, a lot of studies on generating a census of brain cell types have already been done by the US Allen Institute for Brain Science and the EU HBP. The US Human Connectome Project (HCP) started from 2009, has done many works on creating structural maps of the brain, especially on connectomic map of brain white matter, although there is still a far way to go to make the whole brain connectomic map at a synaptic level. As for recording and manipulating neuronal activities, they limited the task at the neural circuit level. Comparing with the EU HBP and the perspective mentioned by President Obama in his speech, their targets are very limited, and their timeline is also much more reasonable. It is suggested that the whole brain activity map of fruit flies could be completed within 10 years, while of mouse within 15 years, and only after that it could be considered to study primate brains. Even so, they did not give any promise when they can get a similar map for primate brains.

Of course, the project has to say something about diagnosis of, protection against, and treatment of human brain disorders to get the support from politicians and the public, but they did not give any timeline. I think there is no ground for blaming such a statement, if it did not give the public an illusive expectation. As a matter of fact, Collins and his colleagues warned:

"*but we must be careful not to overpromise the immediacy of such outcomes.*"
"*Mapping the brain is not as simple as mapping the genome — there will be no linear sequence to decode and no obvious end point. But the lessons learned from this earlier effort — lessons about tool development, ethical implications, and partnerships — can be helpful as we launch a new, even more daunting adventure.*"

In one word, it seems to me that US scientists should be praised by their attitude to their brain project. They have had enough discussion about the goal and approach of the project and focused their target to limited goals which could be predicted to make breakthroughs within a limited period with limited resources. Of course, it is impossible that a project could be agreed on by every person with its every aspect. For example, I am skeptical about the idea that the puzzle of brain mechanism can be uncovered when all the activities of every neuron in the whole brain can be recorded simultaneously. Except for the possibility to do so in technology, the statement sounds like saying that the social problem could be solved at last, if we can record everybody's activity at the same time. As for the problem, what kind of contribution could be made for elucidating the mechanism of a neural circuit if activities of all the neurons in this circuit could be recorded simultaneously, I don't know, let us wait and see.

The letter is already too long, I have to stop here.

Best!

Fanji

17. 08. 2015

Dear Fanji,

Thank you for your kind letter which is, as always, full of interesting thoughts and input.

But before I come to the science news I have to inform you that our daughter Jelka and her husband Jürgen had a baby-boy on July 7th. His name is Alexander and he appeared, quite impressively, in a night full of thunder and lightning. Mother and son are doing well and Adi and I are preparing to, hopefully, become good grandparents. Of course, we are convinced that he's the most wonderful grandson of all, and on top of that, I have a good opportunity to now observe how a young brain develops and how the bootstrapping of consciousness takes place. I will let you know, in case interesting new insights occur. ☺

Although I don't know the precise mind-state of your subjective feelings about our correspondence I'm very happy to hear that we both seem to enjoy our experience of exchanging thoughts. And although we will never know exactly what is going on deep in the other person's mind, my hypothesis is that this subjective pleasure must be pretty similar in your mind and in my mind. ☺

I very much like your expression "*Discussion, … is the best whetstone to sharpen one's mind*". But only open minds can enjoy this process and benefit from it. Discussions between followers of opposed dogmatic schools who are convinced that they possess the one and only truth, are typically of little value. The aim of such debates is often simply to win and to gain victory over another school or ideology. Discussions of this kind often freeze into rituals. The opponents don't listen to each other and the displayed arguments are too foreseeable and therefore such discussions are mostly plain boring. Pretty often established schools of orthodoxy attack new ideas with a similar attitude. Max Planck's autobiography gives an impressive insight into such behavior. When Planck laid the ground for quantum physics as a young man, he had to suffer a lot from the stubbornness and self-centric blindness of his peers. He therefore came to the conclusion that a wrong theory will not die because its advocates realize that they were wrong and subsequently revoke it. A theory will only die, he says with some bitterness, after its protectionists have died. His own experience was probably the reason that, after he himself had become an influential hierarch in physics, he did everything he could to promote the revolutionary ideas of an unknown young Albert Einstein who faced similar resistance from the establishment in physics at the time.

Sometimes we are misguided, not because we are dedicated followers of a specific school, but because we use certain terms, concepts, or views just as thoughtlessly as everyone else. Simply because it is common sense that there is no alternative to this view and that there is no problem involved in such concepts. The role of space and time in the pre-Einstein era is a good historical example in physics, and thinking about the brain as an information processing device, is an example in modern neuro science. I feel ashamed that I, until lately, and also in the course of our correspondence, used this information processing perspective quite thoughtlessly just as if it would be a matter of course. But it isn't, and I simply didn't think about it, like most people don't, and only in the course of our discussion I realized that it is a very misleading, idealized, perspective which logicians and mathematicians imported to a biological subject of which they only had a vague idea.

In this context I have to tell you about two pleasant and encouraging findings which I made, only lately, even though I could and should have made them long ago. One, is that the father of modern computer architecture, John von Neumann, himself already had realized that something was fundamentally wrong with the concept of the computer as an analog to the brain. When he was invited to give the prestigious Silliman lectures in 1955 he chose the subject "The Computer & The Brain". Because of his deadly illness, he wasn't able to hold the lectures and his manuscript was only published, posthumously, after he had died in 1957.

Although the book published under the same title in 1958 is only a fragment, it is very impressive in the way how it displays von Neumann's thinking and his deep analysis of the matter and his comprehension of the problems related to it. It is a delight to read it and one may wish many a modern neuro scientist the ability to express his thoughts in such a clear way as von Neumann was able to do it even on his deathbed.

I don't know whether you have read *The Computer & The Brain*, which is a very small book of merely 80 pages but full of relevant content. In case you haven't, I would highly recommend it to you and, in any event, I would love to hear what you think about it. To me, it makes pretty clear that von Neumann had realized that the brain is not a machine where algorithms are calculated and that it is no digital device but an analogue one. His estimates of the possible accuracy of such an analogue device and the resulting surprise about its, nevertheless, stunning performance are very, very interesting! And so are his speculations about the brain's possible parallel functions. His surprise also comes with the insight that there must have been something fundamentally wrong with the overall theory of the brain at that time. He was very well aware of the fact that what he had designed as the computer architecture carrying his name, had nothing to do with how the brain works, and also that no one at the time had an idea how to solve this very puzzle. To me, it is also clear that he was fully aware that his first try at building a computing machine inspired by the brain and based on the ideas of McCulloch and Pitts about the functioning of neurons and Shannon's thoughts about its logic and information processing fundaments, was a failure. He might have been able to correct it in a second try and probably had wished to do so. But he was not allowed a second try, and so it could happen that modern computer technology developed on the, quite inappropriate and terribly inefficient, principle of a central-time-synchronized digital machine based on logical operations and not on a highly parallel analogue principle. Technical inventions, like biological evolution, sometimes enter blind alleys and it takes a long time to correct such mistakes. The steam-engine and the combustion engine, with their poor energy efficiency and unpleasant side effects, are popular examples. Wilhelm Ostwald, who opted for fuel cells and electric motors, and against heat engines and generators, went furious about those erroneous trends already back in 1894! But it took mankind 120 more years, and the consumption of the major parts of fossil energy reserves, air pollution and what more, to come to the insight that heat engines are not the best technology. It's about 70 years now that we've lived with the von Neumann computer, which is also a monster of wasting energy and the question is whether we have to wait another 50 years to learn how to do better.

In any event, when it comes to the functioning of the brain, von Neumann's insight is about the same as Walter Freeman's and this leads me to my second discovery. You probably came across the Name of Christoph von der Malsburg, the physicist who became a neuro-scientist, often cited as a pioneer of signal processing in the visual cortex. He's also well-known for his contribution to what is called the "Binding Problem" in the theory of cognition.[1]

I knew his name and also that he worked on the problem of face recognition. But I didn't realize how much he shares Walter Freeman's point of view and that he has the same negative attitude against the popular information processing paradigm about the brain.

He too believes that it is mostly analogue phenomena and not digital which control our brains. I'm sure you know more about him than I do and have a better understanding of his role in the neuro sciences. I was happy about this discovery because I like to see that Freeman is not alone with his perspective. And it is no surprise to me that both came to their insight after they worked empirically on two very concrete and well-defined areas of brain functions for many years. One, Freeman, on the olfactory system, and the other, von der Malsburg, on the visual perception system. They must know and appreciate each other because my first research showed that they cite each other. This may not be new to you and you may tell me more about it and whether I may be wrong with my thoughts about the similar approach of the two.

When you say that you have overestimated the role of copying biological mechanisms you probably belong to the majority. There are people however who have an even more distinct opinion and say that we shouldn't be too impressed with what Mother Nature did in the course of evolution. One of them is David Linden an American neuro-biologist who became popular by his book "The Accidental Mind: How Brain Evolution Has Given Us Love, Memory, Dreams, and God". His perspective is refreshingly original and not mainstream at all. Here, a *Newsweek* journalist, Sharon Begley, has nicely summarized his point of view:

"With modern parts atop old ones, the brain is like an iPod built around an eight-track cassette player. One reptilian legacy is that as our eyes sweep across the field of view,

---

[1] https://en.wikipedia.org/wiki/Christoph_von_der_Malsburg

they make tiny jumps. At the points between where the eyes alight, what reaches the brain is blurry, so the visual cortex sees the neural equivalent of jump cuts. The brain nevertheless creates a coherent perception out of them, filling in the gaps of the jerky feed. What you see is continuous, smooth. But as often happens with kludges, the old components make their presence felt in newer systems, in this case taking a system that worked well in vision and enlisting it higher-order cognition. Determined to construct a seamless story from jumpy input, for instance, patients with amnesia will, when asked what they did yesterday, construct out of memory scraps." [1]

This again fits quite well into Freeman's perspective and it also supports the insight that, once you have built an organism (or a machine) on a poor principle it is hard to correct such mistakes in biology (and in machines) and you may be forced to put layer over layer on an ill-fated concept trying to compensate the shortcomings of your earlier concepts.

In many such cases you'd rather start from scratch, which is easier for engineers than for nature, but nevertheless engineers also tend to stay, very long, with inappropriate concepts.

Your question whether Total Turing Tests are appropriate and whether a robot might fail the test because he's too intelligent, is very interesting. What immediately came to my mind as an example was lieutenant commander Data, the famous android at space ship Enterprise, in Gene Roddenberry's Star Trek science fiction series. As you say, he was often too precise and accurate, e.g. when he gave time estimations (*"we will arrive in 2 years, 157 days, 9 hours, 28 minutes, 37 seconds..."*) or cited from encyclopedias word for word in a way a normal person would never do. Being human includes being vague and not precise. So, your question is very smart indeed. I didn't think about it this way before. There is an asymmetry in that a superior intelligence can always pretend to be less smart, while it is impossible to do it the other way around. Turing as a logician must have primarily thought about the problem whether machines can approach human abilities at all.

The other thing is empathy and humor. Those are the real challenge in Data's virtual

---

[1] http://www.newsweek.com/human-brain-marvel-or-mess-97675

life. He knows of this short coming in the architecture of his artificial brain (which is called a "positronic-brain" to indicate the fundamental difference to a human brain) and tries to improve it. So, he goes to the databases and reads all jokes and all humorous stories and watches all comedies ever recorded in history. And although the android realizes that something funny must have happened in a video-recording of a comedian's stage performance because the audience laughs, he doesn't grasp why and tragically fails to detect the real sense of humor. His attempts to imitate humor by mechanically telling funny stories fail miserably because no-one laughs. And when he finally says something funny by accident, he doesn't understand why everybody laughs. Roddenberry obviously had a deep understanding of the difficulty to build a true human-like kind of AI already back in the 1980s. Probably, he understood the problem better than Turing did. Humor and especially irony are really tricky problems. Irony, which needs a second or third layer of perception and consciousness, is the last thing a child learns. And when I look around, my feeling is that many adult humans have never learned the trick. ☺

You are raising an interesting question about empathy and mirror neurons. Is it mirror neurons or mirror neuron systems? I guess it is systems, and in Freeman's theory it would be advisable to look for oscillations over a large area. Freeman makes the point that humans are the only species that have rhythm and can share rhythm. To him sharing rhythm and dancing was the starting point for developing languages. Here I found an interesting article about this by Natalie Geld.[1]

But before we switch to my hobby "language" let me come to your interesting news on the crisis in the big European Human Brain Project (HBP) and about the correction made in the orientation of its American rival BRAIN Initiative.

There are unusual things going on in the HBP. When you informed me last year about the outcry of the European brain researchers against the organization of the HBP and the way it is managed I stated that it looks like there is trouble ahead for Henry Markram. Well, actually he got in very big trouble because the installation of a mediation process alone was mortifying enough. But what the report says, now equals a

---

[1] http://mbscience.org/mbsci/walter-freeman-the-dance-of-consciousness/

condemnation for the proud initiator and organizer. To be criticized and publicly disgraced in such a way is very unusual for a scientist and must feel terrible. I can't think of a solution where he stays with the project in a leading position on the long run. This is really tough and a clear indication that there must have been severe problems within the HBP researchers' community. We know from his dogfight with IBM's Modha that Markram is not the most empathic and sensible man when it comes to arranging with competitors. The organizers and financiers of the project, who have made him king, must have also known it, and they may have run the risk because sometimes such monstrous projects benefit from a strong leader. And, of course, there was a lot of jealousy among those researchers who didn't get their share and who felt that somebody had stolen the governmental money from them and their fields of interest. But when almost all but two members of the mediation committee, including those who participated in the HPB, approved the report, this is a clear signal. Markram must have somehow overdone it with his undiplomatic manner, turning too many people against him. As unpleasant an experience this may have been for the architect and leader of a one billion Euro project, the consequences for the project may be even more important. The loss of face for this prestigious project is enormous because what you have cited from the scientific part of the report, clearly indicates that now everybody agrees that the goals have been too ambitions and the promises were exaggerated. The corrections recommended to readjust the research goals and fields indicate nothing else but that the original concept was a failure.

Well, this isn't nice at all and rather depressing. And although I never liked the concept and always believed that simulating a biological organ which we don't understand on a computer in order to comprehend its function is not a good strategy, I very much regret this development. Especially because the loser in this dogfight is scientific progress and the winner is bureaucracy. I think that in one of my first emails I said that I would rather give a million to each of a thousand PhD students instead of betting a billion Euros on a research monster. In a discipline that is still in its infancy and as long as we have only a vague idea about what is going on in the brain, meticulously planned big research projects in the "man on the moon" style are very likely nothing else but meticulously planned big failure.

At least you should be happy because what the mediation committee states about the concept is exactly what you criticized right after your first examination of the HBP concept and its full-bodied promises! You were very critical about the concept from the very beginning and you were right on the spot with what you published about the ill-fated concept already some time ago. You said that the emperor has no cloths and now — bingo — everybody knows it. And again I'm proud of you!

I wonder what your colleagues say now, especially those who questioned your right or ability to criticize the ideas of such famous Western scientists. ☺

Thank you for telling me about your research around the US BRAIN Initiative. It is an awful lot of work to read all those reports and commentaries and I'm very grateful that you were so diligent and so kind to share your findings with me.

I read, only briefly, about the project so far and your summary is very instructive especially when you compared it with the EU HBP. I think you are right when you are more positive about this second billion dollar brain project and I understand that organizers and scientists have learned their lessons from what happened with the HBP. Although the misery of the latter was only documented in public, quite lately, I'm sure that the US researchers must have been aware of the many problems and, especially, of the discontent of those not participating in the project. The global community of neuro scientists is a small family and I bet that the rolling of thunder from Europe was heard pretty early in the US and elsewhere in the world. Therefore, it is quite understandable that the US managers on both sides (government and researchers) did their best to prevent failures their European colleagues made when preparing the project, laying out the research-plan, and setting up the management. And, of course, the researchers were smart enough to lower the crossbar they promised to jump over. They lowered it a little too much to my taste but quite understandably of course. I would have liked a more competitive and less scholastic approach and a closer link to solving real problems in the real world.

What I find remarkable is the link of the BRAIN initiative with a private foundation, the Allen Institute for brain science. Its chief scientist, Christof Koch, is a very prestigious man and let's hope that he also has the right social skills to deal with his fellow scientists. Do I recall it right that you have translated one or more of his books into Chinese?

So, you may know him and tell me more about his way of thinking.

In any event a lot is going on in the field and we have interesting things to observe.

Good night for today and best to Shanghai.

Karl

Epilogue

August 2018

When our unexpected correspondence began almost 6 years ago, there was no plan and no goal for our journey. Fanji had stumbled across some inconsistencies in the concept of the prestigious and ambitious HBP but he wasn't sure whether his apprehension was justified or not. In the beginning, we were uncertain and simply did what car mechanics do: with much curiosity, we looked under the hood and tried to understand what was going on.

Looking back, we ourselves are surprised at what we discovered and what path our discussion about the brain, the mind, AI and big science projects took.

As we have pointed out in the preface it was more like a random walk; wandering from field to field, stopping to study something in depth whenever we wished. We were guided only by our curiosity, and simply put in more effort when we had the desire to understand things a little more precisely or when we felt the need to fill gaps in our knowledge. And often, we enjoyed following the streams of knowledge back to their sources, including some excursions into the history of our very different cultures. But as chaotic as our journey was, we feel that through our continuous and sometimes controversial debate, we gained insights we would have not acquired had we chosen a more systematic approach.

In retrospect, we are quite surprised at what we thought about certain things, theories and concepts, and in what different light they appear to us today. We have noticed that we have experienced a gradual change in perspective, the extent of which only becomes clear to us in hindsight.

Karl, for example, learned that there are more promising perspectives with which to look at the brain as opposed to the conventional approach, which is to think of it as an information processing machine.

At the beginning of our correspondence, Karl was (and still is) an amateur in neuro science and was simply happy that a seasoned expert like Fanji was willing to discuss such delicate questions relating to the latest state of research with him; while Fanji was (and still is) a layman of information technology (IT), although he is much interested in the rapid progress in recent years in this field. It was difficult for him to find an IT expert who knows brain science as well just like Karl to ask. Experts rarely enjoy discussing their business with outsiders, especially when their opinions may be questioned. Not many of them like to be bothered with the kinds of basic questions children ask, or to express their ideas in a way that children can understand.

We both come from very different fields and our set of professional and cultural experiences couldn't be more diverse. However, we share a passion for curiosity and also for rational reasoning and have a weakness for questioning everything, especially when it comes to established academic insight and the claim for scientific or technological progress.

Although we have already said the following in the preface we would like to repeat it now because of its importance:

We both like to take the perspective of a child who asks simple questions in order to grasp what is happening. Sometimes the child can see that the emperor's new clothes are not as brilliant as they are presented. But we don't want to overdo it with this perspective because it would be presumptuous to say that we are the child in the famous fairy tale "The Emperor's New Clothes" who can see or can't see what others see.

However in the case of HBP, Karl insists that very early on, while others were still praising it, Fanji was able to recognize that there were flaws in this impressively presented project.

We spent a great deal of effort in our early letters demonstrating and assuring ourselves that more than one thing was wrong in the concept of the HBP, and that we shouldn't expect too much from it. It initiated our correspondence, and it became a good model to explore many of the basic elements of the brain and the mind and a possible link with artificial intelligence and computer-technology.

Today, after the facade of this project has been seriously damaged in public such criticism is common and our verve in the old days may seem to some as an attempt to kick a dead dog. Maybe the meanwhile common critique is even too much because in our eyes there are interesting parts in the HBP-concept that have deserved a second try.

Fierce, content-focused debates were once the usual style of scientific disputes, similar to those in sporting competitions. But today this very successful method seems to be going out of fashion, so that one almost has to apologize for using it.

It may also be that this method is accepted more in Europe and may be seen as rude and inappropriate in China.

But whatever the feelings we've triggered in our readers, it's important to remember that we probably owe this harsh method a great deal of the astounding progress made in the golden era of the natural sciences. A glance at publications shows that the debates among scientists at the turn of the 19th and 20th centuries weren't based on polite consensus and harmony, but were often violent disputes. It was often friends who fought as hard as they could for truth and knowledge, but only with the aim of destroying the opponent's theory and not his character. Of course, such fights could hurt and leave wounds, but they also released the energy to come back with better solutions and to fight for ones cause.

In any event, we, and especially Karl, want to make clear that if we sometimes used strong words or metaphors when criticizing a theory or perspective, we did so in this traditional, sportive sense and aimed it only at the argument itself and never with the intention of discrediting an individual.

※　※　※

When Karl received Fanji's first email it was full of questions he couldn't answer and some which he hardly understood. But he was fascinated by the subject and also by the seriousness of the mind he encountered in Fanji, a mind which was trying to understand and fully penetrate the concepts. He was also impressed by the sharp-witted arguments of his counterpart, and his braveness to ask critical questions even when they challenged what great authorities were doing. Even today, Karl wonders whether this unusually determined focus and persistence, not to give way until everything is as clear and transparent as possible, is an original part of Fanji's personality, or a skill he has acquired while translating many of today's most relevant books about neuro science into Chinese.

In any event, it seems to Karl that Fanji has honed his skills in his discussions with many first-class authors to a point that can hurt. Karl got to experience how good Fanji is at forcing an author to make clear what he wanted to say when the time came to translate our English correspondence to Chinese; Fanji came back with questions again and again to make sure that he had caught the true meaning of a statement or sometimes even a single term. This process wasn't always easy and pleasant because sometimes it revealed that the problem was not in the translator's ability to understand, but maybe in the formulation of the idea or, even worse, in the idea itself.

Fanji appreciates Karl's erudition, earnestness and answering every question he raised. He even subscribed to books in order to answer questions. It is difficult to find a friend like Karl who is willing to think, answer and discuss with him the questions that confuse him. Fanji knows that he has found a good friend and teacher.

This continuous process of clarifying what we wanted to say and truly understanding each other's points of view, as well as taking into account subjective and cultural differences, wasn't only a pleasure in its own right, but also had a lot to do with the subject of our mutual interest.

The question of whether and how it is possible to transfer the content of a statement (in the sense of meaning) correctly and completely from one language into another is a recurring theme in our conversation. We came back to it time and again, not only because Fanji is an experienced translator who has often struggled with this problem, but also because it is a well-suited empirical testbed. This translation-problem has a correlation to the general difficulty of transferring content from one brain to another and also to machines. The subjectivity and privacy of the way in which consciousness is composed, as well as how it builds meaning, occupied us a lot. And the question of whether such content can be completely reduced to physical processes and be accessible through objective, rational research is another recurring topic in our correspondence. This of course is relevant to the popular discussion of whether or not, with the help of modern AI, we can build machines that have a consciousness similar to a human's. This fundamental question is the natural link between the two enigmas we are dealing with in our letters. The first is the natural intelligence of the brain, which we understand so little about even after generations of the smartest researchers having spent so much effort to break its secret. The second is the artificial intelligence of man-made algorithms running on computers which are totally different kinds of machines, totally unlike the brain, and that improve their competence with astounding speed.

Actually AI is already superior to human abilities in many respects, something people tend to overlook because the discussion is focused on what machines still can't do.

The key questions we discussed were whether autonomously learning machines will equal and outperform humans in all domains of intelligence, including creativity and other abilities, which many consider to be genuinely human, like emotions and humor. And if so, will they ever be able to develop consciousness and subjectivity at all?

These are issues where we still have different opinions, although our views have altered under the impact of each other's arguments. Before we began our exchange, Karl had thought little about the problem of subjectivity, and under what conditions and how consciousness may emerge. To him, it seemed more like a pseudo-problem that could vanish like the old "élan vital" problem in physics has vanished. Many engineers believe that consciousness may somehow emerge by itself when AI has piled up enough algorithms in high-speed computers to finally perform all intellectual functions our mind is providing. When Fanji insisted that this was not a trivial problem, Karl at first was suspicious whether his opponent may retire to a metaphysical position, and thereby leave the rational turf of the discussion. And Fanji wasn't quite sure whether Karl had grasped the core of the problem that had kept him from sleeping for many years. On the other hand, Karl felt that Fanji may have underestimated the pace with which hardware was still developing, and may have shared the popular belief that Moore's law has to come to an end soon.

We explored these themes together and tried to get an overview of the developments in the various related disciplines in science and technology. And by exchanging arguments back and forth, we finally managed to better understand the issues and what our counterpart's position was. With the difficult case of consciousness, Karl was impressed that Fanji did not escape into metaphysics but presented strong arguments supporting his position.

We finally agreed that a 1 : 1 to copy of a body and brain could contain internal states of subjectivity, and even the mysterious qualia. But Fanji was successful in demonstrating that it's close to impossible to create such a copy. Karl, on the other hand, had to realize that it may be much more difficult than he had hoped to build strong, general AI systems that can develop consciousness and subjectivity. And Fanji had to admit that he had underestimated the speed of AI's development.

So when asked today, Fanji may still be skeptical of whether strong AI is possible at all, while Karl would still opt for a positive answer. But although we never expressed this explicitly in our letters, we both have a feeling that we have managed to plant a seed of doubt in the mind of the other.

In any event, we both agree that mind-uploading is close to impossible, and that it will take much longer to build even mere artificial zombies than the Kurzweil flock hopes. Not to mention, an allegedly near singularity. And we also agree that establishing neural links between machines, the body and the brain is much more difficult than many believe.

Another question that occupied us was how engineers could be so successful with their concepts of computers and AI software, even though both were based on obviously quite inaccurate models of the brain?

And would it help engineers to reach their goals if they would try to make their machines more brain-like? And would neuro scientists be inspired to better understand the brain and the mind by knowledge of AI concepts and getting more help from engineers, mathematicians and logicians? And would bigger and faster computers help in simulating, and thereby improving, neurological brain models?

We both agreed that a computer-simulation of a highly dynamic functional system like the brain which we don't even understand on the physical level, not to mention the higher functional level, doesn't make much sense. However, trying to create one may help in improving the tools needed to achieve understanding.

We also finally agreed that trying to reengineer the biological brain on silicon isn't a promising approach. Karl proposed early on that engineers wouldn't benefit much from the results of brain research, and would be better off going their own way and disregarding the biological model. An idea that seemed strange to Fanji in the beginning, but which he finds less absurd now.

We still can't answer many of the questions that puzzle us, nor can we be sure that what we have agreed on is valid, and of course, we are not claiming that we have gained any new scientific insights.

※　※　※

The point is, that the two of us have made progress in getting a better view of the field in order to gain relevant insight for ourselves. The journey we have made was very rewarding for us, and it may inspire others to explore the field in a similar way. Readers can decide which position suits them more, and some may disagree with both of us.

By publishing our correspondence, it is not our goal to proclaim theories, but rather to propose a method of acquiring knowledge and exploring areas previously unknown to us as individuals.

As an inspiration for those who are interested in this kind of subjective exploration, we would like to reiterate once more what we considered to be the most helpful and inspiring perspective.

There was one man who helped us more than anyone else in discussing all of this, and who gave us a theoretical framework of reference which we found to be very helpful. This was Walter Freeman, Fanji's friend, who sadly passed away during our correspondence. We acknowledge him as one of the most eminent thinkers and researchers in the field. It's a pity that although he's well-known, he hasn't received the recognition in mainstream neuro science to the extent he has deserved in our eyes.

For Karl, being introduced to the work of Walter Freeman by Fanji, which isn't easily accessible, was the most challenging and rewarding experience in their correspondence. Some years prior, when Karl was confronted with the neural system for the first time, he had looked at the brain as an information processing machine, and had believed in the possibility of mind-uploading, and that the paradigms built around logic and mathematics by von Neumann, Shannon and Turing seemed to be the right approach to crack the enigma. However the more he looked into the problem, the more he grew uneasy about the digital nature of this approach and the idea of the brain as a device doing calculations.

Freeman was the first person Karl encountered who understood as much about the brain as one possibly could, and who not only shared this unease, but went even further by calling the principle of "information processing" as a basic misconception of brain function. And via our discussion, even Fanji gained new insight into how helpful Freeman's framework could be, although he had already translated Freeman's work to Chinese.

Trying to understand Freeman's perspective, and especially his point that our brain doesn't just calculate information but is trying to filter meaning in a continuous process of interaction and loops of "circular causality", had a pivotal effect on both of us. Although this perspective may be imperfect, or not applicable for all brain regions and functions, and without a doubt will not be the last word spoken in this debate.

We still do not agree on other things, such as the essence of subjectivity and whether and how consciousness may emerge from matter and be acquired by machines. We also still have a diverging estimation of some authors like Chalmers and Dennett. But our agreement on the value of Freeman's work is much more important than what we still do not agree about.

To us, Freeman's perspective, which he had already presented in "*Consciousness, Intentionality, and Causality*" (1999), might also make a promising theoretical framework that could be applied and tested in research projects like HBP and the BRAIN Initiative.

The idea that our brain is not a digital information processing machine built around logic doing numerical calculations, but is instead, an analogue device that filters meaning from noise, may not sound attractive to mathematicians and programmers eager to solve partial differential equations on ever quicker computers. Nevertheless, we hope that it could be an appealing and inspiring starting point for young, unbiased explorers looking for a different entrance to the magic castle.

If we manage to inspire even just a couple of young students or one or two experienced researchers to try out our idea, our work will have been worth the effort.

In addition to questioning the miracles of the brain, the mind and AI, we wrote a lot about the methods of doing science and how best to make progress in knowledge. Naturally, we don't claim the right to tell others how to do their science, although we have widely criticized how some research is conducted.

The point is, that science itself is subject to rational reasoning, and it is apparent that some methods and research arrangements are more promising than others in the quest for cracking scientific enigmas. First of all, it is noticeable that the treasure of substantial knowledge that we have at our disposal in the natural sciences has been accumulated to a large extent by former generations. And when we look at the curve of knowledge growth, it is amazing how steep the increase was in the phase around the turn of the 19th and 20th centuries. And it is all the more astonishing how small the group of people who prepared the ground for our scientific worldview was, and how little it cost.

If we look at it as a building, we realize that the basement and the first few floors were built by just a few craftsmen in a short time. Today, a whole industry of craftsmen is working at the construction site, yet progress seems to have slowed down and even minor decorations are costing more than those first couple of floors.

Some people argue that the modern science industry has cranked out more knowledge over the last couple of years than has been accumulated in all of history. This may be true if we judge by the amount of papers published. But what if we judge by "creative competence" instead? Some measurement that would indicate what we can do today that our ancestors couldn't. This is quite a trivial and pragmatic perspective, but may be a very helpful one.

We have stated that physicists, and especially engineers, have performed better than life scientists in this discipline, and that engineers seem to be progressing at an exponential rate while the advancement in neuro science and medicine is developing only linearly.

We have been especially critical and have expressed dissatisfaction with the output of many research projects and results published as "interesting findings" where patients are waiting for cures. Maybe we have done this to an extent which was too much for many who are working hard in these fields and may find our critique unfair and depressing. This may especially be the case when they are involved in big science projects that leave them little freedom of choice. But again, we are not aiming at any specific scientists but are simply discussing what caught our attention, and fully understand how difficult the remaining problems are, after low-hanging fruit have already been harvested by our predecessors. Fanji thinks that a big scientific project may be important and feasible, if it is technical in its essence, has a solid foundation of scientific theory and aims at solving a well-defined important task. The multidisciplinary development of new research tools and the large-scale collection of basic brain data may be important for brain research, and need huge investment and team tackling. But this does not solve all the problems in brain research, especially those in which creativity and insight plays key role.

※　※　※

So our point is not that we don't see enough wonders, but that some are promising more than they can deliver which seems to push the whole sector into a mode of overselling. This is intensified by the established academic reward-system where publications and citations are the currency. This favors a tendency in science and research to sell mere leafage dressed up as substantial fruits simply because many published "interesting findings" around a few enigmas may be a safer way to make a career than the risky way of trying to crack a hard enigma.

We don't have an easy solution to this problem but it wouldn't be right not to mention it.

We are well aware that criticizing is easy and doing better is difficult. And although we don't believe that we have a miracle-solution at hand, we don't want to avoid trying to give an answer to the question: "what should we do?"

When we summarized our conversation and prepared for this epilogue, Fanji asked the question below with reference to a comment Karl had made in one of his early emails when he, half-jokingly, said that he would rather give one million to a thousand smart students than pour a billion into a research monster like HBP:

"I agree with what you said about the state of the art in medicine, but there is always a question in my mind, after our discussion, i.e. what should we do in medicine? Cut financial support to these researches? I don't think so, although your questioning is correct, it seems that the way in the past did not work, 'big science' seems also not working, then what should we do? Follow the example of information technologies? Owing to the different natures, this may be difficult. Then what else we can do? Give the money to many post-doctors? Maybe, but what should they do?"

The questions Fanji had raised made us think and discuss this problem once again.

Karl still believes that a seemingly crazy idea like betting on the creativity of a few young, wild and obsessed students may be a valid alternative in some cases. But of course it's not a realistic alternative for all big science, and it may only be an option for small experiments simply because big science is an established, multi-billion dollar industry that can't be revolutionized so dramatically over night. This is even more so in medicine, which is a highly regulated sector with many legal restrictions and ethical boundaries that have been established for good reasons. All this adds to the inertia that has been building up in science for years now, as we can see from Max Planck's complains from over a hundred years ago.

So when we try to give a pragmatic answer to this "what should we do?" question, we don't have the battleship captains or the fleet commanders of big science in mind because they already know all about the calamities we are describing. What we propose may only increase their frustration when they look at how little room there is for maneuverability when it comes to even small course changes in such large structures.

So the most we can expect from them is to hope that they don't take us for naive romantics. Our correspondence suggests that, indeed, we are both a little romantic for determined and self-reliant thinkers who are permanently concerned with a problem and won't let it go until it's solved.

But we are of course aware that modern science is a team sport and that it needs different kinds of people to build a successful team. Besides the individual skills and ambitions, it is the team spirit that is decisive, and it is meaningful that in the history of science, this team spirit, in the sense of a passion focused on a given problem, has been very helpful.

So when we make recommendations we have passionate individuals and small teams in mind, like the captains of small research-speedboats and their crews, and especially young students who already feel that they have a high potential, and are looking for a real challenge and who might profit from a little encouragement.

To them, the very simple and not surprising advice is this:

- Work hard and wrestle intensely with a matter/problem that fascinates you (as Walter

Freeman did), very much as a sportsperson preparing for the Olympic Games.

- Try to become a passionate expert in a field early on, and truly master it in depth.

- Don't mistake good grades and winning prizes for real insight or true mastery (which means avoiding what Feynman describes as the scholastic mistake of the Brazilian university).

- Find/produce real fruit and don't fool yourself into mistaking leaves for fruit.

- Don't dress up negligible results into "interesting findings" just to publish them.

- Don't impress just your peers, but also others in the community who will realize your caliber and cooperate with you to crack a well-defined enigma.

Truly, we don't think that most of those whom we have in mind really need this advice, because they are already operating this way, and they are so busy and obsessed by what they are doing that they wouldn't care about our advice anyway.

But there may be some who are still undecided which way they should go and may get some benefit from our outlook and some encouragement if they are already leaning to the side we prefer.

Karl got an interesting response when he showed his daughter and some of his young friends the draft of this epilogue to see how our advice sounds to the audience in question. They argued that we should not only focus our attention on the genius of exceptional inventors and discoverers, but also on the slightly less gifted that can also make important contributions. Most of us do not fall into the genius category and it doesn't help us to behave as if we do. However less exceptionally gifted people can make very helpful contributions in teams, often by compensating for the typical deficits of the genius-type people.

This is a valid argument and it's especially true when it comes to building teams and companies where scientific inventions have to be transformed into products. And here it is often the case that social skills can carry far greater weight than scientific excellence. We have mentioned previously that much of science is a team-sport where the right orchestration of resources is crucial for success and we should not forget this lesson!

※ ※ ※

There is another, more general, point that stands out when we look back to the golden age of the natural sciences, which is true for everybody participating in the game. It is the average age of a student, who is finally starting out with his or her own research. Typically, they are much older today than a hundred years ago, simply because it takes so long for academic rituals to be completed. The question is whether it really has to take so long in all cases?

There are encouraging examples that this is unnecessary. A hundred years ago, a couple of 19-year-olds were encouraged by their academic peers to redefine the world of physics and they did so quite successfully, leaving their professors utterly baffled. The driving force behind their accomplishment was curiosity, ambition and a burning desire to understand and hopefully crack a fascinating enigma. When Wolfgang Pauli and Werner Heisenberg left gymnasium and arrived at university, they already knew almost everything that there was to know about the theory of relativity and quantum theory at the time. Not because they were super intelligent child-prodigies, but simply because they had been fascinated by a very specific puzzle for a long time. The two were lucky that their passion and skill was recognized by their professor, Arnold Sommerfeld, who accelerated the fulfillment of their academic perquisites. Not everybody in the faculty liked this, but Sommerfeld's brave move was rewarded by the fact that both youngsters did their major work at age 24 which later secured each of them a Noble Prize.

Today, we also have such passionate youngsters in the world. Whatever their fascination is, it keeps them up at night, and the more obsessed they are the more they tend to neglect everything else around them because if it has nothing to do with their primary interest, they find it boring. Traditional academic entry criteria, based on superior school grades or the skill to excel in tests, tends to filter such people out.

So the question is, how do we attract these people and bring them to the playground?

Of course 19-year-olds can't crack all the enigmas in science. But science isn't so different from other kinds of high-performance sports where we often see a decline in performance after the age of 30.

When we look at history, we see how many breakthroughs we owe to extremely young researchers, especially in math, where we have always had a high fruits to leaves ratio. A basic principle of mathematic statistics is the method of least squares. It was developed along with what we call today "curve fitting" by an 18-year-old, Carl Friedrich Gauss, in 1795, and is the core principle around which modern AI deep learning algorithms are built.

Of course we are not suggesting 19-year-old hacker-neurologists should be doing brain surgery on their neighbors in the garage. Not least because there are good reasons for protecting patients by established professional rules and regulatory bodies.

The point is, that other very bright young people have done similarly crazy things in their garages with electronics and software and created entire new industries over the last few decades. Many of them were university drop-outs who found it more compelling to do wonders in the real world than conform to the rituals of academia.

And this brings us to the point of the technological earthquake we are experiencing right now.

We didn't use the term "trade war" in our correspondence but towards the end it could be read between the lines. Yet similar terms made news headlines at the time and everybody talked about it after the strong US reaction to China's announcement of becoming a main world center of AI innovation by 2030.

We were quite surprised to find the technical part of our field of interest in the middle of this debate. It seemed to have come out of the blue after China had made this announcement.

When we started our deliberations over the understanding of the brain, intelligence and AI this was a subject for initiated people and a small group of experts. Today AI is a dominating topic in the media, and is discussed everywhere from science to economics and politics. In 2016, some had predicted the AI hype-bubble and that it would see its next winter soon. But the opposite happened: AI became the most hyped-up technology ever. And now after China's declaration of her AI project and US' strong reaction, interest is ever increasing as the world follows with interest as this showdown develops.

We felt as though a tsunami had hit the small river on which we had been calmly rowing whilst contemplating the topic in our little boat.

Everybody seems to have an opinion on AI now, and to understand what is going on. Of course we don't claim that we have a better understanding of what's going on in the field, or what will happen next, only because we have been dealing with this subject a little longer than some of those who talk and write about it today.

We are even impressed by how well some journalists and analysts have managed to wrap up the essentials of the field and to highlight the crucial points and challenges. That's another example of the enormous help which older versions of AI, like the good old internet with its archaic search engines, provide when it comes to gaining insight into new fields and complicated matters.

And we, too, profited from this increased popularity of the field because many clever people have shed more light on the topic which helped us gain new insights.

So, readers may hear us talk about things that were little known until just recently and have become popular almost overnight. We don't think that we were much off-track with our reasoning, and hope that it may encourage some to have a closer and sober look at the AI-issue.

The major, and maybe game-changing, effect of this AI-hype is that even more money is now flowing into the field. At the time of the big brain research projects one could get the impression that no government of an industrialized country could afford not to have such a project. This is even more true for AI, especially after China revealed its ambitious plan last year. And the price tickets are much higher. A billion dollars seemed to be an breathtaking sum for a research project five years ago, but now, Cambricon in China, a single start-up company working on a hardware-chip to run AI-applications, is valued at a billion dollars (at the time we checked this we found that it had raised to 2.5 billion in June 2018). And there are an incredible number of such start-ups popping up like mushrooms everywhere.

A side effect of this rush is that people who understand how to build AI systems, and especially those who are also competent in the neuro sciences, are very much in demand. Industrial leaders all over the world are trying to recruit these rare experts from where ever they can, especially from universities and research institutions. This is good for skilled people, but a problem for these institutions because they often can't compete financially. At the same time, however, it also opens up opportunities for new forms of cooperation between academia, industry and investors to build start-ups and all kinds of research arrangements and new forms of research institutes. It is amazing to see how many creative constellations are currently emerging and how quickly bureaucratic obstacles are removed. It seems that China is particularly creative and fast in this respect.

It is therefore possible that some of the problems described above, which are related to purely academic research, will also be solved this way. We would not be surprised if part of the research will shift from universities to industry and that we'll experience growth and development similar to the IT industry over the last few decades, where relevant basic research migrated to the laboratories of big companies like Bell-Labs or IBM, which also produced a number of Nobel Prize winners.

Because passion and team spirit is so compelling in such research projects, we see an advantage for our favorites, the small speed boats, over big science battleships, at least in the early phase.

But this is already speculative, and we cannot make guesses to predict the future. Right now we don't see much more than what we have discussed in our letters, the major difference being that we may see change faster.

We left out many things or only mentioned them briefly. For example, the vast field covering the possible social effects of AI, such as people losing their jobs to robots or concerns over the privacy of personal data.

We didn't go deeper into this because we didn't see it. We simply didn't do it because there is already so much speculation and noise in this discussion that we didn't know what we could substantially add to it, and, we also didn't want to trump others with even wilder negative speculations.

However there is one more observation we want to add.

We often cited John von Neumann who was certainly one of the true geniuses of the 20th century and made more contributions to modern science than most. One of these was his work on game theories which he published in 1944 together with Oskar Morgenstern. It was almost as successful as the computer-architecture carrying his name. But as impressive as it was, it also came with some drawbacks. The major one being that it focused on fixed-sum games where the victory of one player was the loss of the other. This perspective became a very popular paradigm, especially among generals of the Cold War era.

It took another scientific maverick and mathematical genius, John Nash, to demonstrate that in economics there are more promising cooperative strategies for improving welfare than trying to defeat a competitor.

And those cooperative strategies surely apply in the process of gaining knowledge in science and technology, and especially when it comes to a war worth fighting, such as the one on Alzheimer's, Parkinson's and similar scourges of mankind.

And thus, the Chinese scientist and the German engineer agree that we better off doing this in friendly competition rather than under the false assumption that we are playing a zero-sum game. And if this idea does not spread easily in people's minds, there is still hope that the next generation of artificial neural networks will find out for themselves.

Scan the QR code for
more information about
the brain research and AI.

This is the first time I have translated a book coauthored by myself. Although translating the part written by myself is easier, as I know what I wanted to say, almost two thirds of the book were written by Karl, which is not so easy for me to translate. This is due to his deep thinking and broad knowledge, especially when the topic involves advanced technology, western philosophy and science history which goes beyond neuroscience. Fortunately, Karl is kind and patient enough to answer my questions in detail, or even re-write paragraphs so that I can better understand them. This makes the translation be true as possible to the original. Therefore, I would like to express my sincere gratitude to my dear friend for his kindness and patience. Without his help, I really could not have translated the whole book in its present form. This reminds me of a problem with current machine translation, which doesn't understand the meaning of the text! Even a poor human translator can know something of the original text that he or she doesn't understand, and if he or she is serious, then he or she may consult dictionaries, encyclopedia, background materials or even ask the original author or other experts, until he understands. However, the present machine translator never asks a question, it even translate a typo as if it were not a typo!

As we wrote in our preface, this book is mainly to inspire readers' thinking rather than tell them what conclusions we came to. The authors do not pretend to know everything and have all the correct answers, and neither do they want to pretend to be prophets and predict the future. They just want to tell readers what they thought based on rational reasoning and hope readers take a similar approach. I have written several popular science books on the brain and mind, and one of my friends once asked me how I could guarantee all that I said was correct. My answer was "No, I could not guarantee that, and I don't think any author could, what I could do is make sure all the material I talked about has a strong foundation and can be justified or rejected empirically." This is also true for this book, where to take things even further, the two authors still have different opinions on some topics, just as Karl summarized in our epilogue. Each of us had their own reasons, but could not persuade the other to believe, while for some topics we may agree but to a different degree. For example, we both think that the simulation of the whole human brain is unrealistic and perhaps even meaningless, but to simulate neural circuits and the like with most parameters within the physiological range is meaningful. To understand the brain one cannot depend purely on big science projects. However, when the subject involves very expensive installations, huge data collections with sound theoretical foundations and requires broad cooperation between experts from different fields, big science project may still be necessary. The problem is the content of the project and how to organize it. And how to have a proper balance between the battleship and the small speedboats. We both agree that technology develops exponentially, while neurology and medicine develop only linearly, but is this difference due to the dominance of "Interest-performers" in the former, and "System-performers" in the latter? Furthermore, in the translator's eye, there is also struggle between "Interest-performers" and "System-performers" in the latter, although the degree is different. Maybe the difference stems from the nature of the two fields, although the different degree of the struggle between the two types of performers may also play role. These are still open problems, and deserve the readers' careful pondering.

Anyway, writing this book and the discussion with Karl has changed many of my ideas, and even the way I think. I hope readers with similar ideas as mine may have the same feeling. Translating gives me a chance to review the whole process and to tidy up my ideas. Now I hope that as I am at a new starting point and can see these open problems from a new perspective, I may see a little clearer and am more sensitive to problems and the rapid progress of these fields. And although these problems haven't been solved yet, and are not expected to be solved in the foreseeable future, it is exactly this point which makes study of these fields so charming.

Fanji Gu

Acknowledgements

The most relevant currency in the world isn't money but time. Our family, friends and colleagues devote to us with their attention and cooperation. We are very grateful for the enormous amount of time others have devoted to us until this book could be presented.

We both enjoy the rare and unlikely happiness of being married to a woman who is an intellectual partner who has been tirelessly reducing entropy in our lives for more than 5 decades. And we both enjoy the good fortune of having a daughter who is also a friend and valuable advisor. Yuemei Ma, Adi, Yimeng Gu, Jelka and Jürgen Seitz deserve our thanks first and foremost. Special thanks to Adi for arranging and editing our drafts in endless loops. And we owe Jelka and her husband Jürgen exciting and controversial discussions about the conclusions to be drawn from our investigations and how they could be implemented. As a now 4-year-old mentor and teacher of his grandfather, Alexander Seitz may not have noticed how much he helped him with the question of how mind and consciousness develop.

With regard to causality, our thanks go to our mutual friend Prof. Hans Braun ("German Hans"). He initiated our contact, our friendship and thus this book. Some chaos-experts say that a butterfly's wing beat in Shanghai can't cause a tornado in the Caribbean. But Hans, a distinguished chaos-expert, has demonstrated that it is possible to make two people work for six years by just redirecting a Shanghai-email from Marburg to Karlsruhe. A performance that should make butterflies, tornados and chaos-experts think!

There is a dear friend and colleague of "German Hans", to whom we also owe a special debt of gratitude. It is Prof. Hans Lilienström ("Swedish Hans") a longtime friend and valuable advisor of Fanji.

The support and advice we received from Fanji's friend Prof. Nelson Y. S. Kiang was especially helpful and we would also like to thank him for his kindness, which allows us to quote from his personal communication.

There were many other friends and colleagues who helped us answer specific questions or expand our research in the course of our conversation. We cannot name them all and would like to mention some of those whose support has been particularly encouraging.

In China it was Yizhang Chen, Tiande Shou, Peiji Liang, Pei Wang, Xiaowei Tang, Aike Guo, Xiongli Yang, Bo Hong, Si Wu, Jintao Gu, Hongbo Yu, Longnian Lin, Ying Pan and Amenda M. Song.

In the West it was Dietmar Harhoff, Gert Hauske, Ehrenfried Zschech, Matthias Bethge, Dirk Kanngiesser, Susanne and Rafael Laguna, Ingo Hoffmann, Serdar Dogan, Ben Hansen, Andreas Bogk, Bernd Ulmann, Hendrik Höfer, Mirko Holzer, Miro Taphanel, Matthäus Paletta, James Wright and Leslie Kay.

Y. Chen, P. Liang, P. Wang, X. Tang, J. Gu, A. M. Song, H. Braun, D. Harhoff, G. Hauske, E. Zschech, and M. Bethge were so kind to read the full manuscript. They made valuable suggestions and comments which helped us to reduce errors and unclear passages before printing.

We are infinitely grateful that they have also taken on the great effort of writing commentaries and reviews, which you find as introductions to some sections of our letters. While we are ashamed that they are generally much too friendly and generous with us, we believe that they perhaps convey the essence of our discourse better than we could have done ourselves.

Thank you to Patrick Hartmann, Agnieszka Paletta and John Broomfield for fixing some of the worst mistakes in our Chinglish-Denglish gibberish without losing the original tonality of our emails.

Very special thanks go to our editors: Mr. Wei Huang and Ms. Mingyue Shen from Shanghai Educational Publishing House for their great support and many helpful ideas on how to make the book better, such as the suggestion to have several tables of contents, and the wonderful form to make the book both in Chinese and in English. They have also invited an artist, Mr. Chuqiao Chen, to draw drawings for the Chinese content and suggested what should be drawn. Of course, we would also like to express our gratitude to Mr. Chen here.

We don't know if Fanji's friend Walter Freeman would have liked what we have to say, but we both feel the need to express our gratitude, admiration and respect by dedicating this book to the memory of this great researcher whom we deserve so much inspiration and insight.

本书由上海文化发展基金会图书出版专项基金资助出版

"科学的力量"科普原创

"脑与人工智能"系列

The Brain and AI

意识之谜和
心智上传的迷思

一位德国工程师与一位中国科学家
之间的对话

顾凡及　[德]卡尔·施拉根霍夫　著

顾凡及　译

上海教育出版社
SHANGHAI EDUCATIONAL
PUBLISHING HOUSE

图书在版编目（CIP）数据

意识之谜和心智上传的迷思：一位德国工程师与一位中国科学家之间的对话：汉英对照／顾凡及，（德）卡尔·施拉根霍夫（Karl Schlagenhauf）著. —上海：上海教育出版社，2019.8
ISBN 978-7-5444-9038-2

Ⅰ.①意… Ⅱ.①顾…②卡… Ⅲ.①意识-研究-汉、英 Ⅳ.①B842.7

中国版本图书馆 CIP 数据核字（2019）第 173648 号

内文插图：陈楚侨（图ⅡF2.1、图ⅡK2.1、图ⅡF6.1、图ⅡK7.1），
　　　　　弗里曼（图ⅡF9.1、图ⅡF9.2）

意识之谜和心智上传的迷思
—— 一位德国工程师与一位中国科学家之间的对话
顾凡及、[德]卡尔·施拉根霍夫（Karl Schlagenhauf） 著　**顾凡及　译**

出版发行　上海教育出版社有限公司
官　　网　www.seph.com.cn
地　　址　上海永福路 123 号
邮　　编　200031
印　　刷　苏州美柯乐制版印务有限责任公司
开　　本　787×1092　1/16　印张 23.25　插页 6
字　　数　370 千字
版　　次　2019 年 11 月第 1 版
印　　次　2019 年 11 月第 1 次印刷
书　　号　ISBN 978-7-5444-9038-2/N·0021
定　　价　66.00 元

如发现质量问题，读者可向本社调换　　电话：021－64377165

作者介绍

顾凡及

顾凡及是复旦大学生命科学学院计算神经科学退休教授。

1961年毕业于复旦大学数学系，1961年至1979年在中国科学技术大学生物物理系工作，1979年到复旦大学生命科学学院工作。在1983年至1985年期间，作为访问研究助理，他在伊利诺伊大学厄巴纳－香槟分校生理学和生物物理学系工作。

他的专业是计算神经科学。他出版过3本专著和约100篇论文。2004年退休后，他于2006年至2011年担任《认知神经动力学》杂志的责任编辑。他还主编了三次国际会议论文集。

之后，他成为脑科学的科普作家。他撰写了6本关于脑科学的科普书籍并翻译了7本书，其中包括弗里曼的《神经动力学》、科赫的《意识探秘》、埃德尔曼的《意识的宇宙》和拉马钱德兰的《脑中魅影》。他的科普读物获得七项奖，包括2017年中国好书、2016年上海科学技术奖（科普书，三等奖），2015年上海科普教育创新奖（科普书，二等奖），他本人还获得了2013年第四届瑞典认知神经动力学国际会议授予的成就奖，2017年上海市科普教育创新奖（个人贡献，二等奖），以及2018年度上海市科技进步奖三等奖（科普人才）。

他现在是上海市欧美同学会留美分会顾问、伊利诺伊大学校友会上海分会名誉会长、全国科学技术名词审定委员会生物物理学名词审定委员会委员和科普公众号"返朴"的编委。

卡尔·施拉根霍夫

　　卡尔·施拉根霍夫是新技术领域的连续创业者，在欧洲和美国的高科技领域创立了许多初创企业。他曾担任应用组织研究所（Institute for applied organizational research, IFAO）和 ADI 软件公司（ADI Software）的首席执行官，该公司专注于工业和银行业的 RDBMS、多媒体和互联网应用，这些都是源自他在 20 世纪 80 年代于卡尔斯鲁厄大学的创业成果。

　　他于 2003 年卸任首席执行官，现任 ADI 创新公司（ADI Innovation）的董事会主席，并经营家庭办公室。他还是他投资的许多公司的董事会成员，并曾在 AP Automation + Productivity（现为 Asseco Solutions），Brandmaker, CAS Software, JPK Instruments 和 Web.de 等多家技术公司的董事会任职。

　　作为一名发明家，他拥有通过互联网安全遥控和用于原子力谱蛋白质分析的纳米机器人方面的专利。

　　他是私募股权公司和政府机构的顾问，也是年轻企业家的教练。

　　虽然他一直关注着前沿技术及其对社会系统和对人力、智力和财务资源的微妙协调的影响，但他也将自己的兴趣从软件、电子学和制造业扩展到了生命科学和人工智能。

　　他拥有卡尔斯鲁厄大学（现为 KIT）授予的经济学／工业工程硕士学位、哲学博士学位以及社会学和科学理论的"特许任教资格"（Venia Legendi）。

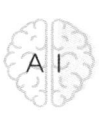
这是一本奇书。对人类洞察力的本质以及我们如何能在对此的认识上取得进展感到好奇的每个人，阅读此书都可以有所得。

——马修·贝特格（德国图宾根大学伯恩斯坦计算神经科学中心主任，教授，deepart.io 联合创始人）

这是一本不同寻常的书，很明显地从许多其他神经科学出版物中脱颖而出。应该向所有希望挣脱传统的主流研究窠臼的 IT 专家和神经科学家推荐此书。

——汉斯·阿尔贝特·布劳恩（德国马尔堡菲利普大学生理学教授，生理学与病理生理学研究所神经动力学研究组负责人）

一位德国工程师和一位中国生物物理学家之间的精彩对话，我们应该感谢他们写了一本有关人工智能、脑活动和意识的好书。

——陈宜张（中国科学院院士，第二军医大学教授）

我如痴似醉地读完了全书，激动得想在有一天会把书拍成电影。

——塞尔达尔·多甘（德国导演兼电影制作人）

我以前只有在很少一些情况下才享受过这样的读书乐趣。它是思想和见解的宝库，时而有趣，时而非常严肃，但总是非常乐观。

——迪特马尔·哈霍夫（德国马克斯·普朗克创新与竞争研究所所长）

中国历史悠久的智慧加上神经生理学的专业知识与西方社会科学和信息技术间的碰撞，"我们需要这样一本书吗？"我的回答绝对是肯定的。

——格特·豪斯克（德国慕尼黑理工大学退休教授，
《生物控制论》杂志前主编）

我真诚地希望由凡及和卡尔撰写的充满原创思考和见地的图书，能够激发科学、政治和产业界之间新的合作方式，以创造下一波"真正的"通用人工智能，从而造福所有人。

——拉斐尔·拉古纳（Open－Xchange 首席执行官，
德国颠覆性创新开发署主任）

作者以理性的视角和客观的思维，带着读者一同探讨智能时代背后的飞腾与迷思。

——梁培基（上海交通大学生物医学工程系教授，
中国神经科学学会理事）

无论你是对人工智能与大脑充满懵懂的好奇，或者是已经在某一个具体科学领域有所涉猎，又或是想给自己的业余生活增加些情趣的读者，相信这本真诚、有趣而深刻的对话录都能带给你丰富独特的体验。

——宋蔓（美国加利福尼亚大学圣迭戈分校博士生）

这是一本发人深思的书，所有对脑科学和人工智能感兴趣的读者都应读一读这本书。

——唐孝威（中国科学院院士，浙江大学教授）

一本如何进行科学讨论的范本，所有对脑科学、认知科学、人工智能、科学哲学等领域有兴趣的读者都会喜欢这种观点和思想的碰撞。我确信在近年出

　　　　　意识之谜和心智上传的迷思　　一位德国工程师与一位中国科学家之间的对话

版的大量关于人脑和人工智能的图书中,本书应该算是最深刻的之一。

——王培（美国天普大学副教授，通用人工智能学会副主席，

《通用人工智能》杂志执行主编，

《哥德尔、艾舍尔、巴赫——集异璧之大成》译者之一）

一本扣人心弦犹如阿加莎·克里斯蒂侦探小说的科学书。

"传统的科学辩论文化"风格，即在书信中严肃地交换意见并基于坚实的基础推进认识，令人耳目一新。我们不应该忘记这些技能，这在一个推特和短信的时代更显得尤为重要。

——埃伦费里德·切希（德国弗劳恩霍夫陶瓷技术与系统研究所系主任，

教授，欧洲材料研究学会理事）

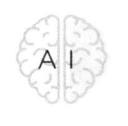

　　在发展人工智能时,不必理会脑机制是否会更好一些? 飞机并不复制鸟类的方法,为什么人工智能要复制脑的方法呢? 计算机和人脑中的信息处理之间是否存在原则差异? 神经科学曾经取得过像物理学在 20 世纪初取得过的那种进展吗(普朗克、爱因斯坦、海森堡等)? 在当代教育和年轻研究人员的晋升问题上,往往鼓励人亦步亦趋地追随权威而非鼓励批判性思维,这种倾向是否存在不当之处? 为了获得数亿美元或欧元的大额资助,人际手段是否比科学概念更重要? 成功申请到了这种巨额资助的人是否真的相信他们的承诺,例如可以用数字计算机完全仿真人脑? 在还不知道人脑中与意识相关的物理机制的情况下,我们怎么能谈论起有意识的人工智能呢? 对脑理解脑有原则性的限制吗? 人工智能能否取得更大的成就?

　　这些只是这一优秀著作中讨论的一些主题和问题类型。它以一种独特、引人入胜和趣味盎然的风格写成,这并不仅仅是因为两位作者采用电子邮件通信的形式,它特别为科学及其有关事务提供了全新的观点。尽管本书向读者全面介绍了脑研究和人工智能活动的现状,但其重点并非是要介绍这两个领域在近年来所取得的成就。相反,它指出了许多悬而未决的问题,揭示了关于脑和人工智能的许多广为流传的神话,并且对当前流行的一些研究概念进行质疑,从而批判性地揭示了一些"皇帝的新衣"。

　　这段电子邮件交流始于 2013 年初欧盟"人脑"计划快要获得批准的时候,当时我的两位朋友几乎不约而同地问我对这个计划的看法。当我的朋友继续讨论下去时,由于许多其他任务,我不得不脱离了这种思想交流。当我现在看到这些讨论的结果时,我当然觉得很遗憾没能继续参与交流,这不仅是因为我美慕我的朋友们出版了这本优秀著作。我认为,作为一名工程师和神经生理学

家,我本可以提出更多的论据。我并不赞同作者的所有观点,这正如两位作者尽管在辩论时态度友好,但他们也远没有完全达成一致一样。

但这正是本书的优点之一:它开启了新的视角,激发了批判性思维。看到两位不仅来自不同研究领域而且来自不同文化背景的作者如何理解对方的论点,并从不同的视角出发以新的形式重新阐发,这一点特别有意思,甚至可以说非常有趣。除了他们的智力之外,作者们还拥有那些人造系统(包括 Siri 和 Alexa)仍然缺失之处,即难以言传的幽默感。

我的德国朋友卡尔经常喜欢用他对科学史和科学理论的广博和深刻理解来强调他的论点。令人印象深刻的是,我的中国朋友凡及,虽然不太熟悉这种主要出自西方研究的背景,他急切地接受所有这些想法并将它们(通常以惊人的新结果)整合到他自己的世界观中,然后巧妙地告诉卡尔他作为一名东方神经生理学家来看的新见解。

另一方面,我的中国朋友凡及尽力为卡尔提供有关脑的各种信息,卡尔反过来又在人工智能设计的背景之下重新进行解释。对于工程师卡尔来说,他很难接受在神经生理学中经常不精确地使用科学术语,也难以接受把知识和推测混为一谈。他专注于近年来在技术上的巨大进步,特别是在计算机科学和信息技术(IT)领域中的飞速发展,这正展现为人工智能新的光明前景。尽管如此,他们都不得不承认,到目前为止,还没有人知道是否可以和如何设计具有可与人类灵活性和创造性(包括意识)媲美的人工系统,以及是否应该这样做。随后,他们又对计算机和人脑中信息处理的异同进行了深入的讨论,这也包括了对古老的心脑问题的讨论,即有关心理功能和意识如何从物理机制中产生的问题。这对他们两人来说又引发了一个同样有意思的问题,即发展人工智能应该在多大程度上基于类似于脑的机制,还是采取模仿心智的策略要更好一些,也就是说,寻求最合适的算法来模拟所需的功能,而不是试图在所有细节上都复制生理机制。

有趣的是,凡及和卡尔从未见过面。我在各种会议上见过凡及,特别是在他发起的"认知神经动力学国际会议"上。他在科学上力求准确、朴实无华的性格,特别是他高度的幽默感给我留下了深刻的印象。本书就充满了这种幽默感,使阅读特别愉快。我从高中时代起就认识卡尔。我们作为卡尔斯鲁厄大学(现为KIT)的学生而重聚。直到今天我仍然感谢他允许我参加他的卡尔斯鲁厄朋友圈。

这本书立刻让我想起了我们往日的学生时代，我们常常在晚上聚在名为"Pschorr - Fässle"或"Bürgerstüble"的两家啤酒屋一起痛饮一番。通常参加者都是跨学科的，卡尔常以他的口头禅开始我们的会议："今天有谁有问题要讨论?"讨论越激烈，我们离开酒吧时就越满意，智力上的收获也就越大。在这本由两位意气相投的伙伴合著的书中，仍然可以觉察到同样的对激动人心的讨论的追求。

有一件事是肯定的：这是一本不同寻常的书，很明显地从许多其他神经科学出版物中脱颖而出。应该向所有希望挣脱传统的主流研究窠臼的 IT 专家和神经科学家推荐此书。特别是，应该推荐给每一位对神经科学和人工智能感兴趣的年轻科学家。他们将学到很多关于脑和人工智能研究现状、关于科学作为一项大产业、关于科学史和许多其他问题的知识。最重要的是，本书中的见解将加强批判性思维，并防止盲目追随科学大师及其教条。仅仅出于这个原因，我就特别要推荐这本书。此外，这本书还提供了许多非常有意思和有趣的阅读材料。

汉斯·阿尔贝特·布劳恩[1]

生理学博士，Dipl. Ing.[2]

生理学教授

神经动力学研究组负责人

生理学与病理生理学研究所

马尔堡菲利普大学

德国

---

[1] 汉斯·阿尔贝特·布劳恩(Hans Albert Braun)博士是德国马尔堡菲利普大学生理学和病理生理学研究所的生理学教授和神经动力学研究组负责人。他对生理学和神经科学方面的许多杰出贡献都记录在《混沌》(Chaos)杂志的一期特刊中，该特刊是 2017 年 6 月为纪念他杰出的工作而举办的一次科学会议的成果。请参阅：

《特刊简介：生命系统的非线性科学：从细胞机制到功能》("Introduction to Focus Issue：Nonlinear science of living systems：From cellular mechanisms to functions")，Epaminondas Rosa, Svetlana Postnova, Martin Huber, Alexander Neiman & Sonya Bahar, Chaos 28, 106201(2018)；由美国物理研究所出版，doi：10.1063 / 1.5065367.

在线查看：https：//doi.org/10.1063/1.5065367

[2] 拥有工程学位证书。

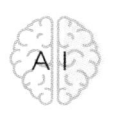

人类的崛起一方面靠的是人的思考能力，另一方面也靠双手的技能。对此的先决条件是由于直立行走而发展出大而复杂的脑，以及可以自由移动的双臂和手。还有一个关键就是有智能的人类行为中的概念和社会因素。本书作者对上述主题深表关注。他们说："我的印象是，大多数神经生物学家和心理学家过分关注单个孤立的脑。他们一般忽视了人脑只有作为有巨大内容的脑网络的一部分才能有意义地工作的事实。"这是一个重要且有价值的信息。显而易见的是，目前的脑研究向我们提出了如此深刻和基本的问题，仅凭一切都要计量的自然科学还无法给出答案。

因此之故，本书非常引人入胜的一点是，两位作者来自不同的专业领域和不同的文化背景，其中一位是中国的计算神经科学家，而另一位是西方的社会科学和信息技术专家。我认识凡及已经很久了，他是一位具有批判性思维的学者，优秀的教师，一些科学期刊的编辑和许多科学论文、科普书的作者。他毕业于数学专业，而后投身生物物理并专攻神经信息处理。他很幸运结识了卡尔这样一位德国的社会学家和信息处理专家，并交流思想。他们的讨论从神经生理学深入到社会问题，从单细胞活动到社会合作，从而大大扩展了他们的视野。这种学科交叉的背景对理解本书中丰富的观点是非常必要的。

本书采取了书信的形式，这使他们的论述更带有个人色彩，比起一般书籍来说，作者们在论述时带有更多的个性。作者们以一种非常带有个人色彩的方式交流思想和发表独创性的看法，这令读者有一种很美妙的体验。作者们恰当地引用了相关文献，这为读者提供了深入探讨相关主题的宝贵线索。因此，其表达就好像是带有练习的讲稿，这从教育学的角度来说非常好。我们知道，科学和科学发展都不是静态的和抽象的，而是与人类及其交流密切相关。他们对

有关脑科学、社会发展方面的介绍都闪耀着智慧的光芒,但也涉及应该怎样做以及在何种情况科学才能健康发展的问题。作者们合理地批评了当代加强脑研究中的某些尝试,不仅因为这些研究消耗了大量资金。某些声名很大和雄心勃勃的这类尝试中的许多概念中的一些矛盾引起了他们的通信讨论。作者说,通常有大量的生物学数据,但这些数据并没有告诉我们脑是如何运作的。

在非常广泛的视角下,跟随作者们从逻辑网络、控制论、感知器、神经网络到人工智能(Artificial Intelligence,简称"AI")的不同角度回顾人们怎样试图理解神经系统功能的过程非常令人兴奋。当然,人工智能扮演着重要角色,作者们对此也比较乐观。然而,一些早期的 AI 承诺被作者们批评为不切实际。某些 AI 鼓吹者要他们的读者相信所有问题都可以通过更强大的计算机能力和应用足够的数学来加以克服,但他们真正需要的是更深入的认识。这些鼓吹者希望读者相信,只要有了正确的算法,那么世界上的一切问题都可以通过计算解决。但是按照作者们的想法,有些重要的现象并不是只靠算法和数值计算就能解决的。

智能依赖于思维过程,这些过程可以是逻辑的、问题求解的或归纳的。归纳性思维具有发现规律的能力,这被认为是人类智能的主要组成部分之一。作者们探讨了人工智能是否可行的问题,其答案取决于人们如何定义它,这显然不是一项简单的任务,因为有 70 种可能的智能定义。作者们也讨论了和意识有关的问题,还牵涉到人工意识是否可能,但作者认为研究人工意识还为时尚早,虽然这并不意味着人工意识就绝无可能。

很明显,与技术系统不同,脑能够产生意图,目标和意义(弗里曼)或预测未来(霍金斯)。为了解决一个特定的问题,脑不仅可以像下国际象棋那样在短时间内处理大量数据,而且还能想出解决问题的适当方法(王培)。因此,他们不仅可以提高能力,还可以创造新的能力。但是从知道脑是如何由一个个细胞构成的,到认识整个系统的复杂功能,这一过程非常困难。作者们说,工程师可以从大自然的方法中获得启发,但不应该在每个细节上都复制生物学。

另一个非常重要的问题涉及人类通过社交的行为能力,即在人际间进行合作和信息交换,也不排除竞争。早在 1940 年,德国人类学家阿诺德·格伦

　　意识之谜和心智上传的迷思　　一位德国工程师与一位中国科学家之间的对话

（Arnold Gehlen）就清楚地指出了这一点。格伦的见解表明，为了生存，人类被迫以社会组织的方式行动、工作和思考。事实上，正是因为人类不能只靠天然的本能来保护自己，这才使他与完全适应自然的动物区别开来。按照格伦的说法，人为自己创造了所谓的"规章"，如婚姻、国家、宗教和经济，最终表现为我们所谓的文化，这一点非常重要。格伦认为，人是一种"深谙生存艺术的机器人"。对于格伦来说，"规章"是文化的缩影，它为人类提供了必要的保障。语言和写作是建立一个超越遗传进化的快速发展世界的重要工具。10万年前发明的语言是一种"智能放大器"，一如不莱梅的脑研究人员格哈德·罗斯（Gerhard Roth）所言。语言也是社会行为的重要工具。

在这种背景下，很有意思的是，昆虫群体早已表现出惊人复杂性的合作和行为特性，而在较低的复杂程度上，甚至连细菌群体也是如此。因此可以假定，人类的自然智能可能是基于各种生物都共有的同样的基本原则。人工智能也许并不像它看起来那么人工，因为它是自然脑的产物，并且和所有各种生命系统一样都要服从相同的规则。要想认识这些规则并应用于人工智能可能仍然需要更多的研究。如果以一种负责任的方式把人工智能融入我们的社会，它或许会成为人类生存的重要工具。

总之，我应该说两位作者是写这样一本书的很理想的一对合作伙伴。看看中国的智慧加上神经生理学的专业知识与西方的社会科学、信息技术如何交融在一起非常吸引人。当我问自己："我们需要这样一本书吗？"我的答案绝对是肯定的。

格特·豪斯克

慕尼黑理工大学退休教授

《生物控制论》杂志前主编

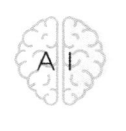

　　仔细算算，认识这本书的作者之一，顾凡及，顾老师，已经是7年前的事情了。那时的我是复旦大学生命科学系的一名大四学生，对人类的意识和心智都非常感兴趣，想申请认知与大脑方面的博士项目，在搜索相关书籍时发现了顾老师翻译的几本关于意识的经典书籍，并由此也幸运地认识了顾老师本人。这些年，承蒙顾老师的指引与鼓励，我开始了在加州大学圣迭戈分校（UCSD）认知科学系的博士生生涯。这些年，我的兴趣渐渐由意识与大脑转向了人工智能和社会科学，脚步渐渐从学术界迈向了工业界，也能够更加从多个角度来看待当初感兴趣的心智问题。非常幸运，能够在第一时间读到顾老师的这本关于脑与认知的佳作。意识与心智是一个学科高度交叉的问题，它吸引着来自神经科学、心理学、社会科学、哲学与计算机科学等领域的学者。这本书的两位作者来自不同的背景，但双方都被共同的问题所吸引，从不同的角度求索着关于大脑与智能这一终极难题的答案。

　　阅读这本对话录，是一次奇妙的旅行。在沿途，你可以感受到一如顾老师以往的科普作品一样的平易近人和引人入胜，但不同的是，这份旅途里融入了两位思想者的碰撞，迸发出了别样的"1+1>2"的智慧火花。这本书不仅仅能带你领略关于脑和人工智能方面的最前沿研究，对各个方向的重要问题与研究方法手段有全局的清晰认识，更重要、可能也更有趣的是，它还是两位学者的思想与生活的珍贵刻录本。

　　你会跟着他们一起，逐步领会一个想法是如何引出了新的想法，一个问题是怎样触发了共鸣，继而又引发了新的思考。你可以看到他们对研究方法与方向的方法论上的反思，对"权威"的不盲从和挑战，对人工智能方面最新进展的热切关注与解读，在滚滚向前的人工智能浪潮中对未来大趋势的预测、思考和

判断。你可以看到两位理性而富有好奇心的学者是如何取长补短，用友好而富有建设性的方式为对方补充自己擅长领域的知识，像海绵一样持续吸收和补充新的知识，并激发出有深度的思考。你得以窥见两位知识渊博的学者富有幽默又包容开放的日常，看他们如何信手拈来，娓娓道出科学史与技术发展史上的趣闻轶事，又对自己不了解的事物和领域保持谦逊与坦诚。你还能见证两人的友谊是如何在一轮轮的讨论中逐渐加深，成为彼此思想精神上的挚友。

　　无论你是对人工智能与大脑充满懵懂的好奇，或者是已经在某一个具体科学领域有所涉猎，又或是想给自己的业余生活增加些情趣，相信这本真诚、有趣而深刻的对话录都能带给你丰富独特的体验。希望你也和我一样，有幸跟随两位作者，一起开启这场奇妙探索之旅，领略沿途关乎科学、关乎人文、也关乎作者本身的美丽风景！

宋　蔓

加州大学圣迭戈分校博士研究生

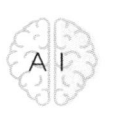

　　本书即使不说是绝无仅有，也是很独特的。两位作者是在不同的文化氛围中成长起来的，从未谋面。卡尔是一名德国的工程师和企业家，而凡及则是一名中国的脑科学家和科普作家。我们是在6年前通过一位共同的朋友，神经生理学家汉斯·布劳恩（Hans Braun）教授的介绍结识的，此后一直就脑研究和人工智能（Artificial Intelligence，简称"AI"）方面的问题进行通信。

　　我们成了好朋友，通信频繁，甚至超过了与一些多年老友的通信。我们对脑、心智、意识和人工智能之谜的共同兴趣维系着我们的友谊，我们以极大的热情共同关注着这些领域中的迅速发展。

　　我们都喜欢理性思考的方法，而且我们总是渴望追究事物的原因和理由，而不是随大流或囿于学究式的思维。由于我们经历的不同，我们的观点也有明显差异，卡尔在产业界工作，而凡及则在学术界工作。凡及的工作主要是创造知识和传授知识，而卡尔则致力于如何通过技术应用来利用知识。

　　在跨学科领域和多种技术行业中工作几十年后，我们都到了法定的退休年龄。然而，科学家和企业家是永不"退休"的，因此我们以更大的热情去利用自己的时间和经验。我们享受由此得到的自由，我们不用再为前程操心，也不用考虑要给同行留下深刻印象。不再受到这些约束而只凭自己的兴趣行事真是妙不可言，我们都非常享受这一点。

　　就像卡尔经常说的那样：我们就像两只自由而快乐的鸟儿，可以待在喜欢的任何一棵树上，讨论感兴趣的东西。但这并不意味着我们就漫无目标或要求不高。

　　首先，我们总是要求自己尽可能好地了解我们感兴趣的复杂领域中所发生的事情，并评估它们将如何发展。

其次,凡及有许多对此深感兴趣和挑剔的读者,卡尔有很多人(年轻的科学家,工程师,企业家以及工业和政府部门的管理成员)在这个问题上征求他的意见。所有这些人都有理由要求我们所说既非信口道来也不肤浅。

我们的讨论是从一个问题开始的。凡及在考察了后来名满天下的欧盟人脑计划(Human Brain Project,简称"HBP")的技术概念之后,向汉斯·布劳恩问了一个和特定类型神经元有关的问题。汉斯把这个问题转给了卡尔,我们小小的旅程就从2013年1月正式开始了。这样就有了一系列电子邮件,我们的讨论从神经元开始,延伸到人工智能的最新发展以及某些人所谓的中美之间的技术和贸易战。

本书就是将我们的通信经过重新组织以后的结集。其中的信件都是按照昔日的辩论文化传统写成的,按照这种传统,科学家们在精心思考的信件中交换看法并进行有争议的讨论。当然这并非我们的发明。事实上,这是一百年前科学的黄金时代科学家们进行交流和完善他们的想法的常用方式。在推特和短信服务大行其道的当今,这看起来有些过时,现在所有内容都必须以标题表达,几秒钟内即可读完,讲得快也忘得快。

对于更习惯于达成共识的年轻科学家来说,我们信件中的对抗性语气可能读起来有些奇怪。然而,应该提到的是,对抗方法是目前在最先进的人工神经网络应用中引入的一种非常有前途的技术。以老式的对抗方式进行交流可能非常耗时且要求很高,但对于那些喜欢深入探究以便彻底了解真相的人来说,它也可以非常高效和有益。

如今人们已不太习惯写长信了,但信件比普通出版物有一个很大的优势。它们不那么正式,为创造性甚至猜测留下了更多空间;它们使说话的人更容易改变立场,从而向对方学习。你还可以用更平易的语气提出更为尖锐的问题,并直抒某个想法。我们发现这种方法对于我们感兴趣的、内容迅速变动的领域非常有用,在这里没有什么东西已有定论,而且还流传着种种迷思和概念滥用。

特别是在中国宣布将在2030年成为世界主要人工智能创新中心的雄心之后,许多人对我们已经关注了很长一段时间的那些问题感兴趣起来了。

为了清楚起见,需要对相关的科技现状以及有可能实现的前景和极限进行

意识之谜和心智上传的迷思　一位德国工程师与一位中国科学家之间的对话

清醒分析。

这是我们在这次热潮开始之前就已经在尝试做的事情,如果也有人想了解这些领域中正在发生的事,并将事实、流行观念、现实希望、梦想和营销噱头区别开来,那么我们愿意和他们分享我们的见解。

你现在读的既不是一本科学教科书,也不是典型的科普书,更不是对这两个学科的系统或完整的介绍。我们所做的更像是随意漫游,从一个领域转悠到另一个领域,随着我们的意愿不时停下来深入探究。我们只是受到好奇心的驱使,当我们想要更准确地理解事物或者当我们觉得需要填补我们的知识空白时,我们就会加倍努力。通常,我们喜欢对知识追根究底,也包括我们不同文化的历史回顾。但是,尽管我们的探索看似无序,我们觉得,通过我们持续的、有时甚至是有争议的辩论,我们得到了如果选用了更系统的方法得不到的见解。

我们都喜欢从孩子的视角来看问题,他们会提出简单的问题,以了解真相。有时孩子可以看到皇帝的新衣并不像所说的那样华丽。但是我们也不想过于夸大,因为说我们就是著名童话故事《皇帝的新衣》中那个勇敢说出看不到别人"看到"的东西的孩子,就未免太自以为是了。

然而就 HBP 而言,卡尔坚持认为,从很早开始,当其他一些人还在赞扬它的时候,凡及就认识到这个令人印象深刻的计划存在缺陷。

我们在早期的信件中花费了大量的精力来说明并使自己确信在 HBP 的概念中有多处错误,我们不应该对此计划期待过高。由此开始了我们的通信,它成为探索脑和心智及其与人工智能和计算机技术的可能联系的许多基本方面的良好试验田。

今天,在这个项目的名声在公众面前已严重受损之后,这种批评很常见,而我们过去的批评在一些人看来似乎有点像在打"落水狗"。也许现在一般性的批评甚至过多了,因为在我们看来,HBP 概念中也确实有一些有趣的部分值得再作尝试。

除了讨论有关脑和人工智能的各种迷思之外,我们还讨论了理性思维和意识问题。在此期间,在脑科学和人工智能研究中都发生了若干重要事件,例如

美国先进创新神经技术脑研究（Brain Research through Advancing Innovative Neurotechnologies，简称"BRAIN"）倡议的启动，阿尔法狗（AlphaGo）击败前世界围棋冠军和自动驾驶汽车上路等。我们跟踪了这些令人印象深刻的事件，并讨论了如何评价它们的重要性。一些新进展支持了我们的推测，并鼓励我们进一步讨论。一些进展甚至超出了我们最好的期望，我们不得不重新考虑我们的观点并从错误中吸取教训。所有这些都激发了我们的讨论，重新聚焦要讨论的问题，引发新的争论，有时这会使我们改变想法。在某些问题上，我们达成了共识，对于另一些问题我们仍然存在分歧，还有某些问题我们从未找到过任何答案。

我们并不指望所有读者都会阅读所有的信件。一些人可能是神经生理学方面的专家，他们只想知道可以指望从"仿神经结构"芯片中得到些什么，而另一些读者则可能是熟练的人工神经网络编程人员，他们希望更好地理解把生物神经元和人工神经元相提并论有什么问题。还有些人可能对脑的生理细节或计算机架构都不太感兴趣，而是对一些我们两人都仔细考虑过的问题感兴趣，这些问题包括意识的涌现、自由意志、模因的意义、自组织、循环因果关系或是在生命科学和工程学科中的研究组织问题。我们曾请关心科学组织的一些朋友审阅过本书的草稿，他们觉得我们有关官僚主义大科学弊病的讨论对他们很重要。还有些人可能会忽略技术细节，并喜欢像看旅行纪录片那样，观看两位作者在生物学和计算机技术这一困难的交叉领域中摸索前行的故事。

我们提供了两种不同的目录，那些只对特定问题感兴趣的读者，可以由此找到包含这些主题的信件。

把我们之间的通信出版成书是后来才想到的。开始想到的读者群是那些对科学和技术感兴趣的雄心勃勃的外行人，他们希望更好地了解在这两个热门的领域中的事实和迷思。但是当我们的讨论在一些地方深入到许多细节之后，如果没有适当的背景知识就很难理解。因此，本书的中文部分用专栏、脚注和插图进行解释，给予读者辅助的材料，从而帮助读者知道我们所讨论的问题的有关基础知识。增加这些材料需要花费大量时间，但丛书内容也因此而丰富起来。为了方便阅读，我们把本丛书设计为三卷。

对于那些对我们的推理总结以及我们所得到的结论感兴趣的人,我们增加了一个比通常要长得多的跋,总结了我们最重要的发现。对于好奇的读者,就像在读侦探小说时一样,急不可耐地想及早知道凶手是谁,那么可以先阅读这个跋。

当然,我们并不声称已对所有问题都给出了答案。事实上,我们的探索甚至还没有完成,许多问题仍然没有得到回答,而新的问题又产生出来了。我们也并不声称我们比其他人懂得更多。在某些部分中我们表现出来的自信和坚定的语气不应该被误解为对自己立场的绝对把握。这只是老式辩论文化中惯用的方法。这样做只是要把某种立场尽可能清楚地表达出来,这并非是为了捍卫这种立场,而是为了请对方对它进行反驳。

我们的许多假设、结论和评论可能都是错误的或不完整的,需要尽快予以纠正。问题在于我们不知道哪个是错的。无论如何,在我们经常大肆批评别人之后,我们也准备好接受读者的批评。

如果我们的看法最后被证明是错了,我们可能并不会因此感到高兴,但如果我们想要取得进展,这是不可避免的。我们都认为,在不断通过实证研究挑战理论的过程中,理性思考是增加我们知识的最佳方式。但是理性思考本身并不能代替在现实世界中的实践活动,而在这个过程中,理性思考也不能代替研究人员和工程师的好奇心、勇气和雄心壮志,尤其对年轻一代来说更是如此。

对于一些人来说,看到我们对自己心智之谜所知之少,以及我们在认识心智问题上进步速度之慢,可能会感到失望。而引人注目的是,在该领域的技术方面,则进展要快得多。还有些人可能会把这当成是进入一个非常有前途的工作领域的机会。在这个领域中,对于那些准备打破传统的人来说,可望收获令人难以置信的有价值的发现。

我们都确信我们并没有做出任何重要的发现。但是我们都希望能激发某些人才来试试运气,并找到魔法城堡的新入口。并非所有人都会成功,但我们希望许多读者能够发现我们的见解是有帮助的,并且会像我们在过去 6 年里所做的那祥,享受在这魔法城堡中的漫步。

顾凡及,卡尔·施拉根霍夫

丛书内容概览^[1]

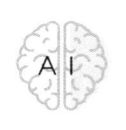

本套丛书共包含 31 对书信,讨论了有关脑和人工智能的一系列开放性问题。这些讨论和争辩贯穿整套丛书,但是在各个分册中侧重面又有所区别。其中第一册《脑研究的新大陆》包含 11 对书信(编号 I 001 - 011),重点是讨论脑研究中的开放性问题;第二册《意识之谜和心智上传的迷思》包含 10 对书信(编号 II 001 - 010),重点是讨论对意识研究的不同见解和心智上传的可能性问题;第三册《人工智能的第三个春天》包含 10 对书信(III 001 - 010),重点讨论人工智能的潜能和前景。在所有三册中也讨论到科学方法论和科学组织的问题。三册图书的书信内容是按照时间进行排序的。为了便于读者从整体上了解整套丛书的内容,我们提供了如下列表:

1. 引言(I - 001 凡及)

2. 科学方法论

兴趣派与规矩派(I - 002 卡尔,I - 003 凡及,I - 003 卡尔,I - 008 卡尔,II - 003 凡及)

自然和工程采用不同的方法(I - 002 凡及,I - 003 凡及,I - 005 凡及,II - 004 卡尔,II - 009 卡尔)

不同学科的不同思想习惯(I - 003 凡及)

科学家之间的竞争和合作(I - 003 卡尔)

学术争论(I - 004 凡及,I - 004 卡尔,I - 005 凡及,I - 006 凡及,II - 002 卡尔)

---

[1] 这是一个按信件内容进行分类的目录,也可以说是索引。括号内的数字表示信件的编号,表示在此信件内有这方面的内容,但并非只有这方面的内容。

理论、模型和仿真（Ⅰ-001 卡尔，Ⅰ-002 凡及，Ⅰ-004 卡尔，Ⅰ-006 卡尔，Ⅰ-010 凡及，Ⅰ-011 凡及，Ⅱ-003 卡尔，Ⅱ-008 卡尔）

循环因果关系和线性因果关系（Ⅲ-002 卡尔，Ⅲ-003 凡及，Ⅲ-005 卡尔，Ⅲ-009 卡尔）

3. 脑研究中的一些未解问题

脑研究是万里长征（Ⅰ-001 凡及，Ⅰ-005 凡及，Ⅰ-005 卡尔，Ⅰ-006 凡及）

神经元（Ⅰ-002 凡及，Ⅰ-009 卡尔，Ⅰ-010 凡及）

功能柱可能并非皮层的标准模块（Ⅰ-002 凡及）

记忆（Ⅰ-007 凡及，Ⅰ-008 卡尔，Ⅰ-009 凡及）

情绪（Ⅰ-007 凡及，Ⅲ-001 卡尔）

连接组（Ⅱ-009 卡尔）

脑是一种信息处理系统还是创造意义的机器？（Ⅱ-009 凡及）

社会的脑（Ⅲ-001 卡尔，Ⅲ-002 凡及，Ⅲ-002 卡尔）

4. 人工智能中的争论问题

智能（Ⅰ-001 卡尔，Ⅰ-002 卡尔，Ⅱ-005 凡及，Ⅲ-003 凡及，Ⅲ-003 卡尔，Ⅲ-004 凡及，Ⅲ-005 凡及，Ⅲ-005 卡尔，Ⅲ-006 凡及，Ⅲ-006 卡尔，Ⅲ-007 凡及）

"图灵测试"和"中文屋"思想实验（Ⅱ-005 凡及，Ⅱ-010 凡及，Ⅱ-010 卡尔）

神经网络和深度学习（Ⅱ-001 卡尔，Ⅲ-004 卡尔，Ⅲ-008 卡尔）

机器翻译（Ⅲ-001 凡及，Ⅲ-001 卡尔，Ⅲ-002 凡及，Ⅲ-003 卡尔，Ⅲ-008 凡及，Ⅲ-008 卡尔，Ⅲ-009 凡及）

阿尔法狗（Ⅲ-003 卡尔，Ⅲ-010 凡及）

人工智能领域正在发生范式转换（Ⅲ-004 卡尔）

奇点（Ⅱ-002 卡尔，Ⅱ-003 凡及，Ⅱ-004 凡及）

摩尔定律（Ⅱ-004 凡及，Ⅱ-004 卡尔，Ⅱ-005 卡尔）

脑启发计算和脑样计算（Ⅰ-010 凡及）

仿神经结构系统（Ⅰ-010 卡尔，Ⅰ-011 凡及，Ⅲ-007 凡及，Ⅲ-007 卡尔）

逆向工程（Ⅱ-003 卡尔，Ⅱ-004 卡尔）

弱人工智能和强人工智能（Ⅲ-004 凡及）

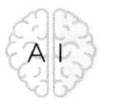

目录[1]

---

[1] 本目录给出了书信列表和专家点评信息，其中包含了每封信所讲到的主题和关键概念，并列举了一些与主题相关的专栏资料供读者参考。

性;恰当的问题是意识从什么样的脑活动中涌现出来的,以及涌现意识的充分必要条件是什么;科学家致欧盟公开信批评 HBP;美国 NIH 顾问委员会工作组提出对 BRAIN 倡议的建议

意识之谜和心智上传的迷思　　一位德国工程师与一位中国科学家之间的对话

意识；困难问题；简单问题；主观性；私密性

2013-11-20

亲爱的卡尔：

　　非常感谢你的美言。在我的那篇文章中，应该写上致谢，以感谢和你、布劳恩和其他同事的讨论，由于编辑给我篇幅上的限制，我很遗憾不得不省略了这一部分。

　　如果我在投寄那篇文章之前就已经读过亨利·马克拉姆（Henry Markram）的公开信，我很怀疑自己是否仍然可以如此"勇敢"地发表我的文章。你对此事的深刻分析一定是对的，我就做不到这一点。

　　你的话让我想起 HBP 有关意识或觉知的话题。在《人脑计划——向欧盟委员会递交的报告》（2012 年 4 月）中，有标题为"意识和觉知"的一节，其中宣称：

　　"HBP 平台将提供机会，通过用现有模型，或开发新模型，并在硅片上做实验以测试这些模型。这样的实验可能会引发根本性的突破。"

　　而在 HBP 网页的常见问题答复（FAQ）一栏中，它也宣称：

　　"仿真人脑将为我们提供对脑工作方式的深刻见解：我们的知觉、思想、情绪直至我们的意识的来源。"

　　然而，早些时候在蓝脑计划（the Blue Brain Project，HBP 的前身）网页的

"有关蓝脑计划问题的回答"一栏中,对"脑模型中是否会涌现出意识?"的问题,其回答是:

"我们真的不知道。如果意识是由极大规模的相互作用产生的话,那么这也许有可能吧。但我们真的不明白意识究竟是什么,所以很难说。"

又一次是以前的回答更为可信,而后来的说法则可能会给人一种虚假的希望。虽然在发表这两段陈述之间有一段时间,但在这段时间里,在意识研究上并没有发生什么突破性的进展可以解释这种变化。

意识是我非常感兴趣的话题之一,它困扰着一代又一代的哲学家和科学家,而且还没有得到解决。在关于意识的基本问题上也还存在许多不同的意见。幸运的是,我碰巧读到英国科学作家布莱克莫尔(Susan Blackmore)的一本书《意识对话》[1](*Conversations on Consciousness*)。为了探索意识研究的现状,她采访了 21 位顶尖科学家,并向他们提出了许多有关的基本问题和关键问题。虽然此书是早在 2005 年出版的,有些旧了,但它至少能让我知道直到 2005 年为止有关意识的主要不同观点。

在采访一开始,她总是问同样的问题:"为什么意识如此特别? 这种研究的关键难点是什么?"大多数科学家都认为问题在于如何解释物理的脑产生主观意。澳大利亚哲学家查默斯(David Chalmers)把这个问题称为意识研究的"困难问题",而把如何通过脑活动解释由此产生的行为或可观察的功能称为"简单问题",尽管解决这些"简单问题"一点也不简单。⊗在阅读了这 21 次采访内容后,我发现也许可以把被采访的科学家分成两个阵营。第一个阵营的科学家同意"困难问题"的看法,他们认为意识研究的特殊困难就在于意识的主观性和私密性。在意识研究开展之前的科学史上,科学家们总是用第三人称视角

---

[1] Blackmore S J. Conversations on Consciousness[M]. Oxford:Oxford university press,2005.有中译本:苏珊·布莱克莫尔. 对话意识:学界翘楚对脑、自由意志以及人性的思考[M]. 李恒威,等译.杭州:浙江大学出版社,2007.

来研究客观现象,应该避免任何主观性。然而,在意识研究中,主观性本身就是研究的主题! 因此,有些人认为现有的理论不能解决意识之谜,需要发展新的理论。他们中的一些人认为意识是一些有高度组织的系统的基本属性,就像所有实体的空间、时间、质量等一样,不能用更基本的概念来解释。这是不可还原的。相反,来自第二个阵营的一些科学家甚至否认有这样的"困难问题"。他们认为研究意识就跟其他科学研究一样。他们认为意识和脑活动是同样的,就像硬币的两面。因此,尽管他们从来没有使用过"简单问题"这个术语,但对于他们来说只存在查默斯的"简单问题"。但是,我对后一个论点表示怀疑,并将在后面再作解释。

在我看来,主观性和私密性是意识研究困难的核心,否认"因难问题"只是回避或忽视意识的主观性和私密性。但是,它们仍然在那里! 在可预见的未来,我看不出有什么可能性可以解决这个难题。具有讽刺意味的是,虽然上述两个阵营看似持相反观点,但实际上都是在说目前不要去研究意识如何从脑中涌现出来的问题。

美国著名神经科学家萨克斯(Oliver Sacks)讲述了一个有关美国神经科学家巴里(Sue Barry)的故事[1],这个故事确凿无疑地表明主观体验并不等同于这种体验背后的神经机制。2004 年,萨克斯第一次见到了巴里,他们有一次有趣的谈话。巴里告诉他,她从小就是对眼,所以一次只能用一只眼睛看世界,眼睛迅速而无意识地交替着看。因此,事实上她没有双眼视觉,可能无法直接看到深度,但她仍然可以通过单眼线索来判断深度。所以她也可以像其他人一样开车、玩垒球等。作为神经科医生的萨克斯急于想知道她是否有任何立体视觉的观念。她有深度知觉的主观体验吗?巴里的回答是"是的",因为她是神经生物学教授,她读过休伯尔(David Hubel)和维泽尔(Torsten Wiesel)的论文,她几乎对视觉信息处理无所不知,包括双眼视觉和立体视觉的脑机制。萨克斯把她的想法形容如下:"她觉得这些知识使她对她之所失有了特别的了

---

[1] Sacks O. The Mind's Eye[M]. New York:Alfred A. Knopf, 2010.

解——她知道立体视觉必定像是什么样子,即使她从来也没有体验过体视[1]（stereopsis）。"这听起来就好像是如果你解决了"简单问题"——知道双眼视觉的大脑机制,那么"困难问题"也就将消失——你也可以知道体视是什么样的体验！但是,在你阅读下一段后,你就可以知道上述的话只是一种错觉！

几年后,萨克斯收到她的一封来信。她在信中说道:"你问过我是否能想象得出用两只眼睛看世界会是什么样。我告诉过你我以为我可以……但是我错了。"她说这是因为她进行了一次手术,并在术后做了各种训练,最后成功了,现在她有立体感了！她描述她的经历如下:

"我回到我的车里,随意看了一下方向盘。它从仪表板处"跳了出来"。我闭上一只眼睛,然后闭上另一只眼睛,然后再双眼同时看了一下,方向盘看起来就是不一样。我断定这一定是夕阳的光线作弄了我,于是开车回家。但第二天起身,我做了做眼部练习,坐进车子开车去上班。当我看后视镜时,它从挡风玻璃处跳了出来。

……

"我以前一直不知道我失去的是什么。普通的东西看起来都非同寻常了。灯具飘浮在半空,水龙头插入空中……这有点像身处游乐场,或者吸毒后忘乎所以。我盯着东西看……世界真的看起来不一样了。

……

"我注意到我办公室敞开的门的边缘似乎向我伸来。虽然我一直知道门打开时是向我伸过来的,这是由于门的形状、透视和其他单眼线索使然,但是我从来没有见过它有深度感。我双眼看一下,然后用一只眼睛看,再用另一只眼睛看,以便使自己相信看起来确实不一样。门绝对是伸在那里。当我吃午饭的时候,我低头看一碗米饭上方的叉子,叉子在碗前面的空气中静静地停着不动。叉子和碗之间还有空的地方。我以前从来没有见过这样的事……我一直望着我叉起的一颗葡萄。我可以感到它的深度。"

---

[1] "体视"一词一般专指由双眼视差所引起的深度知觉。——译注

通过巴里的故事，我们可以知道：即使知道了产生立体视觉的条件，你可能仍然没有体验过深度知觉！如何从双眼神经活动中产生这样的知觉依然不得而知！尽管计算机仿真有可能帮助我们揭示行为或某些可观察的功能的脑机制，但是计算机仿真不能说明客观过程如何产生主观体验。你无法确认你的计算机上运行的任何程序具有主观体验，即使这个程序声称它有主观体验也不行！图灵式的测试不能告诉你这一点。IBM 的"沃森"（Watson）可能会告诉你它是有意识的，但是，它绝对没有意识！在我看来，巴里的故事告诉我们不能简单否定所谓的"困难问题"，确实存在主观体验！虽然我们可以研究各种"简单问题"，这对于我们理解意识很有价值，但是，如果不解决"困难问题"，我们似乎就不能宣称我们已经彻底认识意识了。

由于篇幅限制，我不能再引用巴里的故事了。如果你有兴趣，你可以自己阅读萨克斯的书《心灵之眼》（*The Mind's Eye*）[1]。

除了 HBP，现在我们也可以不时地听到所谓的"人工意识"，甚至都已出版了以此为题的国际期刊。但是，由于上述原因，我认为研究人工意识还为时过早。我不是说人工意识是绝对不可能的。无论如何，无论人脑有多么复杂，它仍然是一种物理系统，因为我们知道人脑可以产生意识，这就没有理由断定没有任何其他物理系统也可以有意识。但是，我们不知道什么样的物理系统才可以有意识。是复杂的分层系统吗？但是"复杂"是什么意思？即使对于脑，我们也只是知道人脑可以产生意识。至于其他物种，灵长类动物甚至脊椎动物都可能是有意识的，鹦鹉和乌鸦也有可能，但昆虫并没有意识。边界在哪里呢？为什么？如果我们连哪种脑可能有意识都无法判断，那么我们如何来判断一个人造物体是否有意识呢？就像霍金斯（Jeff Hawkins）指出的那样，仅根据机器的行为来判断机器是否具有智能是不合理的。仅根据其行为来判断一件人造物体是否有意识也是不可能的。因此，在我看来，讨论人工意识的问题还为时尚早。建模和仿真不太可能引发在意识研究上的"根本性突破"。

虽然意识很可能是某种高度组织起来的系统的一种涌现性质，但它是由系

---

[1] Sacks O. The Mind's Eye[M]. New York：Alfred A. Knopf, 2010.

统内部复杂的相互作用中涌现出来的，而且就如同许多涌现性质一样，你无法解释这种性质是如何涌现出来的。但是，正如我上面所说的那样，我们还不知道什么样的"高度组织的系统"能涌现出意识，我们也不知道对这样的系统，究竟要在什么条件下才会涌现出意识。我们可以测试一个模型，看它的行为是否与有意识的生物的行为类似，但是由于意识是一种内在的状态，所以不可能根据它的行为来判断这个主体是否是有意识的。为了彻底揭开意识的奥秘，核心难点——主观性和私密性是不可避免的！我说得对吗？

到现在为止，我主要讨论了由于主观性所引起的困难，而邮件就已经太长了，私密性问题我就放到下一封信中再谈吧。

一如既往的良好祝愿。

凡及

计算和信息处理;神经元和神经网络;单层感知器和多层感知器;理论和实际

2013－12－04

亲爱的凡及:

感谢你所说应该在《脑-心智杂志》(*Brain-Mind Magazine*)那篇有关 HBP 的文章中向我们表示感谢的话,那是完全用不到的。你敏锐地看到了 HBP 在其基本层面有些不对头之处,而其所做出的承诺也过高了。我能提供的唯一贡献就是说:是的,你是对的。

我为你有勇气发表这篇文章感到骄傲,即使你不知道这有多么危险。人的许多英勇壮举都是由于不知道危险才做出来的。☺要是一个人知道某件事很危险,通常他们就不会去做了。

我们有一篇专讲这一现象的著名文学作品《穿越康斯坦茨湖》(*Ritt über den Bodensee*)。故事情节是,在深冬,一名骑马人急于乘渡船渡过康斯坦茨湖。然而,湖面已经结冰了,这是很少有的事,但是骑马人并不知道,就从薄冰上疾驰而过。当他到达湖的另一边,人们告诉他他做了什么的时候,他感到非常震惊,吓昏了从马上跌了下来。

缺乏信息或无知往往是壮举的来源,其他的原因则是惊吓、苦难和绝望。奇思妙想和发明也是如此。☺

不管怎么说,意识和理性认识在许多情况下都是有用的,但在有些情况下却没有什么帮助。在某些情况下,关于某个问题的信息太多似乎反倒会降低创造性。这也许就是为什么科学上的重大突破往往是青年人做出的原因之一,他们根本不知道或不在意在某门学科中的种种清规戒律。

从蓝脑年头到 HBP 时代,你所发现的参与者对意识和觉知研究的期望或许诺所发生的变化非常有趣。虽然在此期间意识研究并没有什么突破,但在那么短的时间里其观点就从现实变为过分乐观,这是令人吃惊的。

意识是你很感兴趣的领域,你一定很了解这个研究领域的情况,因此我相信你的专业见解。我对意识也很感兴趣,但是对这个问题的学术讨论只有些零星知识。对于我来说,这个问题所涉及的层次太高了,就像你的朋友江渊声指出的那样,其下还有很多我们不了解的脑功能层次。在我们得以认识这些更低的层次之前,我总是有点怯于去谈论这样复杂的事情。

但是你提到的这些书听起来很值得一读,而且你对主观性问题所说的话表明这背后似乎真有一个谜题,所以我很高兴你激励我冲动起来也去考虑这个问题!

你给了我一个克服惰性的理由,这很好,但是要拿到书、阅读、思考、对你的论点给出有道理的意见,我还需要一些时间。

今天我还做不到这一切,让我借此机会回过头来谈谈你以前的电子邮件中所提出的一些尚未解决的问题。

在你 8 月 15 日的邮件(Ⅰ-010 凡及)中,问了一些非常有趣和具有挑战性的问题:

a. "这是由于'脑样'或'脑型'这类表达意思比较模糊,谁能清楚地告诉我脑是如何进行计算的?"

b. "……我怀疑脑是否总是像计算机一样计算来执行其功能。我不太确定对于脑来说计算究竟是什么意思,这里所说的计算是不是就是冯·诺伊曼(John Von Neumann)计算机或图灵机中计算的意思呢?或者它只是信息处理的同义语?对我而言,前者似乎不大可能,否则所有的脑功能就都应该可以通过数字计算机来实现了,我对这种可能性持怀疑态度。"

c. "尽管'信息处理'这个术语本身的概念也不是那么清楚,我依然总是把'计算'看作为'信息处理'的同义语。无论如何,使用'信息处理'而不是'计算'一词可避免将'计算'误认为'四则运算'或在数字计算机中执行的操作。

当然,可以提出下面的问题:除了在现代数字计算机上进行的操作之外,是否还有任何其他计算? 如果有,那么这种计算是些什么样的操作,特别是那些在脑中进行的所谓计算究竟是什么意思?"

d. "麦卡洛克-匹茨(McCulloch-Pitts,简称"MP")神经元模型可以执行逻辑计算,因此原则上,麦卡洛克-匹茨神经网络可以执行冯·诺伊曼计算机可以执行的任何计算"?

你的问题 a 和 d 在我看来很容易回答。问题 a 的答案是"没有人能说出脑是如何进行计算的",问题 d 的答案是"是"。

对问题 b 和 c 的回答则更为复杂。但是在我回过头来讨论这两个问题之前,让我再就你的问题 d 做些相关评论,这是因为只有在句子的后半部分加了"网络"一词之后,对这个问题的回答才是肯定的。

正如坎德尔(Eric Richard Kandel)的"圣经"[1]附录 E(P.1585)中所指出的那样,单个 MP 神经元可以执行 AND 和 OR 的基本逻辑功能,也可以通过突触抑制执行"非"运算(逻辑 NOT)。

然而,单个 MP 神经元不能执行逻辑异或(XOR)功能。在有两个输入线的情况下,逻辑异或门只有当其中一个输入为真,但不能两个都为真的情况下,其输出才为真。而对 OR 函数来说,只要任何一个输入为真时,输出都为真,而且两个输入都为真时,输出也为真。

不熟悉布尔逻辑的人在第一次碰到这个问题时,很难理解为什么这个 XOR 函数看起来会如此棘手。事实上,专家们最终发现如何通过 MP 神经网络执行 XOR 也确实是经过了一段时间才做到的。这顺便又牵涉到另外一个科学上两个对手之间恶斗的故事。这个故事没有像马克拉姆和莫德哈(Dharmendra S. Modha)之间的争斗那样粗鲁,但是也带来了严重的后果和悲惨的结局。

1957 年,罗森布拉特(Frank Rosenblatt)提出了"感知器"(Perceptron)的概念。这是一个模仿视觉的神经元网络。在坎德尔的"圣经"的附录 E 中也讲了

---

[1] Kandel E, et al. Principles of Neural Science[M]. 5th ed. New York:The McGraw-Hill Education,2013.

这个概念,以此作为例子说明如何在有许多神经元的脑中进行并行数据处理,从而使得我们的视知觉如此之快。但是这种说法没有切中要害,因为它没有把罗森布拉特的"单层"感知器和后来的"多层"感知器区分开来。不同之处在于单层感知器不能实现逻辑 XOR 操作,而多层感知器则可以。罗森布拉特的概念是最早的前馈神经网络之一。

这项研究得到了美国海军研究办公室的资助,发明者夸大宣扬了其成果。在一次新闻发布会之后,《纽约时报》(*New York Times*)报道称感知器是"海军所期望的能够行走、谈话、看、写、自我复制并能意识到自身存在的电子计算机的雏形"。[1]

正如你所看到的那样,即使是在人工智能早期,在为获取基金和政府资金的竞赛中,也不能表现谦虚。

这种声调和夸张听起来很熟悉,与此同时在这场争取资助和成名的竞赛中,竞争对手的反应也如出一辙。

在这场竞争中,罗森布拉特的对手是明斯基(Marvin Minsky),后者后来成为 AI 的巨匠之一。明斯基碰巧也是罗森布拉特的同学。罗森布拉特于 1962 年在康奈尔大学出版了一本名为《神经动力学原理:感知器和脑机制理论》(*Principles of Neurodynamics*:*Perceptrous and the Theory of Brain Mechanisms*)的畅销书。明斯基研究的也是类似的问题,他以前的同学的成功和名望似乎令他很不好过。不管怎么说吧,他觉得有必要告诉全世界罗森布拉特有关感知器的思想有多么糟糕。1966 年,明斯基和他在麻省理工学院的同事帕佩尔特(Seymour Papert)[2]一起出版了一本名为《感知器》(*Perceptrons*)的书,在书中揭露了罗森布拉特感知器的重要缺陷。其中之一就是单层感知器不能执行逻辑 XOR 功能。这两位麻省理工学院(MIT)重量级教授的攻击非常成功,不仅使感知器几乎在一夜之间就跌落神坛,而且对神经网络和人工智能的投资也江

---

[1] https://en.wikipedia.org/wiki/Perceptron
[2] Seymour Aubrey Papert(1928—2016),在南非出生的美国数学家、计算机科学家和教育家,也是人工智能的先驱之一。——译注

10   意识之谜和心智上传的迷思   一位德国工程师与一位中国科学家之间的对话

河日下。这本书促成了第一个"人工智能的冬天",并使神经网络的研究停滞了大约十年。罗森布拉特没有机会回应明斯基的攻击,因为他在一次划船事故中于1971年四十三岁生日时不幸身亡。

不无讽刺的是明斯基本人后来也把兴趣转向了神经网络[1],而多层网络和引进"反向传播"的概念则解决了早期单层感知器的缺陷。

麦卡洛克和匹茨早就讨论过其神经元有关布尔逻辑的完备性问题。匹茨曾在罗素(Bertrand Russell)那里学习过,他非常清楚形式逻辑和布尔逻辑的规则问题。而且他早就发现,当你把多个人造电子神经元组合起来后,就能解决异或问题。现在大家都知道,当你有一个可以执行基本布尔函数 AND、OR 和 NOT 的器件时,你可以通过组合两个或多个这样的基本元件来执行所有更高级的逻辑函数。这个领域的先驱们早在 70 年前就已经在这方面有了这样的远见卓识,真是令人惊讶和印象深刻。

在技术计算上,把数字电子学中的基本元件、触发器和锁存器(latch)组合起来构建更复杂的门,可以执行更复杂的功能,这使用的也是同一种方法。当我们考虑这种逻辑与时间无关(组合逻辑,combinational logic)或是与时间有关(时序逻辑,sequential logic)时,还有一个因素要起作用。关键是时序逻辑有存储器,而组合逻辑没有。对于时序逻辑(换句话说,就是带有存储器的布尔逻辑),我们可以构建"有限状态机"(finite state machines),这是几乎所有现代电子设备的基本构建模块。从原则上说起来,无论是用继电器、二极管、电子管、晶体管,还是用流体、光学,甚至生物元件来实现这种"有限状态机"并不重要,只要可以应用布尔逻辑并具有存储器就行。

然而,不应忘记,人造电子神经元或门,非常像它们的生物亲属,也是抽象的理想模型,而当它们真的在物理世界中实现时也不可能百分之一百相同。

程序员经常会忽视这个事实,尤其是那些习惯于用高级语言编写代码来解

---

[1] 事实上,明斯基早年就对神经网络有兴趣,他在普林斯顿大学的博士论文的题目就是《神经-模拟强化系统的理论,及其在脑模型问题上的应用》(Theory of neural-analog reinforcement systems and its applicaition to the brain-model problem),这其实就是一篇有关神经网络的论文。

决远离物理世界的抽象问题的人。当一个逻辑设备不像它应该做的那样运行时，他们往往会感到惊讶。首先，他们总是怀疑程序设计中有错误，实际上却是因为在用设备进行物理实现时和程序的逻辑起了冲突。

编写控制诸如机器人或物联网（IoT）应用程序代码的程序员，需要有一个不那么理想化的观点。为读入信号的微处理器编制程序的人首先需要注意信号的物理现实。典型例子就是机械开关的"跳动"。与神经元中的动作电位非常相似，电压不会无需时间就从 0 跳到 1 或从 1 跳到 0。这总需要一点时间，逻辑开/关信号在实际上看起来并不会像在理论上那样干净和理想。当你将示波器连接到开关的输出端时，你可能会因看到它在打开和关闭时所显示的噪声而感到惊讶。

噪声来自开关的"跳动"，这和系统中的电感和电容有关。这可以用硬件方法（例如将一个电容器并联到开关）或通过软件进行补偿，后者可以在读取状态之前在程序中设置几毫秒的延迟时间。对这种情况不熟悉的工程师常常会因为开关状态的波动而感到困惑。

我之所以要提到这样明显的问题，是因为这能说明一个重要的问题。我们似乎对电动力学原理有完善的理论知识，而麦克斯韦方程组则是我们在物理学中所有的最为坚实的基础之一。然而，将这些理论知识应用到现实世界却可能非常棘手。机器并非像数学上定义出来的那样纯粹是的封闭系统，生物机体就更不用说了。事实上，理论与实际脱节在 19 世纪末期跨大西洋电报有线通信的建设中就造成了很大的问题。当海维赛德（Oliver Heaviside）仔细思考电脉冲穿越电缆时就碰到了这样的问题，这时不仅需要考虑电阻而且还得考虑电容和电感也要起作用。事实证明，这需要一种新的数学方法，这使得他努力创造这种方法，而与此同时他也重新表述了麦克斯韦方程组，并赋予它们以现代形式。他还发明了一种称为海维赛德阶跃函数[1]的数学工具，后来证明这对于描述电报电缆中的信号传输是非常有帮助的，对神经动作电位也是如此。海维赛德实现了个人所能创造的最令人印象深刻的智力成就之一。不管怎么说，令

---

[1] 阶跃函数就是当时间小于 0 时其函数值为 0，而当时间大于或等于 0 时其函数值为 1 的函数。——译注

人称奇的是,19世纪的数学家和逻辑学家已经取得了长足的进展,并且能够提供构建现代信息处理机器所需的几乎所有基础。他们都远离生物原型,但在某些工程领域却非常成功。

顺便说一句,世界上最先专门用于计算的是人而不是机器[1]。我们的祖先已经发明了一些相当巧妙,也很省时的方法来进行一些有难度的计算。这就像他们以前发明过的其他工具一样,例如刀、锤子、轮子、衣服、语言或所得税。这样的计算工具被称为算法,并且在我们有机器运行它们之前就已经在人脑中运行了。

因此,我同意你在问题c中所说的那样,谈论"信息处理"而不是"计算"是个好主意。顺便说一下,"计算机科学"这个词也有同样的问题。爱丁堡大学教授,贝尔实验室前成员瓦德勒(Philip Wadler)[2]是我上面讨论过的问题中世界顶尖的专家之一,他也喜欢用"信息学"这个词,他是这样说的:

"除了'计算机'和'科学'这两个字之外,'计算机科学'这种表达方法并没有什么错。首先'计算机科学'与计算机无关。其次,一门真正的科学从来也不把科学这个词作为自己名称的一部分。"[3]

虽然关于计算机的上述说法听上去很滑稽,它却代表了对一个问题的深刻理解,这个问题经常会在讨论计算机和人脑之间的相似性时引起混淆。

瓦德勒不仅富有幽默感,而且还是该学科中最敏锐的思想家之一。如果你愿意,我还可以告诉你更多关于他的事情,因为在回答你的问题的其余部分时,他会非常有帮助。但是现在,当我意识到这封电子邮件已经写了有多长,而且对你提的问题我回答了还不到一半时,我想还是停下来的好,下次我要提高效率!

一如既往地祝你一切顺利。

卡尔

---

[1] "computer"一词在计算机发明之前是用于指办公室里专职的计算员。——译注

[2] Philip Lee Wadler(1956—),美国计算机科学家。——译注

[3] 卡尔记得瓦德勒在他的讲话中说过这样的话,但是未找到文字记录。他的下列著作值得一读:Wadler P. Propositions as Types[J]. Communications of the ACM, 2015, 58(12): 75–84.

II-002　凡及

意识的私密性；人工智能的冬天

2013 - 12 - 23

亲爱的卡尔：

非常感谢你一如既往的美言和评论。也感谢你有关骑手的故事，这让我松了口气，无论如何，即使是在你告诉了我有关马克拉姆的公开信后，我也没有失去知觉。☺

你是对的，意识太复杂了，从自然科学的角度对意识的认真研究要一直到20世纪80年代末才开始。1989年，萨瑟兰（Stuart Sutherland）写道：

"意识是一个迷人而又难以捉摸的现象：不可能说明意识究竟是什么，它起什么作用，或者为什么要进化出来。已经发表的有关意识的著作没有一篇是值得一读的。"[1]

虽然此后情况发生了巨大的变化，正如布莱克莫尔在2005年采访了这个领域的21位顶级科学家后所指出的那样：

"但是我现在是否就了解了意识呢？当然我对许多意识理论的认识要比以前深得多，但是关于意识本身（如果真有意识的话），我恐怕还是不了解。"[2]

---

[1] Sutherland S. Consciousness[M]. Macmillan Dictionary of Psychology. London：The Macmillan Press，1989.

[2] Blackmore S J. Conversations on Consciousness[M]. Oxford：Oxford University Press，2005.

虽然将近十年过去了,基本情况恐怕还是如此。[1]

如果考虑到十年在科学史上只是一瞬间,那么这种情况对于像意识这样的千古之谜也就不足为奇了。正如我在上一封电子邮件中所强调过的那样,意识研究面临的主要困难是意识的核心属性——主观性和私密性,这是其他科学分支所从未面对过的问题。

在我的上一封信中,我比较详细地讨论了主观性问题,并简要地谈了一些关于私密性的问题。这两个概念密切相关,但强调的要点有一点点不同。关于私密性的问题是人的意识只有主体本身才能体验到,而不能为其他人所分享。然而,最近当我读到 V. S. 拉马钱德兰(Vilayanur S. Ramachandran)的经典著作《脑中魅影》(*Phantoms in the Brain*)时,我读到了他关于私密性起源的一个非常有趣的论点。他认为,私密性来源于两种语言之间的翻译障碍:不同人之间交流的书面语言,以及主体体内神经系统所使用的神经语言。书面语言不能携带神经语言中的所有信息。因此,他提出了一个绕开这一屏障的思想实验:用电缆或神经束把两个对象的相应脑区连接起来。他认为这样这两个主体就可以分享他们的体验了。他甚至建议说,如果有人能够把人和电鱼的相应脑区连接起来,那么人也就可以体验到电鱼的电感觉了。但是,即使只是从原则上来说,我也对他的这一论点深表怀疑。

我的主要论点是有三点。首先,没有两个脑是完全一样的,脑总是随时间而变化。其次是接收者的脑,特别是连接到的那个脑区决定了体验的内容,而不是发送者发出的锋电位序列。想想拉马钱德兰本人在他自己的书中多次提到的话吧。他解释了为什么当被截肢者在他的脸被碰触时会感觉到他已被截去了的手受到了触碰,这是由于脸和手的代表区在躯体感觉皮层中比邻而居,在这种情况下从面部出发的传入神经入侵到了现在空无所用的手区,以致被截肢者把来自面部的刺激误认为来自断手!第三,脑是电化学机器,神经递质、神经调质等激素在体验中也起着不可或缺的作用,它们都不能通过电缆传播。因此,在我看来,根本不可能准确地分享他人的体验。

---

[1] https://en.wikipedia.org/wiki/Consciousness

人脑里没有任何电觉区,即使其脑中的随便哪一个区域受到来自电鱼电器官所发出的脉冲序列的刺激,他也不会有电的感觉!他的感觉一定还是受到刺激的那个感觉区原来的感觉!

这个问题让我想起了一个中国古代典籍《庄子·秋水》中的故事:

庄子与惠子游于濠梁之上。

庄子曰:"儵鱼出游从容,是鱼之乐也。"

惠子曰:"子非鱼,安知鱼之乐?"

庄子曰:"子非我,安知我不知鱼之乐?"

惠子曰:"我非子,固不知子矣;子固非鱼,子之不知鱼乐,全矣!"

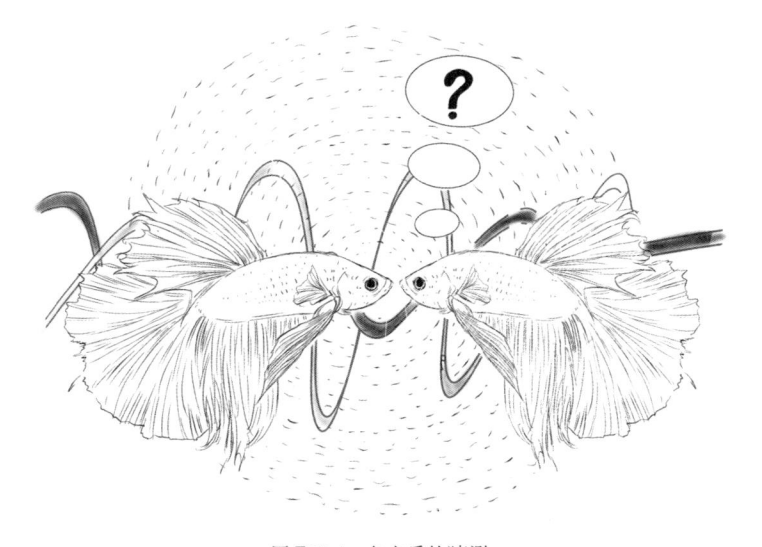

图 Ⅱ F2.1　鱼之乐的猜测

让我回到你上封信中的论点。似乎我们都同意,在脑中,除了在数字计算机或图灵机上所执行的顺序逻辑计算或操作之外,还有另外一些"计算"或"信息处理"。我们也同意,对脑来说,信息处理是比计算更好的一种表达。似乎很难把脑中的"计算"究竟是什么意思讲清楚:也许我们可以用另一个同义词"变

换"来强调计算的本质,变换就是把一个状态或一组元素变化到另一个相应的状态或元素集合。问题是这样的计算究竟是什么样的,我们能否清楚地描述这样的计算,以及如何在人工系统中进行这样的计算。

从你上封信中所讲的人工神经网络的第一个冬天的故事,以及人工智能的前两次冬天[1]可以得出下面的教训,这就是:破坏某个科学领域的声誉,使其陷入寒冬的最有效的方法就是过度夸张,或是给出不切实际的宏伟前景。1965年,AI 的创始人之一司马贺(Herbert Alexander Simon)宣称:

"在今后二十年内,机器就能做人所能做的任何工作。"[2]

明斯基(在 1970 年)也宣称:

"在三到八年之内,我们就将拥有一台具有普通人通用智能的机器。"[3]

然而,近半个世纪过去了,即使在今天或可预见的将来,他们的预言也没有成为现实。马尔可夫(John Markoff)在 2005 年说道:

"在其低谷,一些计算机科学家和软件工程师都避免使用人工智能这个术语,因为害怕被别人当做是在做白日梦。"[4]

宣传得越过分,其破坏性影响也就越强。另一个著名的例子是日本的第五代计算机计划(1981—1991)。他们的目标是编写程序,建立起可以像人一样进行对话、翻译、解释图片和推理的机器。但到了 1991 年,这个计划依然未能达

[1] https://en.wikipedia.org/wiki/AI_winter
[2] Simon H A. The Shape of Automation for Men and Management[M]. New York: Harper & Row, 1965.
[3] Darrach B. Meet Shaky, the First Electronic Person[J]. Life Magazine, 69(21): 58 - 68.
[4] Markoff J. Behind Artificial Intelligence, a Squadron of Bright Real People[N]. The New York Times, 2005 - 10 - 14(A10).

到其主要目标,其中有些目标一直到现在都还没有实现。对它的期望远远高于实际可能。

在总结经验教训时,从事 AI 研究的科学家莫拉韦茨(Hans Moravec)说道:

> "许多研究人员陷入了一张越来越夸张的怪网,他们最初对美国国防高级研究计划局(DARPA)的承诺过于乐观,当然,他们到时候交账的东西远远不及他们最初的允诺,但他们觉得他们在下一次申请时的承诺不能低于第一次承诺,所以他们就给出更多承诺。"[1]

结果是公众和有关部门失去了对人工智能的信心,人工智能研究在一段时期内难以找到资助,这就成了人工智能的冬天。

不幸的是,人们容易忘记教训,特别是当前方似乎有丰厚的报酬在等着你的时候!你的话"在为获取基金和政府资金的竞赛中,不能表现出谦虚。"是绝对正确的!当局应该记住从申请中挤出水分!

你的故事告诉我们,当问题涉及巨额基金和其他资源时,科学争论可能会变得多么残酷。尽管明斯基和帕佩尔特对仅有一层的感知器的批评是正确的,但他们的下列说法却是误导性的:

> "我们认为,阐明(或驳斥)我们的直觉判断是一个重要的问题,也就是把感知器进行推广[2]并没有多少希望。"[3]

这暗示对人工神经网络的进一步研究只会浪费时间。由于他们对单层感知器的分析是严谨的,所以虽然他们并没有仔细分析多层感知器的问题,人们

---

[1] Crevier D. AI: The Tumultuous History of the Search for Artificial Intelligence[M]. New York: BasicBooks, 1993.

[2] 指推广到多层感知器。

[3] Freedman D H. Brain Makers: How Scientists Are Moving beyond Computers to Create a Rival to the Human Brain[M]. London: Touchstone, 1995.

还是很容易地相信了上述判断。他们对人工神经网络研究的打击是如此之沉重，以至使其在此后十年陷入了寒冷的冬天！几乎所有涌入神经网络研究的人员都回到了符号人工智能的领域，同样是为了得到经济支持！虽然帕佩尔特多年后写道：

"明斯基和我是不是试图要消灭连接主义呢？……不错，在撰写《感知器》研究报告的动机中是有一些敌意……我们没有想到我们的工作会有杀伤性的作用，我们只把它看作是一种认识的方式。"[1]

这句话不禁让我想起了"此地无银三百两"的老话。[2]

是的，请告诉我有关瓦德勒的故事！你可以想象，我多么急于知道你对我其他问题的回答。

圣诞快乐，新年快乐！

凡及

---

[1] Freedman D H. Brain Makers: How Scientists Are Moving beyond Computers to Create a Rival to the Human Brain[M]. London: Touchstone, 1995.
[2] 在英文稿中，为了帮助西方读者的理解，有一段有关"此地无银三百两"的故事，但是由于国内读者都很清楚这个成语，所以就把有关故事删去了。——译注

II-002　卡尔

奇点;基金争夺战;多前沿学科融合发展计划;纳米组装器和纳米材料;学术争论

2014 - 01 - 15

亲爱的凡及:

谢谢来信。也祝你和家人新年快乐!

节日期间和往常一样忙碌,所以我过了段时间才给你写回信。按照宗教传统,一年的这个时候应该是专注于反思,但实际上却恰恰相反。也许这是因为只有少数人才真正信教,也可能是因为不知道出于什么原因,现在沉思默想已经不再时兴了。传统上,在十二月的最后几天里要和商界朋友以及员工共进午餐和晚餐,然后在连续三天的假期里阖家团聚,而在一个星期之后又有新年的庆祝活动!

包括我们在内的每个人都在全国各地旅行。因此,虽然几乎所有的公事都停了下来,但节日那段时间却是一年中压力最大的时候。☺

我听说你们在农历新年里也有类似的传统,至少在旅行方面是如此。

好吧,你知道我这样讲节日活动只是为自己的懒惰找个借口,说明为什么我不能在这段时间里做好有关意识和心智问题上的功课。我读了你推荐的那些书,发现它们非常有启发,特别是布莱克莫尔对该领域中许多顶尖思想家进行采访的那本书。这些人中有一些我是知道的,例如 C. 科赫(Christof Koch)和克里克(Francis Crick),而另一些人则从来也没有听说过。这个领域和哲学非常接近甚或有重叠,而正如你在哲学中一直看到的那样,一些非常聪明的人总是雄辩地捍卫自己这派的观点。他们的论据听上去很有说服力,即使他们的论点与对立一派的论点有矛盾也罢,而这些对手的论点听上去也言之有理。这有时让我产生了一种不愉快的感觉,感到这些才华横溢的人可以用同样的口才来

　　　　意识之谜和心智上传的迷思　　一位德国工程师与一位中国科学家之间的对话

表达完全相反的观点。

当谈到要站到哪一边,或采取哪种基本立场,比如是采取行为主义立场还是建构主义立场时,我总是感到不安。从某个有明确方向的坚实起点出发,就像在航空无线电信号的导航下安全地飞行,这很有诱惑力。这种通常以公理系统的方式出现的框架的缺点是,你很容易身陷其中而不能自拔。而一些特定理论或学派的代表人物之间的辩论,常常是徒劳无益的,因为他们并不是以求知为目的,而只是为了表明自己一派理论的正确性。

因此,我更喜欢以实用的方式提出关键问题。对所有这些原教旨主义者[1]来说,最重要的问题是:

"要什么样的经验证据才可以让你放弃自己的立场?"

如果答案是"我想不出有任何这样的证据",那么你最好和这样的人保持距离。☺

我同意你对拉马钱德兰有关把两个脑用电缆连接起来以分享彼此体验的思想实验的怀疑,或让一个人体验电鱼的电感觉的怀疑观点。事实上我觉得这个想法是完全愚蠢的,但是要是他真能用实验证明有可能建立这样的联系的话,那么我就收回所有表示怀疑的话。

我懂得你的那些论点,但是我还没有完全想好,我需要再做些准备,再读些书和做进一步的思考,然后才能更好地回答你关于意识和心智的问题,也才能对你独特的主观性和私密性的论点发表意见。

但请不要期望太高,不要忘记你对这个问题已经思考多年了!

你所引用的那些有关人工智能严冬背景的言论,以及那些有关极度夸大的承诺导致进一步夸大承诺的恶性循环的话,都是非常有启发性的!在过去30年里,每个敢于给出实现这种 AI 承诺截止期的人,实际上在事后都被证明为只是让自

––––––––––––––

[1] 卡尔在本书中专指那些在没有任何支持性的证据的情况下就将其称为真知灼见,或者在理性的伪装下作出类似宗教一样的预言的人。——译注

己出丑罢了。但是有些人比别人更受伤害。AI 中最大胆的传道士之一库兹韦尔（Ray Kurzweil），在他 1999 年出版的一本书《灵魂机器的时代》（*The Age of Spiritual Machines*）[1] 中就发表过如此超级乐观的言论。正如你所引用的司马贺和明斯基的话一样，他承诺在 25 年内 AI 将达到人的智力水平。然后，当机器和人类的智慧融合为一时，我们将会有一种叫做"奇点"的知识超新星，而在不到 100 年后，那时占据统治地位的机器人将把残存下来的智人放到动物园里展示，就像我们现在对待猩猩一样。这确实是一个令人印象深刻的预言，并引起了一场激烈的辩论，但辩论的并不是这种说法是否可信，而是科学家和政治家应该采取什么行动来防止这种情况发生。[2]

好吧，25 年即将在 2024 年到期，不管现代计算机设备的性能给我们留下多么深刻的印象，但有一点是很清楚的，如果要想使预言成真，库兹韦尔所承诺的奇点就得抓紧时间了。库兹韦尔的作品也必须放在当时开始的争夺基金的背景之下来加以考察，莫拉韦茨也参与到了这场竞争之中。莫拉韦茨也曾经发表过一本类似的但知名度远没那么高的书，书名为《机器人：从普通机器到超级心智》（*Robot: Evolution from Mere Machine to Transcendent Mind*），这本书早在 1998 年末就讲了人工智能的惊人未来。当时莫拉韦茨和库兹韦尔都是克林顿（Bill Clintion）总统的总统信息技术咨询委员会（President's Information Technology Advisory Committee，简称"PITAC"）的委员，并帮助制定了后来被称为"提高人类业绩——把纳米技术、生物技术、信息技术和认知科学融合在一起"（Converging Technologies for Improving Human Performance — Nanotechnology, Biotechnology, Information Technology and Cognitive Science，简称"NBIC"）的科学和工业发展计划。

这种想法的目的是要以某个像 20 世纪 40 年代建造原子弹的曼哈顿计划，或肯尼迪的"登月"计划那样的计划，带动一种科学和工业的"新政"（New

---

[1] 中译本：库兹韦尔.灵魂机器的时代：当计算机超过人类智能时[M].沈志彦，等译.上海：上海译文出版社，2002.——译注

[2] https://www.wired.com/2000/04/joy-2/

Deal）。所以这些书既可以看作是对这种行动的说明，也可以看作是基金申请书。不过这一计划从未像克林顿和他的 PITAC 顾问所希望的那样实施起来。首先是因为计划推出的时机不巧，当时媒体对莱温斯基（Monica Lewinsky）丑闻的细节更感兴趣。其次是因为新总统乔治·W.布什（George W. Bush）不愿推动他前任的想法，这一主张带有民主党人的印记。

NBIC 背后的智力推动力之一是美国副总统戈尔（Al Gore），他是德雷克斯勒（Eric Drexler）[1]的坚定支持者，后者的想法是从纳米级开始建设一个新的工业世界。今天差不多已经没有人知道德雷克斯勒了，但是在他出版了《创造的引擎：即将到来的纳米技术时代》（*Engines of Creation: The Coming Era of Nanotechnology*）（1986）和《纳米系统：分子机械制造与计算》（*Nanosystems: Molecular Machinery Manufacturing and Computation*）（1992）之后，他曾经是整整一代技术爱好者的导师。早在 1959 年，费恩曼（Richard Feynman）在他著名的讲稿《底层大有可为》（*There's Plenty of Room at Bottom*）[2]中就猜测过原子和分子水平上的技术，与此类似，德雷克斯勒提出了一种基于纳米尺度上的分子组装器[3]（molecular assemblers）的新型工业。

德雷克斯勒的思想直到今天读起来也依然是一种真正的智力享受，他的想法听起来非常有说服力，当时很少有人怀疑这一切有朝一日都能实现，因此把纳米这一方面纳入"新政"的概念之中似乎是合乎逻辑的。但在布什政府上台以后，情况发生了变化。其他人走上了前台，全盘修改了这些好想法，尤其是那些由戈尔的门徒德雷克斯勒提出的纳米装配器（nano-assemblers）的想法。

随之而来的是另一场科学混战，它的调子并不比马克拉姆和莫达哈之间的

---

[1] Kim Eric Drexler（1955—　），美国工程师。《纳米系统：分子机械制造与计算》是根据他 1991 年在麻省理工学院的博士论文改写而成的。——译注

[2] 该文全名为"*There's Plenty of Room at the Bottom: An Invitation to Enter a New Field of Physics*"。这是 1959 年 12 月 29 日费恩曼在美国物理学会年会上作的报告。其后一年内在一些著名杂志和报刊上发表。——译注

[3] 按照德雷克斯勒的定义，分子组装器是他设想的一种装置，这种装置"可以以原子的精度移动参加化学反应的分子，从而引导化学反应的进行"。分子组装器是一种分子机器。某些生物大分子，例如核糖体就符合这种定义，它可以按照信使 RNA 的指令把特定序列的氨基酸组装成蛋白质分子。不过一般只用这一术语指称理论上的人工装置。——译注

图 II K2.1 巴克敏斯特富勒烯
的分子结构

好,而后果甚至更大。

　　这一次是斯马利(Richard Smalley),他是诺贝尔化学奖获得者,在 1996 年因发现巴克敏斯特富勒烯[1](buckminsterfullerenes)而声名鹊起。他起而宣称德雷克斯勒的大部分想法都是无稽之谈,而且从根本上来说是永远无法实现的。在刊载在《科学美国人》(Science American)(2001)上的一篇文章中,他严重质疑德雷克斯勒分子组装器的可行性。在接下来的三年中,两人以公开信的方式进行了一系列的反驳和争论。虽然大多数人觉得很难理解这些论点,但是至少在政府应该把钱投到哪里的问题上,诺贝尔奖获得者赢了。斯马利成为政府顾问,而纳米组装器几乎在一夜之间就销声匿迹了(就像明斯基和帕佩尔特攻击罗森布拉特的感知器之后,神经网络研究所处的情况一样)。纳米技术的新焦点是纳米材料。尤其是碳纳米管,斯马利不仅在研究方面专长于此,而且还创立了一家公司(碳纳米技术公司,Carbon Nanotechnologies Inc.)专营于此。虽然碳纳米材料的研究领域非常有前途,也富有成果,但斯马利没有太多时间来欢庆胜利。不幸的是,他在 2005 年赢得胜利后不久就去世了。纳米组装器这一领域一直没能从这次沉重打击中恢复过来,仍然处于纳米严冬之中,尽管纳米计量学家在抓住和移动单个原子方面已经取得了很大的成功。我从来都不相信斯马利的话是对的。首先是因为当时我自己也参与纳米技术工作,因此知道我们可以用原子力机器人(atomic-force robots)做什么,其次也是因为我相信只要工程师努力工作,斯马利的反对意见是可以克服的。

　　他指出的那些问题,让我想起了 20 世纪 80 年代那些用以反对在瑞士吕许利空的 IBM 研究实验室工作的两位物理学家的论点,他们试图看到单个原子。每个人,特别是权威的理论物理学家都告诉他们,根据量子物理学的规则和海森堡(Wernerkarl Heisenberg)的不确定性原理等,他们试图做的一切都是徒劳

---

[1] 巴克敏斯特富勒烯,分子式是 C_{60},富勒烯家族的一种,呈球状分子。——译注

的。但是他们并没有放弃,有一天连他们自己都惊讶不已地发现他们的扫描隧道显微镜(scanning tunneling microscope)确实有效。这使这两位物理学家罗雷尔(Heinrich Rohrer)和宾宁(Gerhard Binning)赢得了诺贝尔物理学奖(1986),他们的发明为下一代纳米级计算机芯片的研发铺平了道路。它也使得人们能深入到纳米生物学的世界之中,20世纪的研究成果中很少能有影响如此巨大的工作,并且还对工业产生影响。

我非常高兴能见证罗雷尔向德国顶尖的计量学人士发表讲话,然后还和他谈了谈。他知道有些重要人物(至少在他们自己的眼中是如此)就坐在听众席中,他们中的一些人曾经宣称过他想要做的事是不可能的。现在,他凭借刚获得的诺贝尔奖有机会可以取笑他们了,不过是以一种非常亲切、轻松和友善的方式。他是你可能想得到的最谦虚、最低调的人。他很幽默地说到,当他和宾宁在准备发表他们所取得的突破性成就时遇到了一个问题,就是他们没有新的理论,特别是没有"公式"。于是他们终于找出一个使得这个发明看起来更科学的公式。他把它展示在一张幻灯片上,它看起来确实令人印象深刻,其中包含所有数学上必不可少的成分。虽然每个人都对此印象深刻,并试图理解其意思,但罗雷尔笑着并以其可爱的瑞士口音说道:"这个公式不错吧? 不过请相信我,有没有公式都一样。"

由此得到的经验教训是,我们不应该对正统的学说亦步亦趋,不要为我们的理论偏见所蒙蔽,也不要为大理论所设定的界限所吓倒。而且,我想补充一点,年轻的科学家不应该怯于打开贴有"禁止入内"标签的大门。我很喜欢罗雷尔的清新和务实的观点,我也很高兴能够在随后的谈话中告诉他这一点,他也告诉了我更多他不得不克服许多学究式的反对的故事。然而,并不是所有的听众都像我一样接受这种经验教训,这是我在事后从组织者那里听到的。有些人甚至觉得受到了冒犯,为了是罗雷尔而不是自己获得诺贝尔奖而愤愤不平,他们觉得罗雷尔不能像他们自己那样理解他们的奇妙理论。天哪,我不禁想,聪明人怎么能如此心胸狭隘? ☺

德雷克斯勒的《创造的引擎》中有一章直到今天依然值得一读,它涉及计算机器的问题。显然他对有关图灵机的概念想了很多,他提出了一种机械的信息处理机器,其中的逻辑门就像乐器里的阀门一样。这清楚地说明了信息处理机

器可以由任何一种结构构成。我不知道还需要多长时间，才有新的一代重新发现德雷克斯勒，并启动一个新的纳米之春。不管怎么说吧，德雷克斯勒—斯马利之争是另一个值得拍成电视肥皂剧的故事。[1]

我们还可以从这个故事中学到另一个教训。德雷克斯勒的科学生涯就这样结束了，尽管他的承诺就其大胆性来说还不及库兹韦尔的一半。然而，对于库兹韦尔来说，他的科幻般的幻想成了真正的地位促进剂，不仅作为作家是如此，而且作为一个行业的相关参与者也是如此。他是一位世界著名的作家，现在有了大量的追随者。大约一年前，他被谷歌聘为"工程总监"。虽然我不是库兹韦尔奇点预言的忠实拥趸，并且对他所宣传的一切深表怀疑，但我还是认为谷歌再也找不到比他更好的 AI 布道者了。

在 NBIC 提出的十年之后，我们又看到了这种把各种科学技术结合起来的"新政"思路以某种方式复活了起来。现在的重点是工业制造、智能机器人、3D 打印机和能复制自己的计算机，这在现在被称为工业 4.0。当然，AI 再次处于这一切的中心。这一次是一位记者启动了这一切。《连线》杂志（*Wired Magazine*）前主编安德森（Chris Anderson）出版了一本名为《创客：新工业革命》[2]（*Makers：The New Industrial Revolution*）的书，这本书在美国和欧洲影响巨大，其中也讲了许多与中国有关的事。其实它讲的是我们的工业在如何生产商品方面的全球变化趋势，以及科学、技术、机器人和人工智能在其中所扮演的角色。我不知道它在中国的反响如何，但你可以告诉我。

但是，我又一次写得太长了，所以得搁笔了。

一如既往致以最良好的祝愿。

卡尔

---

[1] 如果你有兴趣的话，可从下面的网站知道更多的细节：
https://www.wired.com/2004/10/drexler/
https://en.wikipedia.org/wiki/Drexler%E2%80%93Smalley_debate_on_molecular_nanotechnology
[2] 中译本：克里斯·安德森.创客：新工业革命[M].萧潇，译.北京：中信出版社，2012。——译注

不要听从你教授的每句话；奇点临近

2014‑02‑22

亲爱的卡尔：

　　非常感谢你的节日问候！虽然我们中国人没有圣诞节的宗教传统，但是现在越来越多的大城市，特别是年轻人也庆祝这个节日。对所有华人来说，类似的节日是春节，也就是农历新年。全国有数亿人流动以阖家团聚！农历新年的日期不固定，但一般来说是一月底到二月中旬。事实上整个庆祝活动可能会持续近一个月！所以你看，这也给了我延迟回复的借口。☺

　　除了我的懒惰和节日之外，另一个借口是由于缺乏有关人工智能的知识，我不得不急于阅读一些材料，以便更好地理解你的论点。我不能说现在我就可以完全理解你说的话，因为有很多东西要学，所以不能指望几天的阅读就能解决问题。尽管对我来说很困难，但我觉得这个努力是值得的，现在媒体上有这么多关于人工智能的消息，这个领域的快速发展给了人们很深刻的印象。人们津津乐道国际象棋的计算机弈棋系统"深蓝"（Deepblue）在1997年战胜国际象棋世界冠军卡斯帕罗夫（Га́рри Ки́мович Каспа́ров），以及2011年知识竞赛抢答的计算机系统"沃森"击败了知识竞赛"危险！"（Jeopardy！）的前获胜者拉特（Brad Rutter）和詹宁斯（Ken Jennings）。同时，有些人正在谈论智能机器战胜人类并成为地球上下一个统治物种的"迫在眉睫"的危险。有人估计，这种风险可能在21世纪中叶之前就会发生。我对这样的预言非常怀疑。然而，合理的怀疑并不能仅仅基于直觉和信念。

　　我根本不知道关于德雷克斯勒和斯马利之间的争论，这个故事跟你告诉过我的其他故事一样有趣。你在纳米技术方面的专业知识和经验使你的话更具

说服力。你下面的一段话很有启发性："我们不应该为我们的理论偏见所蒙蔽……年轻的科学家不应该怯于打开贴有'禁止入内'标签的大门。"这让我想起了拉马钱德兰在他的经典著作《脑中魅影》中所讲的一个故事：

20世纪60年代，医科生都知道，不仅吸入玫瑰花花粉会引起哮喘发作，有时仅仅看到玫瑰，甚至只是塑料的玫瑰也会引起哮喘。这也就是说，暴露于真正的玫瑰花和花粉的环境之下，会把玫瑰的外形和支气管收缩在脑中建立起一种"习得性"联系。

当时拉马钱德兰还是一名医学院学生，他想道："如果仅仅给患者看一下塑料玫瑰就可以通过条件反射引起一次气喘发作，那么是不是也可以通过条件反射来中止或者缓解发作呢？"比如说，当患者哮喘时，在给患者支气管扩张剂的同时也向他/她展示一朵向日葵。那么在经过一段时间的训练后，是否有可能只需要向患者显示塑料向日葵就可以缓解哮喘？当时正巧有一位牛津大学的生理学教授来访，所以他就和这位教授讨论，教授认为他的想法纯属异想天开，这样拉马钱德兰至少在相当长的一段时间内放弃了这种想法。然而，20世纪末一位美国生理学家阿德尔（Robert Ader）研究了老鼠对食物产生厌恶的问题。为了引起恶心，他给老鼠一种引起恶心的药物环磷酰胺，与此同时他还给老鼠服用糖精，他想知道以后当只给老鼠糖精时，它们是否也会出现恶心的迹象。结果果然如此。然而，令他感到惊讶的是，老鼠也生了重病，受到各种感染。人们知道药物环磷酰胺除了引起恶心外，还会深深地抑制免疫系统。这提示只要将无害的糖精与免疫抑制药物配对，就会导致小鼠免疫系统"学习"该关联。一旦这种关联建立起来之后，每当老鼠遇到糖精时，它的免疫系统就会崩溃，使其容易受到感染。拉马钱德兰提的问题一点也不是什么异想天开。他很后悔自己轻易地放弃了自己的想法，并引以为训，"不要听你教授的话——即使他们是来自牛津的也罢[或者如我的同事泽基（Semir Zeki）所说，特别是来自牛津的教授]。"☺然而，他的结论可能是一个悖论。拉马钱德兰教授当然是一位教授。根据他的上述说法，我们不应该听他的话，包括上面的话。☺因此，我把上面的

话稍加修改：“不要听你教授的每句话——即使他们是来自牛津的也罢。”也就是说，你应该根据自己的经验和专业知识，用批判性思维作出你自己的判断。把真理引向极端就成了谬误！

我很遗憾，在收到你上封信之前，我从来也没有读过库兹韦尔的书，尽管我经常听到有人说到他。这一方面是由于我的懒惰，而且我也读过一些作出类似预言的书，而我对这些预言并不相信，所以我也就不急于读他的书。前几天，我从复旦大学图书馆借了一本中文版的库兹韦尔 2005 年出版的《奇点临近》（*The Singularity Is Near：When Humans Transcend Biology*）。正如我告诉过你的那样，我很懒，所以当我读一本非我专业的书时，我以为读中译本会容易些。读这本书时，我碰到了很多来自纳米技术、机器人技术、基因工程和信息技术的术语，这些都是我所不熟悉的。我很高兴我做出了一个正确的决定，否则，对我来说读原版可能会困难得多。但我错了！昨天我刚刚找到了一本原版，这才发现翻译中有这么多的错误，其中很多我读过的译文并不是作者原来的意思！我只是浪费了我自己的时间。☺问题是许多地方译者自己没有读懂，只是做了直译。不幸的是，正如许多年前著名中国语言学家吕叔湘所说的“英语不是汉语”，这两种语言之间没有一一对应的关系！中文版如果由机器来翻译不知道会成什么样子？对不起，我已经远离了我们的主题。也许我不应该过于抱怨翻译。也许就像弗里曼（Walter Freeman）在他的《心智是如何在脑中产生的》（*How Brains Make Up Their Minds*）（1999）一书中所说的那样：“事后回顾，如果我们成功了，我们可以归功于己，如果失败了，则可以责怪他人。”☺

我的主要问题是缺少有关技术问题的知识，所以我不得不说，我无法对与这些技术有关的话题作出令人信服的判断。幸运的是，你是这些领域的专家，所以我急于听你的分析。同时，我觉得也许我应该读一下他的书中和脑有关的一些章节或段落，并以此作为样本，这样我就有可能对他的话作出判断。于是我跳到了“恼人的意识问题”（*The Vexing Question of Consciousness*）这一节，并找到如下的论述：

“到二十一世纪的二十年代末，我们将完成对人脑的逆向工程，这将使我们

得以创造出在复杂性和精巧性方面(也包括情商在内)足以与人相匹敌甚至有过之而无不及的非生物系统。"

这是一个与马克拉姆非常相似的预言,甚至更加大胆!无论如何,马克拉姆所承诺的只是在 2023 年制造出一个人工人全脑,还只能匹敌而不是"超越人类的复杂性和精巧性"!由于我才刚开始阅读他的书,这对我来说有点难,而且我已经把我的答复推迟了太久,所以我不得不在这里停下来,有一天可能会再次回到这个话题上来。

至于说到安德森的著作《创客:新工业革命》,这本书有一个很好的中译本。正如你可能知道的那样,现在电子商务在中国很受欢迎,但我没有听到在中国有什么该书中所描述的制造业的事。我不知道。

一如既往的良好祝愿。

凡及

猜测和宗教式预言；计算；仿真；"当我们拥有足够强的计算机时，我们就可以计算一切"只是一种误导；混沌；"只有在认识清楚脑（特别是通过找到构建它的算法）之后，才能进行逆向工程，然后再创建接近脑的人工制品。"是另一个核心误导

2014－03－20

亲爱的凡及：

谢谢你极好的来信！

我不知道拉马钱德兰和牛津大学教授的那一段经历，但这非常富有启发性，事情真是如此。另外，我也非常喜欢泽基的那段警告，不要听教授的话，"特别是来自牛津的教授"。☺我把这段话告诉了我的女儿，她毕业于牛津大学，她立即领会并笑了，她同意这句话。

我也不知道阿德尔和他对小鼠免疫系统建立条件反射的巧妙的工作，尽管我应该知道，因为这确实是一个重要的工作。如果这种效应也适用于人类，我一点都不会感到惊讶。现代的流行性过敏，你几乎可以称之为一种时尚，可能也是由类似的影响造成的。无论如何，我感谢你提供了关于脑和边缘系统之间的联系的信息，这些想法非常有趣。身体只有一个，脑是它的一部分。你不能把这两者截然分开！

我明白你并不太喜欢读库兹韦尔的作品。读他的书，你随时都得准备好碰到许多相当疯狂或立论不足的想法和大胆猜测。而且他所持的，肯定不是你所喜欢的那种精确、谦虚和脚踏实地的科学态度。

我早些警告你一声就好了,让你花了很大力气去读他的书并深感头痛,为此我觉得有点内疚。

但是,你灵敏的嗅觉再次发现这位盛名在外的专家和精神导师有些什么地方不对头! 你所发现的问题并不是你在技术方面的短板或理解能力上的缺陷。你所不喜欢的是库兹韦尔世界观上的真正问题。简而言之:没有哪个炼金术士能够欺骗凡及,无论是马克拉姆还是库兹韦尔都不行。我再次以你为荣,我亲爱的朋友!

如果你发现很难欣赏库兹韦尔的作品,你并不是唯一的一位,你有许多伙伴。如果你看一下维基百科中有关库兹韦尔的条目,你会读到霍夫斯塔特(Douglas Hofstadter)的以下评论:"垃圾和好主意混在一起,你很难把这两者分开,因为这些人都是聪明人,他们并不愚蠢。"

还有一位生物学家迈尔斯(P. Z. Myers)把库兹韦尔的预言批评为:"基于'新时代教派[1]通灵术'(New Age spiritualism)而非科学。"他还说道:"库兹韦尔不懂基础生物学。"[2]

我认为这两位评论家的话都切中要害。我很理解你作为一位清醒的、有自我批评精神的和严肃的科学家,当你遇到的论点或预测仅仅只是一些猜测、准宗教式的预言或干脆就是垃圾时,很可能会感到沮丧。

也许你对库兹韦尔稍后的书《如何创造心智——人类思想揭秘》(*How to create a mind — The secret of human thought revealed*)[3](2012),失望会稍轻一点。该书中有一段围绕意识和心智问题讨论的总结。但是,也许你不用浪费时间去阅读该书的其余部分或他的其他著作,维基百科上有关库兹韦尔的文章

---

[1] 新时代教派实际上并非一种宗教,而是许多宗教和哲学思想的混合体,它反对理性和科学,崇尚情绪。——引者注

[2] https://en.wikipedia.org/wiki/Ray_Kurzweil

[3] 有中译本:库兹韦尔.如何创造思维:人类思想所揭示出的奥秘[M].盛杨燕,译.杭州:浙江人民出版社,2014.——译注

就已经足以让我们理解这位技术和未来学"大师"所要传递的信息。我们有句老话:"如果你想知道瓶子里是酒还是醋,不必把整瓶都喝光"。☺

然而,正如霍夫斯塔特所说的那样,库兹韦尔的问题是瓶中有许多层酒和醋。☺使用大量技术和科学术语,创造重重迷雾,这是炼金术士打动人心的传统手段,库兹韦尔是这门艺术的当代大师。

我不得不承认,如果他是作为一位科幻小说作家的话,那么我喜欢他,因为他似乎在年轻时也读过我在青年时代读过的同样的科幻小说,他和我当时对科技的未来有过类似的猜测,这些猜测激发了我儿时对科学和技术的兴趣。通过心智上传实现永生,把生物与技术融为一体,对于一个 16 岁的男孩来说,还能想出比这更令人兴奋的想法吗?世界受到科学家和工程师的控制并加以塑造,将来人类演化成超人,那时工程师们可以为所欲为,而博学多才的人成为宇宙的主人。这些就是当时我最喜欢的科幻作家范福格特(A.E. Van Vogt)以及他所提出的"交叉主义"(关于这些我在以前的信中都提到过)令我浮想连翩的事。库兹韦尔似乎也一直有同样的梦想——不过他从未从梦中醒来。☺

猜测本身并不像严肃的科学讨论中常常认为的那样是什么坏事。实际上所有重大的科学发现都是从假设、推测或猜测开始的。只要猜测还没有得到经验证据的支持,那么一位谨慎的科学家就会称之为假设。但是,当猜测能启发发明家和其他研究人员去检验或修正它们并设法寻找经验证据时,猜测就是有帮助的。猜测就是前面提到过的"Einfall"(想法),也就是韦伯(Max Weber)所说的提出科学理论的起点。所以,只要不像原教旨主义者那样武断,或者在理性的伪装下作出类似宗教一样的预言,那么猜测就没有什么错。

正是这种宗教预言式、不容置疑的语气使库兹韦尔很难为严肃而理性的科学家所接受。如果谁真的想要理解和检验他的预言,而不是成为他的奇点教的信徒,那么这些人就更难以接受。

正如我说过的那样,我对他的话总是非常怀疑,而不只是一点点怀疑。所以我从来没有花太多时间去阅读他的任何东西。当这本《如何创造心智》出版后,引起了我的兴趣,想看看里面究竟有什么新东西,以及他是否改变了立场或方法。

然而,尽管我试图以一种轻松而不抱成见的方式来读库兹韦尔的这本新书,我以此心态只读到引言的第 5 页,就为他一系列的胡说八道所激怒了。他和马克拉姆一样对计算机和仿真的作用抱有非常乐观的观点,他的这些观点与我们一段时间以来所讨论的问题有很大关系。我本人对计算机和仿真的作用也持乐观态度,但他对我们用计算机和仿真所取得的成就的不实之词,激起了我的恼怒。书中一开始声称:

"随着计算资源继续呈指数式增长,我们可以仿真其折叠的蛋白质的复杂性也不断增加。我们还可以仿真蛋白质在复杂的三维原子力的舞蹈中如何相互作用。对生物学越来越深入的理解将大大有助于发现进化赋予我们智能的秘密,然后利用这些生物启发而得的范式来创造更加智能的技术。"[1]

然后他继续称赞马克拉姆的蓝脑计划(HBP 的前身)为"有充分的理由可以认为是人机文明(human-machine civilization)史上最重要的努力"。而在说了人类可能是宇宙中唯一的智能物种之后,他继续说道:"从这个角度来看,逆向工程人脑可以看成是宇宙中最重要的项目。"

就这样,一位炼金术士赞美另一位同行,马克拉姆很可能会喜欢这种赞美,也许一些决定资助 HBP 的政治家也会对此留下深刻的印象。然而仔细一看,几乎所有的论点都是误导性的或是错的。

我们可以仿真其折叠的蛋白质的数量非常少,而我们可以仿真的蛋白质相互作用的数量就更小。实际上这些数字都小得可怜。但主要的问题是我们所谈论的都只限于仿真,而不是真正的折叠或真正的相互作用。这些仿真都是基于数学模型的,这些数学模型的真实程度就像我们脑中的神经网络模型一样。

在你向我推荐的书中,丘奇兰(Patricia Churchland)在接受布莱克莫尔采访时说道:

[1] Kurzweil R. How to Create A Mind: The Secret of Human Thought Revealed [M]. New York: Viking Penguin, 2012: 4.

"蛋白质折叠的问题被认为是一个简单的问题,但我们仍然不知道蛋白质是如何折叠的。"[1]

这一点在 2005 年当丘奇兰说这段话的时候是对的,今天依然如此。

当说到对蛋白质如何相互作用的认识时,情况就更糟糕。我们只是对于某些小的蛋白质或大的蛋白质的某个局部如何相互作用有一些模糊的想法。许多人不知道蛋白质可能有多么大而复杂,有多少蛋白质相互作用以及它们传输了些什么信号,我相信库兹韦尔也是这些人中的一分子。你通常在教科书中看到的是一些蛋白质漂浮在细胞中或镶嵌在细胞膜上。这种简化是出于便于教学的原因,免使学生茫然不解。在现实生活中,一个细胞中充满了令人难以想象数量之多的蛋白质,细胞膜上也满是蛋白质,就像春天的草地上遍布着草和花朵。除了其他小分子之外,它们所涉及的通信路径和信号级联有着天文般的复杂性。对于新陈代谢在其最低层次上我们已经有了某些认识,也破译了一些信号级联。对 DNA 的认识是一项重大的成就,但即使在 60 年以后的今天,细胞间通讯的主要部分,特别是事关脑内以及脑与肢体之间的通讯仍然神秘如故。

正如你所知,我参与过一项建造仪器的项目,其目的是借助于原子力波谱法(atomic force spectroscopy)将蛋白质解折叠,一些优秀的生物物理学家和数学家都试图以此理解折叠和结合现象。库兹韦尔以为只要有更强大的计算机,这些问题就很容易解决,并几乎已经解决。

是的,确实有了一些进展,特别是在原子和分子水平上的成像和计量技术正在变得越来越好。但是我们对这个"原子力的舞蹈"还远不了解。而且我们也不了解在更高层次上使我们体内的新陈代谢得以维持运转、控制免疫系统或执行这种神秘的体内稳态的信号和代码。更不要去说记忆、语言或意识等更高层次的功能了。

我们甚至对一些基本问题的认识还很少,这就是为什么尽管我们发动了向

---

[1] Blackmore S. Conversations on Consciousness[M]. Oxford:Oxford University Press, 2005.

癌开战、脑的十年、阿尔茨海默病项目等，并在这方面花了好几十亿美元，但在医学方面依然进展甚微。当问题牵涉到认识我们身体的功能时，我们正面临你在有关脑仿真问题上所指出的那些相同的问题。我们还没有一个有关这种"舞蹈"的理论，如果你对现实世界中的某个对象还不了解，那么你又如何能在计算机上仿真它呢？

库兹韦尔希望他的读者相信，所有这一切都可以通过更强大的计算机并应用数学来加以克服。作为一个例子，他谈道：

"伯努利原理（Bernoulli's principle）说明，气流沿着曲面流动所产生的气压要比沿平面流动的气压略小。虽然伯努利原理如何产生机翼升力的数学问题在科学家中尚未完全解决，然而工程学就是根据这种还不那么牢靠的见解，集中力量，创造了整个航空业。"[1]

这让我摇头，因为这是你能想得出来的最不恰当和误导性的例子。首先是因为机翼并不必须弯曲才能使飞机飞起来。其次是因为并不是伯努利方程或任何其他数学算法指导工程师创造了航空业。他的话中唯一正确的部分是"伯努利原理如何产生机翼升力的数学问题在科学家中还没有完全解决"。

甚至在物理教科书中，你也会读到这种弯曲机翼是飞机得以飞行的主要原因的胡言乱语，它是不加思考只照搬书中所说的学究式教学方式的典型例子，可以把它添加到费恩曼在谈到他在巴西所遇到的学究式大学系统中令人沮丧的经历时抱怨的例子之中。

在《认识飞行》（"*Understanding flight*"）一文中，安德森（David Anderson）和埃伯哈特（Scott Eberhardt）对这个看似简单的机翼升力问题的困难和微妙之处有很深刻的见解。[2]

---

[1] Kurzweil R. How to Create A Mind：The Secret of Human Thought Revealed［M］. New York：Viking Penguin，2012，4.

[2] Anderson D，Eberhardt S. Understanding Flight［M］. 2nd ed. New York：McGraw-Hill，2010. See also：https：//en.wikipedia.org/wiki/Lift_（force）

这清楚地表明了,数学家必须对物理现实做出不符合实际的理想化(如气体不可压缩或气流中不存在涡流)才能够对气体流过某个物体这样相当普通的现象进行计算。

航空工程并非受数学的启发而建立,而是梦想飞行的结果,经过了长时间的反复试验才得以实现的。这些总是伴随着物理假设,结果证明这些假设是不完备的,往往只是在实际上已经行了之后才想起要试着进行数学计算。

库兹韦尔要他的读者相信,只要你有了正确的算法,那么世上万物都可以被计算出来。实际上很多人都相信这一点。我的印象是,对自然科学和数学了解越少,就越容易相信这种神话。

事实上,正如你所知道的那样,有一类现象不能用算法和数值计算来研究。在关于脑和心智研究的典型教科书中,你常可找到这类例子。然而,这往往归之于由于量子力学和测不准原理所导致的非决定论问题。

我们不知道这后一个论点究竟有多少关联,但是物理学家和数学家早就知道,即使在古老而完美的牛顿力学中,也存在另一个使完全决定论系统不可计算和不可预测的问题。像亨利·庞加莱(Henry Poincaré)这样的19世纪的物理学家和数学家也认识到,要想计算太阳系中行星的长期运动及其位置是非常困难的。就像著名的三体问题所表明的那样,即使是像太阳、地球和月球这样仅有三个物体的小型系统也很难计算。只有在与实际情况不符的理想化条件下,我们才可以对只有点质量的极度简化的三体系统进行计算,而对于较大一些的系统,就像 n 体问题所证明的那样,并没有解决方案。行星围绕太阳以完美的椭圆运行着,开普勒(Johannes Kepler)和牛顿(Isaac Newton)笔下的这种干净世界,只存在于纸上和我们的头脑中。在现实世界中,每个天体都通过重力互相干扰。这使得所有的天体都有一点摆动,并使其轨迹不那么完美,这些轨迹很难描述,但随着时间的推移可能会产生巨大的影响。即使在有"点质量"的理想化假设的模型中也早就发现了这样的现象。在现实世界中并不存在理想化的点质量,就像并不存在理想化的点神经元一样。每个天体都有一定的大小,既不是完美的球形,也不是均匀分布的。如果考虑到行星中固体或像熔铁这样

的液体分布不均匀,那么对于这种质量分布不均匀并有复杂弹性的天体,即使像地球和月球这样的两体系统也无法长时间计算。实际上,这就意味着我们不可能计算出何时和何处会出现某个星座,我们也不可能算出太阳系中的某颗行星何时会被甩出或碰撞到另一个天体上。

庞加莱对这些问题的研究促进了混沌理论的发展。混沌系统可能很难计算。正如松德曼(Karl Sundman)[1]和你的同胞王秋冬(Qiudong Wang,音译)[2]所展示的,在加了限制条件的情况下,三体问题存在收敛幂级数形式的全局解析解。但是这些解没有多大实用价值,因为这样的级数的收敛速度非常缓慢。如果你想用松德曼级数进行天文观测,"计算将至少要包括 $10^{8,000,000}$ 个项"。[3]

所以你需要一台大的计算机和大量时间。☺

对这个无法计算的确定性系统的问题,人们已经知道了有 100 多年了,但许多人宁愿视而不见。这样的系统并非奇特也不罕见。纳维耶-斯托克斯(Navier-Stokes)方程所描述的混沌和湍流-涡旋世界无处不在,从机翼周围的气流到气象和股票市场的振荡莫不如此。我们的身体中也充满了混沌系统。

试图通过数值计算来解决这些问题,也仅仅只在很短一段时间里才有帮助。在我们有更强大的计算机来进行数值计算之后,天气预报在 1 到 3 天的时间段内已经有了相当大的改进。但如果要想作 2 到 3 周的预测,其结果仍然和扔骰子不相上下。而且事情也不会有多大改进,因为模型中的事件空间会随着时间的推移而爆炸性增长。我们确实也会看到有所改进,但是如果你想预测一年中所有星期一上海的天气,那么即使有像地球一样大小的计算机也不会有什么帮助。其原因就在于(就像每个数学家都知道的那样)存在着各种各样尺度的无穷大:小的无穷大、大的无穷大和无限大的无穷大。☺

---

[1] Karl Frithiof Sundman(1873—1949),芬兰数学家。他用分析方法证明了三体问题存在收敛的无穷级数解。——译注

[2] Qiudong Wang,1982 年毕业于南京大学,现为美国亚利桑那大学的数学教授。1991 年他把松德曼对三体问题的结果推广到了 n 体问题。——译注

[3] https://en.wikipedia.org/wiki/Three-body_problem

请记住,我们正在谈论的还只是古老的牛顿物理学和仅有几个元素的简单系统。值得一提的是,通过迄今为止人类创造出来的最精彩和最有用的工具——薛定谔方程,量子力学可以在分子水平上进行更好的计算。这就是为什么像策(Dieter Zeh)[1]这样的物理学家主张干脆抛弃牛顿物理学而完全转向量子力学的原因之一。但这完全是另一回事了。

　　长话短说:当库兹韦尔先生告诉我们"计算资源……继续呈指数式增长"时,这对他想解决的问题没有任何意义。不幸的是,"当我们拥有足够强的计算机时,我们就可以计算一切"并不只是库兹韦尔观点中唯一的概念错误。

　　另一个错误的信念是,只有在认识清楚生物原型比如说脑(特别是通过找到构建它的算法)之后,才能进行逆向工程,然后再创建接近原型的人工制品。如果在库兹韦尔的世界观中有什么垃圾的话,这就是它的"核心"。

　　但现在我很惊讶地觉察到,这封电子邮件早就太长了,而我对库兹韦尔的批评还只说了一半。我不知道你到底是否对这些问题感兴趣。所以如果我所讲的一切使你感到厌倦,那么请原谅。多年来,我一直不喜欢广为流行的库兹韦尔哲学,现在我只是利用这个机会把对它的想法写了下来。所以如果我万一被公共汽车撞倒了,你还会知道我的论点是什么——或者至少我的50%的论点。☺如果你对剩余的50%也感兴趣的话,请告诉我。

　　不管怎么说吧,在此期间我在意识、心智和主观性方面都做了更多的功课,虽然我还没有做完,当然也不能和你的专长相提并论,所以我非常期待与你讨论,也许在讲完库兹韦尔的问题之后。

　　晚安,向上海致以最好的祝愿。

卡尔

---

[1] Heinz-Dieter Zeh (1932—2018),德国理论物理学家。——译注

背景专栏 II K3.1

蛋白质的折叠与解折叠[1]

各种蛋白质都有其特定的三维结构(即三维空间上有规则的折叠),一旦解折叠,伸展的直型肽链因其不少疏水性氨基酸侧链暴露在外而彼此聚集起来,成为不可逆的变性聚集体状态而丧失了功能。一个常见例子就是,生鸡蛋煮熟实际上就是蛋白质解折叠发生聚集结块成变性蛋白的过程。所以,在细胞内生物刚合成的多肽链会马上被"分子伴侣"蛋白质保护起来,然后运输到特定位置上再正确折叠成有功能的蛋白质分子。此举就像西方上流社会中一个未成年的少女在参加社会活动时必须要有老妇人(分子伴侣)陪同一样。那么,蛋白质如何正确折叠的呢? 内因是该蛋白质分子的氨基酸序列,什么样的氨基酸序列就包含了各自正确折叠的所有信息。但细胞内不同的环境(亲水性、疏水性等)也是个因子,另外还要有分子伴侣的帮助。蛋白质折叠的过程很复杂,要经历多个构象状态,走能量最小化的途径,这也是分子动力学家、数学家、计算机模拟专家热衷的研究课题。除了基本理论问题外,还有很重要的应用价值。几乎所有的神经退型性疾病都是关键蛋白质解折叠成聚集态(如 β-淀粉样蛋白的聚集体)所造成的。所以研究蛋白质的折叠、解折叠非常有意义。那么,研究中用何方法能将蛋白质从三维卷曲状态拉伸成解折叠状态呢? 一个新的方法是用原子力(atomic force)显微术,用"力"将蛋白质分子内的原子间结合打开。

---

[1] 这一专栏由复旦大学张志鸿教授撰写,由于译者不懂,向张教授请教,他回信详细说明,译者就把他信中的相关内容摘录了下来,作为一个专栏,帮助读者理解。为此特向张教授表示由衷的感谢。

奇点临近;加速回报定律;摩尔定律;"把指数增长趋势作无限期的外推是错误的"

2014‑04‑20

亲爱的卡尔:

我非常喜欢你极好的来信,并衷心感谢你的美言。你洞察库兹韦尔的问题所在,在阅读你的信之前,我对此只有一些模糊的感觉。

是的,我必须承认,我不喜欢《奇点临近》这本书,尽管我曾以为我会喜欢,因为它是许多媒体评出的"2005 年度最佳图书",并受到包括盖茨在内的许多名人的赞扬。这让我再次想起拉马钱德兰博士的话。

虽然我确实感到他的书里一定有什么不对头的地方,但是我一开始并没有意识到其主要问题所在。我最初的印象是,他说话的口气总是好像他所说的每一句话都一定是真理,他使用了大量包括我在内的许多读者不熟悉的各种领域中的科学术语,正如你正确指出的那样:

"使用大量技术和科学术语,创造重重迷雾,这是炼金术士打动人心的传统手段,雷[1]是这门艺术的当代大师。"

人们容易敬畏他们理解不了的貌似深奥的话。他常常在没有充分证据的情况下就跳向结论,就好像这是不言而喻似的。如果读者读不懂,他们或许会认为这一定是由于自己的无知,这样一位学富五车的聪明人怎么可能会错! 你

---

[1] 库兹韦尔的英文名 Ray Kurzweil,此处简称其为"雷"。——译注

下面的话又说对了："对自然科学和数学了解越少，就越容易相信这种神话。"另一关键点就是你所引用的霍夫斯塔特的话："垃圾和好主意混在一起，你很难把这两者分开……"这让我想起了一部中国武侠小说《鹿鼎记》中的一段故事。小说的主人公韦小宝非常善于撒谎，让对手相信他所说的谎话。他的诀窍就是在关键点上撒谎，同时在大量有关的次要问题上却提供真实的细节。看来，无论西方还是东方，古代或现代，所有的大师都采取类似的策略！

我非常喜欢你下面的这句话："如果你想知道瓶子里是酒还是醋，不必把整瓶都喝光。"你的话以及你推荐给我的维基百科全书上有关他的条目，可以帮助我免受进一步硬着头皮读他的书之苦，尽管如果我有时间的话，也许我会在某天阅读其中的某些篇章。无论如何，我已经阅读了《奇点临近》中的几个章节，所以我已经从他的瓶子里喝了一些，并且能够理解你批评的内容。你说得对，使他的作品如此难读的关键问题不仅仅是由于我的技术知识或理解能力所限，更是由于他的世界观和方法。对我来说，痛苦来自你所说的我试图"理解和检验他的预言"，结果却只发现了"像原教旨主义者那样武断，或者在理性的伪装下作出类似宗教一样的预言"。这就是为什么我不喜欢他的书。当然，我不想"成为他的奇点教的信徒"。

你正确地指出了，他的核心论点"当我们拥有足够强的计算机时，我们就可以计算一切"是错误的！你的信已经详细解释了这一点，我不再重复。你摧毁了他和其他类似大师的预言的基石。非常好！

至于库兹韦尔的超智能狂想曲的另一块基石，也就是通过找到脑工作的算法从而逆向工程创造人工脑，在讨论欧盟 HBP 的可行性时，我们早就指出过了，在可预见的将来，不可能实现这个目标。因此，现在我明白了库兹韦尔和马克拉姆确实是同类人。

我认为库兹韦尔所谓的"加速回报定律"（The Law of Accelerating Returns）是他预言的第三块基石，据此，技术将呈指数式增长。[1]

这是对摩尔定律的推广，摩尔定律声称集成电路的功能呈指数式增长。尽管自 1971 年以来摩尔定律已持续了大约半个世纪，但它不是自然规律，而是对

---

[1] https：//en.wikipedia.org/wiki/Accelerating_change#The_Law_of_Accelerating_Returns

数据的经验性拟合。它不能永远维持正确，现在制造半导体芯片的线宽尺寸已经接近 10 纳米，进一步减小将使其线宽尺寸与几十个原子相当，因此散热和量子力学效应将成为一个严重的问题。摩尔（Gordon Moore）自己在 2005 年也说："它不能永远持续下去。指数的本质是推到极端（you push them out），最终就发生灾难"。[1]

库兹韦尔认为，当传统半导体行业等这样的技术领域遇到某种障碍时，人类都会发明出一种新技术来让我们克服这个障碍。他从历史数据中只选择了支持他想法的数据来支持他的猜测，这与加尔（Franz Gall）选取数据支持他的颅相学非常相似。我们知道，人口虽然在 20 世纪前呈指数式增长，但现在增长速度已急剧下降。在许多发达国家，甚至有趋势为零或负的情况。奥运纪录也趋于接近饱和水平。这些例子表明，指数趋势不会永远持续下去。库兹韦尔不能否认上述事实，他在《奇点临近》[2]一书中问自己："那么我们目睹的信息技术的指数趋势是否也有类似的限制？"

他的回答是："答案是肯定的，但在本书所描述的深刻变革发生之前并不会受到这样的限制。"

然后他只是说，使用他所想象出来的技术就可以将能源消耗节省到几乎可以忽略的水平，这样问题就消失不见了。虽然他知道对他的"定律"的批评意见：

"把指数增长趋势作无限期的外推是错误的，因为这不可避免会耗尽资源而难以为继。而且，我们也没有足够的能量来为极其密集的计算平台提供动力，即使我们做到了，它也会像太阳一样热。"

他只是用模糊的承诺作为回答：

［1］ https：//en.wikipedia.org/wiki/Moore%27s_law

［2］ Kurzweil R. The Singularity Is Near：When Humans Transcend Biology［M］. New York：Penguin Books, 2006.

"指数趋势确实会达到某个渐近线,但每次计算和每一位上计算和通信所需的物质和能量资源都非常小,因此这种趋势可以持续到非生物智能比生物智能强大数万亿兆倍的程度。可逆计算可以将能量需求和散热降低许多个数量级。即使将计算限制在'冷'计算机上,非生物计算平台也依然可以达到大大超越生命智能的阶段。"

即便如此,他也只回答了后一部分的批评,而并没有回答前一部分批评——最为关键的一部分! 他假装好像已经回答了所有问题!

事实上,人口增长服从 S 曲线而不是指数曲线这一事实并非偶然,这意味着我们的生存资源是有限的,当人口数量达到一定规模以后,人口数量越多,增长速度就越慢。因此,为了我们物种的生存,人口数量迟早会达到饱和水平。这是技术进步速度的一个重要制约因素:更多的人意味着更多的潜在发明者,更多的发明者意味着更快的技术发展。这可以解释过去几个世纪技术进步的速度,现在随着人口数量接近饱和水平,这种速度能否还永远保持下去呢?

总之,从经验总结中加以外推,而不考虑收集数据的条件就给出预言是危险的。

哦! 你真是一位现代的莎赫札德,总是在黎明来临时讲到紧要关头,并问苏丹是否想知道故事的结局。当然,他希望! 尽管我不是苏丹,而且你也不是莎赫札德,还是请告诉我你的另一半批评吧!

至于意识的主观性问题,由于这封电子邮件已经太长,所以学你的榜样,我也要在我的下一封电子邮件中再说。☺

一如既往的良好祝愿。

凡及

摩尔定律;"工程上通过复制生物原型(例如脑)来创造人工机器的想法"不可行;技术发明与自然进化;"要想对一台像脑那样的你不懂得的机器进行逆向工程,是你所能做的事中最愚蠢的事情";"计算机将会在某一天在几乎所有各种应用中在信息处理方面胜过人类"但很难给出时间表;人机融合

2014－05－18

亲爱的凡及:

你不会相信我多么喜欢你的上一封电子邮件!特别喜欢你所讲的韦小宝的方法:"他的诀窍就是在关键点上胡说八道,同时在大量有关的次要问题上却提供真实的细节。"更重要的是你所做的总结:"看来,无论西方还是东方,古代或现代,所有的大师都采取类似的策略!"这完全适用于你从库兹韦尔的论点中举出的那些例子。

我也喜欢你关于指数增长和摩尔定律的讨论。然而,这是一个非常微妙的领域,属于霍夫斯塔特所说的"垃圾和好主意混在一起,你很难把这两者分开……"它的垃圾部分是简单地根据过去的发展过程就归纳得出未来也将像过去一样发展,你已经对此进行了恰当的批评。然而,在技术上确实有可能使摩尔定律不是在芯片层次上而是在系统层次上持续很长时间,甚至还可能会加速。但这一切都不是显而易见的,也不容易下断言。事实上,要就此下断言的难度就像要判断一个数学级数究竟是收敛还是发散一样难。通常这并非一目了然。有时需要花很长时间才能找出答案,有时候根本就不能下断言。我想更详细地讨论这个问题,因为它在库兹韦尔的论证中至关重要。但我不认为这是

他理论中的主要缺陷。

因此，为了不引起另一个"黎明到来"的中断，请允许我稍后再来讨论这个问题，让我接着履行我以前的许诺，即概括地说明我对库兹韦尔的立场的剩下一半的批评意见。非常感谢你接受了我的提议，但也许是你没有机会拒绝。☺但是我会尽力去证明它对于你感兴趣的意识和心智问题也是相关的。

在我上一封信的结尾处，我说过，如果在库兹韦尔的研究方案中只有一处是垃圾的话，那么这就是在工程上通过复制生物原型（例如脑）来创造人工机器的想法。

人类的发明、工具和人工机器的历史确实令人印象深刻。库兹韦尔强调这是在一个令人惊讶的短时间内做到的，他的这些话是正确的。自从大自然经过几十亿年的进化之后，工程师确实在几百年内就设法改善了几乎所有人类天然得来的感觉和能力。但重要的一点是，这从来都没有首先要通过理解生物原型才来完成。相反，正因为工程师并没有把时间浪费在毫无结果的尝试中，去详细了解鸟类怎样飞行，眼睛如何看到，或耳朵怎样听到声音，这才使这一切成为可能。以上这些事是科学家和研究人员所想做的，而非工程师。工程师从梦想和理想化的想法开始，并在机器中加以实现。有时候他们也受到生物学的启发，不过大部分都是错误的，就像脑的情况一样，然后还是根据与生物世界毫无关系的一些原理继续研发。

机器所用的力超过了生物肌肉可以提供的力的好几个数量级，这是通过杠杆，活塞和液压装置，爆炸物或电磁体实现的。这些原理与肌肉的生物学机理完全无关。当我们将"视力"提高到可以看到原子的程度，或是看到远处的星星，并看到人眼难以察觉的频率范围内的波时，它们都是基于与生物所用的方法完全不同的技术完成的。听觉和嗅觉也是如此，工程师提供的最令人印象深刻的功能甚至没有生物模型。电信和互联网是其中翘楚。而当工程师发明了快速交通工具，甚至最终实现人类自古以来的梦想——飞行时，他们都没有向马匹或鸟类学习。达·芬奇（Leoaardo da Vinci）试图认识鸟类飞行的原理，并按照这些原理建造飞行器。他像所有试图这样做的人一样悲惨地失败了。问题在于功率重量比。那时没有足够坚固的轻质材料，并且最重要的是没有合适

　　　　意识之谜和心智上传的迷思　　　一位德国工程师与一位中国科学家之间的对话

的能源。只是在有了内燃机之后才开始了航空时代，而内燃机根本就没有任何生物模型。

工程师以全新的方式实现了他们的想象和他们的机器，他们几乎在所有方面，也包括脑的许多功能在内，都超越了人类的表现。

当我们将计算机和人工智能与我们的脑功能进行比较时，我们习惯于将重点放在那些计算机和人工智能仍无法完成的脑功能上。这方面确实还有一个明显的差距。然而，我们也不应该忽视在记忆和信息处理的许多方面，机器可以比生物做得更好。事实上，在搜索我的笔记本电脑和互联网找到费恩曼有关巴西的文章后，我坐在这里，键入我的想法，以便通过电子邮件发送给你，这就是最好的例子。我们的祖父母甚至不会相信有这样的可能性。在这里，机器不仅能够做生物模型可以做的事情，而且它们通常在这方面做得更好。

人脑在进行数值计算和正确记忆方面非常糟糕。事实上人的记忆很不清楚。不仅对人的记忆还缺乏认识，而且其工作也很不可靠。搜索和寻找东西对脑来说是一件挠头的事，而我们的记忆还往往不完整、误导甚或完全错误。我们不仅忘记事情，我们的脑甚至错误地以为记住了从未发生过的事情！没有哪一位头脑清醒的工程师会设计或想要复制这样一个不可靠和有缺陷的系统。计算机在存储信息（记忆）方面要好得多，并具有令人难以置信的可靠性。它们在多种信息处理方面也更好。事实上，我同意库兹韦尔的说法，可以说明机器赶不上其发明者的地方已经所剩不多了。

关键的一点是，所有这些机器都是按照一些公理系统造出来的，这些系统与生物系统不同，我们对它们有透彻的认识，因为它们都是由发明家先理性地想出来的，然后按一些明确无误的规则设计出来的。

在充满了像乐高积木一样的逻辑构建模块的工程师工具箱中，你可以找到只有在数学中存在，而在现实世界中并不存在的所有那些纯净而理想的东西。工程师在按照蓝图设计时，应用了各种如何把这些模块组装起来并且使其如何一起工作的规则。整个宇宙中并没有完美的圆形。但它存在于数学、工程师的头脑和蓝图中。当然，机器的这些构建模块并不完全是虚拟的和理想化的，因为它们是由真实的物质组成的，并与真实世界相互作用，机翼升力就是这样的

例子。由于材料疲劳,机械轴可能会断裂,晶体管可能会因过压而烧毁,但一般来说,这些构建模块会按照设计的要求工作,并且可靠地按规则运行。它们代表理性和有目的设计出来的系统,不管怎么说,这些系统都比由曲折和随机的进化过程所塑造的生物系统世界更易于理解和预测。

工程师取得了很大的成功,这是因为他们并不关心生物学的混沌世界,而是用纯净和可计算的构建模块建立了一个新的世界。当你想在开放系统的现实世界中总结出自然法则时,归纳法(inductivism)[1]是一种很差且不可靠的原则,但在工程师按照自己的意图建造起来的机器世界,也就是在一个理想化的封闭系统世界中,它是非常成功的。这对信息处理机器尤其如此,这种机器从使用真空管作为逻辑元件的第一代计算机,直到最近用CMOS电路实现的"仿神经结构"芯片都是如此。

从麦卡洛克和匹茨到莫德哈(Dharmendra Modha),计算机工程师和科学家们常常对生物脑着迷,并多少受到生物脑的启发。但是他们的技术创造与生物神经元相差甚远。首先神经细胞就像我们身体中的其他细胞一样也是一个细胞。这意味着它包含一个核,细胞核中有DNA和所有其他的细胞器,如线粒体、核糖体等。即使单个神经元也是比核电站或炼油厂复杂得多的整个工厂。它不仅沿着轴突产生和传送动作电位。它还可以复制,改变形状,连接并与其他细胞进行通信(不仅通过数千个突触和树突),还可能有许多其他我们仍然不知道的功能。

因此,把网络中的一个电子开关称为人工神经元,就像将蜡烛称为人造太阳一样不恰当。这确实很荒谬。人造神经网络与我们脑中的神经网络的唯一共同之处的就是它们的名称。

然而,令人惊奇的事实是,一个如此原始且完全不同于生物原型的技术装置如何能够提供和原型类似、甚至更好的信息处理能力?(顺便说一下,你的朋友弗里曼对脑是否是信息处理机器提出了质疑,我相信他的质疑是绝对有道理的!但这完全是另一回事。)

---

[1] 在这里,意指明日世界的行为会和今日世界完全一样。——译注

但是,当我们看其他机器性能优于人类能力的例子时,这一点就没有那么令人惊讶了。按逻辑原理设计机器,并且忽略掉生物在进化过程及其无数歧途中累积起来的种种疵瑕似乎是一种很好的设计策略。

信息处理机器仍然无法完成某些我们的脑和心智所能完成的重要任务,于是就产生了下列问题:这究竟是由于人造"神经元"和网络简化不当所引起和所限呢,还是可以用工程方法进一步改进现有技术平台来加以克服?像库兹韦尔这样的人相信后一点,而另一些人则相信(或者似乎是希望),始终会有一部分神秘的功能只保留给我们人类,使我们的心智和意识优于机器。

这是一个悬而未决的问题,工程师正在为此而竞争,努力减小上面所说的后一部分。他们迄今做得相当好,问题在于在改进人的能力时,为什么他们只在脑这方面提出应该采用一种与众不同的策略,而在其他领域他们从来也不采用复制生物模型和原理的方法?

机器的设计和改进通常是在经验和启发式过程中不断摸索完成的。但是当这些机器终于可以工作,并且可以按照蓝图进行复制时,它们就成了应用逻辑和数学的代表。它们成了基于理想的元件和规则之上工作的楷模。要想知道机器的功能,可能并不需要完全认识其各个元件的所有细节,对这些元件只要知道它们确实能以某种方式工作,除此之外不需要知道其他更多的东西。例如,你不必知道在场效应晶体管电极上要加上多少电压才能打开或是关闭门电路,从而可以表示一比特信息这样的具体细节。

当然,这些机器并不能像哥德尔(Kurt Gödel)、图灵等人笔下的机器那样,不受物理或逻辑上的约束。而他们的元件也并非像柏拉图(Plato)意义下那样的完全理想,因为一旦离开图纸并用真正的材料制造出来就不可避免地带有真实世界所有的"不纯净"。但比起活的生物体来,它们在可预测、可计算、可理解、可修理和可改进方面还是要高出好几个数量级。

这就是为什么我们在技术和工程方面的能力取得某种爆炸性的增长,而在理解生物系统和医学方面所取得进展却如此令人失望地缓慢!

所以这正好和库兹韦尔所讲的情况相反!

要想对一台像脑那样的你不懂得的机器进行逆向工程,是你所能做的事中

最愚蠢的事情,就像你、我亲爱的朋友凡及早就在你有关 HBP 计划的文章中所认识和表达出来的那样!

这就是为什么像莫德哈这样的人喜欢说他们的计算机并非是"脑样"的,而只是受到了脑的启发。在此领域中工作的大多数工程师都非常清楚我在这里指出的内容,但他们也足够聪明,称他们的计算机芯片为"仿神经结构的"。他们只是通过贴上一个时髦而神秘的标签就可以卖得更好,尤其是当"皇帝"、政治家和基金资助者似乎痴迷于我们应该复制脑以获得最佳计算机和通用 AI 的想法时。

然而,工程师们应该对生物信息处理的有一个特性感兴趣,那就是其低能耗。这清楚地表明了,和我们已有的 CMOS 芯片上运行的冯·诺伊曼机器比起来,还有许多其他更有效的信息处理途径。但自从齐拉特(Leo Szilard)[1]和兰道尔(Rolf Landauer)[2]那时以来,他们早就已经知道这一点了。实际上,我们很清楚要写入或删除一比特的信息所需要的最少能量是多少。人脑在这方面做得非常好,比基于 CMOS 平台的计算机要经济好几个数量级。但是,在其他各种各样的平台上也可以做得更好一些,很可能可以做得比脑更好。找出如何做到这一点的方法只是努力和时间的问题。

到此为止,就库兹韦尔预言计算机将会在某一天在几乎所有各种应用中,在信息处理方面胜过人类这一点上,我一般说来是同意他的观点的。不过在我看来,当谈到时间表时,他是太过乐观了。我不认为能在 20 到 50 年中做到这一点,我宁愿计划 100 到 500 年的时间尺度,其实这个时间尺度还是相当快的。但我知道你不能许诺"皇帝"或硅谷的大亨要等 500 年才做到永生。☺

然而,由于上面提到的那些问题,我很怀疑我们能活着看到把技术和生物融合成某种混合系统的一天。基于其他原理构建比脑更好的信息处理机器,比起把机器和脑这样一个我们还不认识的系统连接在一起要容易些。因此,我也

---

[1] Leo Szilard (1898—1964),匈牙利裔德国和美国物理学家和发明家。他在 1933 年提出了原子核链式反应的思想,1939 年起草了爱因斯坦等致罗斯福总统的信,最后导致曼哈顿计划和造出原子弹。——译注
[2] Rolf William Landauer (1927—1999),德裔美国物理学家,他在信息处理热力学、凝聚态物理等多个领域都作出了重要贡献。——译注

看不出把心智下载到机器中去的可能性。

我不会说这个令我儿时着迷的奇妙想法就完全没有可能,但是拉马钱德兰的电缆问题比起库兹韦尔所认为的要困难得多。

根本问题在于究竟是否有这种需要。未来把许多计算机连在一起的高度发达的 AI 系统,总可以涵盖世界上所有的知识和可用信息,至少对非主观的信息是可以做得到的。几百万种书籍和电影描述了几乎所有的人类感受和情绪。人们将把所有他们之所见、所说、所闻、所写、所喜的历史都记录在从电子邮件到推特,从脸书到微信或者将来任何一种记录人类活动的系统中。

所以剩下的就只有个人的非常主观的知识,以及被称为"主观体验特性"[1](qualia)的那个东西,不管它究竟是什么。所有无法用观察或语言交流的东西都很难,特别是当它必须与肉体的体验联系起来时就更是如此。例如,你很难用言语讲清楚为什么当你抚摸一块丝绸围巾或抛光过的金属会有一种愉快的感受,特别当对方没有过相应感受时就更是如此。这对个人来说可能是非常重要的,但从更高的角度来看,这也许对人类来说并没有太大的意义。在库兹韦尔所说的奇点之后,这可能就像蝴蝶对以前的毛虫阶段不再感兴趣一样。

我不想贬低主观感受、记忆和主观体验特性对个人的重要性。但我在问自己:那些想要在死时深度冷冻起来而在 100 年后又解冻复活的人究竟想要做些什么呢? 难道就是为了和其他像他这样的人一起建立一个俱乐部,来谈论 21 世纪有多好吗? ☺

好吧,当有人问希尔伯特(David Hilbert)如果他能在 100 年后活过来,那么他的第一个问题是什么时,他可能给出了一个冬眠一世纪的绝妙理由。他的回答是"黎曼假设证明出来了没有?"。我喜欢这个问题,还要加一句:"是否已经揭开纳维耶-斯托克斯方程之谜了?"☺

库兹韦尔的奇点预言当然也带有某些宗教成分。他和他的追随者通常从

---

[1] 指我们对红色物体在主观上所感到的"红",或牙周发炎所感到的"痛"等在主观上感受到的特性。也有译为"感质"的。不过译者以为"质"容易让人以为是有某种客观的"质"在那里,而与本义相悖。——译注

严格的唯物主义立场出发,相信达尔文的进化论,并相信只有信教的不接触现代文明的人才会相信"智能设计"(intelligent design)(一种认为世界是由神设计和控制的思想[1])。

但是当人类设计出可以与人融合在一起的机器,并且这个新物种主导着自身的进化过程时,这样,你就有了一种新的智能设计。

我们成了我们自己的神,而雷就是我们的先知!☺

当然,这个想法并不新鲜,每个思考自组织、自我复制、意识演化和目的论问题的人最终都会思考这个问题。

毫不奇怪,这些世俗预言在美国的技术精英中特别能引起共鸣,他们忙于通过科学知识和技术来重新塑造世界。他们相信,几乎所有的梦想都可以通过理性和技术来实现,而且他们能够做到这一点,在理想情况下甚至在他们的有生之年就能做到。这么说吧,你知道硅谷的生活感受有多乐观和热烈,在这方面没有人能比库兹韦尔表达得更清楚。

阅读库兹韦尔的书并不总是很有趣,但即使他错了,或者说特别是当他错了的时候,他能给人以启发。每次当我来到加利福尼亚州时,我不禁会想如果他说对了的话,那该有多么迷人。因此,尽管他满口大话,但我还是在某种程度上喜欢他,因为他绝不枯燥乏味,万一要建造一条新的诺亚方舟的话,我要建议把他带到船上作为某种知识弄臣为船员解闷。☺

这种耀眼的光芒再加上他的预言,看起来能确保他拥有大群追随者。我们不难看出库兹韦尔有关技术救赎的奇点的话,与摩西带领他的子民到达《圣经》上允许的牛奶和蜂蜜之地的承诺之间的相似之处。

其实我从来都没有搞清楚库兹韦尔究竟是真的不能区分科幻小说和梦想与现实之间的区别的呢,还是他只是一位优秀的推销员。

但正如我所说,无论如何谷歌无法找到一位比他更好的布道者,我很好奇,看看他到底能在现实世界中做出些什么来。

你有关阿德尔的例子使我们回到你关于意识以及私密性和主观性方面的

---

[1] https://en.wikipedia.org/wiki/Intelligent_design

问题上。在我先前的一封电子邮件中，我说我对这些问题还没有想好。我现在仍然不能说我已经想清楚了，但我现在确实更清楚你的观点究竟是什么。我发现布莱克莫尔对该领域的一些领军人物所进行的采访非常有用。她设法让他们以简短明确的方式表达自己的立场。我知道你对通过研究大脑的物质基质来发现意识之谜的可能性持怀疑态度。

当你第一次提到私密性和主观性时，我是用一种不同的、更技术性的方式来解释这一点的。在我阅读了这许多不同的立场之后，我在看法上有一点改变。但我仍然不确定自己是否真的掌握了这个问题。

为了更好地理解你的观点，请你告诉我，在布莱克莫尔采访的人中，哪些人最接近你的立场，而你不喜欢哪些人的观点或抱有怀疑，这会对我很有帮助。

感谢你的耐心，日安！

卡尔

又及：谷歌公司最近收购了深心（DeepMind）[1]公司，这是一家和深度学习有关的公司，它声称拥有最先进的神经网络技术并代表着人工智能前沿。我不知道库兹韦尔是否参与了此举，也不知道这方面的计划是什么，但我会尽力找出答案。

---

[1] 多译为"深度思维"，不过在科学上"心智"（mind）（或简短一些就称为"心"）和"思维"（thinking）并非同一个概念。所以笔者建议作今译。——译注

II-005　凡及

脑科学和技术;人造智能机器能做我们的脑能做的一切吗? 心智上传;主观性和私密性;光看行为不能判定智能;内心世界是认知科学和人工智能两者的真正瓶颈;意识是某些复杂系统的不可还原的基本涌现特性;意识研究的真正问题是意识在哪些条件下才能涌现出来;"图灵测试"和"中文屋"思想实验

2014－06－12

亲爱的卡尔:

很高兴听到你说喜欢我上一封电子邮件。至于说到我对指数增长和摩尔定律的评论,我必须承认我并没有太大的把握。不管怎么说吧,我不能否认技术进步加速的事实。我只是想知道是否一切都会按指数规律增长,并且没有任何限制。技术似乎很可能在可预见的未来是以这种方式发展的,但它能否永远如此发展下去? 经验规则可以无限外推吗? 我很高兴你将进一步讨论这些问题。我非常赞赏你对库兹韦尔的想法的态度,你并不否定他提出的每一个想法,只是因为它是由库兹韦尔提出的,而是把垃圾和好主意区分开来! 我应该效仿你的榜样,有一个开放的头脑! 人很容易有成见。

你在上一封电子邮件中所讲的故事非常精彩,这清楚地说明了库兹韦尔的下列想法是错误的:"工程上通过复制生物原型(例如脑)来创造人工机器。"事实上,我想库兹韦尔并不是第一个提出这种想法的人,它已经流行了很长时间。我必须承认,在相当长的一段时间里,我也是类似想法的信徒,我的许多同事也都这样相信。这些人大多数出身物理学、数学或信息学,只懂得一点点生物学。他们为生物的巧妙而惊叹不已,并相信由于生存竞争,能存活下来的生物的机

制必定得到了优化,或至少要足够好。这甚至几乎成了一种信念。孙复川教授是我国内朋友圈中最聪明的同事之一,他批评我们是"仿生教的信徒"！你的信解释了为什么这个"教"的主要教条是错误的。我想不出比你更好的论据了。当然,这并不意味着我们不应该从生物中寻求任何灵感。如果一种物种能够在生存斗争中生存下去,它的机制必定有一些优点,至少大自然给了我们某种"存在定理",告诉我们什么样的功能是可以实现的。此外,正如克里克所说,把一只手臂绑在自己背后绝不是一种好的战斗策略,从大自然中寻求灵感是有好处的,但要把它作为技术的主要的甚或唯一的途径,特别是想复制每一个细节,那就荒谬了。

你的以下论点非常有力:

"人类的发明、工具和机器的历史确实令人印象深刻。库兹韦尔强调这是在一个令人惊讶的短时间内做到的,他的这些话是正确的。自从大自然经过几十亿年的进化之后,工程师确实在几百年内就设法改善了几乎所有人类天然得来的感觉和能力。但重要的一点是,这从来都没有首先要通过理解生物原型才来完成。相反,正因为工程师并没有把时间浪费在毫无结果的尝试中去详细了解鸟类怎样飞行,眼睛如何看到,或耳朵怎样听到声音,这才使这一切成为可能。以上这些事是科学家和研究人员所想做的,而非工程师。工程师从梦想和理想化的想法开始,并在机器中加以实现。有时候他们也受到生物学的启发,不过大部分都是错误的,就像脑的情况一样,然后还是根据与生物世界毫无关系的一些原理继续研发。"

你的话让我想起了弗里曼的以下评论:

"在 20 世纪 40 年代,神经生物学家和逻辑学家重新概念化了神经活动的功能。他们将神经元看作是执行布尔代数和亚里士多德逻辑的二值开关。神经元的动作电位不再被解释为电脉冲,而是作为信息单位,表示开或关、是或否、1 或 0 的二进制数字。这很快就导致可编程数字计算机的发展,这一切是从

对神经元如何工作的一种错误认识开始的。"[1]

坦白说,当我第一次读到他在其《神经动力学:对介观脑动力学的探索》(*Neurodynamics: An Exploration of Mesoscopic Brain Dynamics*)一书的中文版序言中的下列论点时,是很不理解的(我正好是该书的译者之一):

"随着对神经动作电位在神经元之间通讯中所发挥的作用的发现,细胞神经生物学兴旺了起来……这个概念被动作电位作为信息单位的想法所取代,动作电位被当作成二进制数字。这种推论是不正确的,……然而,尽管这个新想法在脑科学中是错误的,它却导致了神经网络和数字计算机的诞生。这些装置为信息时代提供了国际社会的技术基础,这表明理论不一定要正确才能取得成果。"[2]

只是在经过我们的长时间通信之后,我才清楚了这些话的含义,并且还不仅限于计算机的特殊情况!

在你的信中,你提出了一个非常有趣且重要的、远未解决的问题:人造智能机器能做我们的脑能做的一切吗?对我而言,原则上我不能排除这种可能性,正如我上面所说的,大自然已经给出了关于我们的脑所有功能的"存在定理"!由于脑也是一种物理系统,所以没有理由否定建立可以执行相同功能的其他系统的可能性。因此,在我看来,预测"计算机的信息处理将在几乎所有类型的应用方面胜过人类"并不是问题。问题是什么时候?我同意你所说的20到50年大概不行。100至500年似乎更实际些,但我不能确定。要估计究竟多长时间,对我来说是太难了!

至于第二个问题,我得承认我以前也曾认为仿真脑是人工实现脑功能的唯一途径。然而,你的论点令人信服地表明了事情并非如此。正如你所说的那样"要想对一台像脑那样的你不懂得的机器进行逆向工程,是你所能做的事中最愚蠢的

[1]弗里曼.神经动力学:对介观脑动力学的探索[M].顾凡及,等译.杭州:浙江大学出版社,2002.
[2]弗里曼.神经动力学:对介观脑动力学的探索[M].顾凡及,等译.杭州:浙江大学出版社,2002.

事情"! 我们已经对工程师和大自然所采用的非常不同的方法做过讨论。

顺便说一句,我一直不太明白"把心智下载到机器里去"究竟是什么意思。心智不是一种状态,而是一个随时间变化的动态过程,它随着内外环境的变化而变化。正如我在 2013 - 11 - 20 和 2013 - 12 - 20 的那两封信(Ⅱ - 001 凡及和 Ⅱ - 002 凡及)中所主张的那样,心智和意识一样是主观的和私密的。在我 2013 - 12 - 20 的信中,我认为,即使使用电缆或神经束把两个脑连接起来,接收的脑所感受或体验到的仍然是它的感受或体验,而不是发送者的感受或体验! 只有当两个脑在各个方面和细节上都完全相同,而不管这些细节有多微小时; 也只有当两个脑可以点对点连接,而且不仅传输电活动,同时还传输化学物质时,接收者的脑才有可能完全感受到或体验到发送者的感受或体验。但是,在这种情况下,前者本身就是后者! 如果一个像脑这样的湿件机器都不能下载另一个脑的心智,那么一个人造物体如何能做到这一点呢?

让我再回到主观性和私密性的问题上来。我认为它们是意识问题的核心,甚至也是智能、心智、感受之类问题的核心。这些中的任何一个都不能仅由主体的行为来定义而存在于主体的内心世界中。

在科学史中,行为主义者试图否认意识,他们认为只有行为可以被研究,所有的脑功能只是一系列反射,但他们失败了,认知科学这样一门新学科已经发展成为现代科学的前沿领域之一。在工程和技术方面,行为主义也试图根据行为的相似性来开发相应的人工装置。事实上,著名的图灵测试就是一种基于行为的测试。虽然到现在也还没有机器真能通过图灵测试,但我们真能说通过图灵测试的机器就具有智能吗? 塞尔(John Searle)用他著名的"中文屋"思想实验给出了否定的答案。霍金斯在他的书《智能论》(On Intelligence)中说道:

"在我看来,大多数神经网络最基本的问题和 AI 程序的问题是一样的。它们都过于重视行为。无论它们将这些行为称为'应答''模式'还是'输出',人工智能和神经网络都假定智能取决于程序或神经网络在处理给定输入后所产生的行为。计算机程序或神经网络的最重要的属性是它是否给出了正确的或所期望的输出。正如图灵所暗示的那样,智能等于行为。

"但智能并不只是一个行事或行为聪明的问题。行为只是智能的一种表现,但并不是智能的核心特征或主要定义。只要想一下下面的情况就可以证明这一点了:你可以躺在黑暗中,思考和理解……,那时你依然是有智能的。"[1]

上面的评论表明,人类智能中有某些人工智能、人工神经网络或人工意识所缺乏的东西:发生在主体的头脑里的某些东西,这些是主观的和私密的! 当然,从功能应用的角度来看,也许这样主观和私密的方面并不那么重要,但从伦理的角度来看,这将是一个大问题。真正的困难在于,我们不知道如何解释从物理的脑活动中涌现出这种主观体验,甚至不知道该如何定义我们头脑中的内心世界,我想这一点是认知科学和人工智能两者的真正瓶颈。

至于"在布莱克莫尔采访过的人中,哪些人的立场和你最接近,而哪些人你最不喜欢或怀疑?"的问题,我很难给你一个明确的回答,因为布莱克莫尔就意识的许多方面提出了许多问题。我可能同意 A 的某一个观点,又同意 B 的另一个观点。如果一定要我选择一个我最同意的观点的话,那么我想选择查默斯的观点。我同意他的以下评论:

"主观体验不能还原为脑过程……如果最终表明,关于意识的种种不能从我们已有的基本物理属性如空间、时间、质量和电荷中得出,那么一个合理的回答就是说:'好吧,那么意识就不能被还原。它是不可还原的。它是基本的。这是世界的某种基本特征。'

"所以当我们谈到意识时,我们所必须采取的态度就是承认它是世界的一种基本特性,就像空间和时间一样不可还原。那么我们需要做的是寻找它所遵循的法则,寻找主观体验的第一人称数据和第三人称客观物理属性之间的关系。最终我们也许可以提出一套关于这种关系的基本定律,就类似于我们在物理学中发现的简单的基本定律一样。"[2]

[1] Hawkins J, Blakeslee S. On Intelligence[M]. New York:Levine Greenberg Literary Agency, Inc., 2004.

[2] Blackmore S. Conversations on consciousness[M]. Oxford:Oxford University Press, 2005.

然而,虽然我同意意识是某些复杂系统的不可还原的基本属性,但并不是每个物体都有这种属性。一个有待解决的问题是:哪类系统才会有这样的属性?在什么样的条件下,这种特性才会从这样的系统中涌现出来?

C.科赫虽然在许多观点上不同于查默斯,但他在他的经典著作《意识探秘》(*The Quest for Consciousness*)中也宣称:"我认为意识的物理基础是神经元及其元素之间特定相互作用的一种涌现特性。虽然意识和物理定律完全能够相容,但要想从这些定律中预测或理解意识却是不可行的。"[1]在这里,我想强调,意识是全局性脑活动的一种宏观涌现性质,很可能还是混沌的。这是一种性质,而不是一种物体!作为一种复杂系统的涌现性质,它不能还原为其元素的属性。至于意识和脑活动之间的关系,我更喜欢用"涌现"一词,而不是用"产生"。对于前者来说,它意味着意识是脑活动的一种涌现性质,它同时伴随着某种形式的脑活动;而后者可能意味着两者是不同的实体。对涌现特性来说,我们很少问及它是如何从下一层次元素的活动中产生的,而是问这些性质是在哪些条件下涌现出来的。因此,在我看来,虽然查默斯的"困难问题"很难,但对科学研究来说,这并不是一个合适的问题。这有点类似于数学公理,数学家从来不问为什么这个公理是真实的或公理是如何来的。所以对我来说,问题是:意识在哪些条件下才能涌现出来?另外,我想强调,意识的核心是主观性和私密性,我认为这是一种不可还原的涌现性质,这可能是我们内心世界的核心。到目前为止,所有现有的人造物都没有主观性,因此它们没有意识或没有真正的智能。但是,这里也存在一个大问题:由于主观性和私密性本身,我们怎么能知道它们有没有主观性呢?

在上面,我刚刚讨论了一些关于意识的主观性和私密性的问题。我还没有讨论很多其他问题,比如,意识是全局性的呢还是局域性的?意识能被测量吗?等等。然而,要在一封信中讨论所有这些话题就太长了。仿照你的榜样,我想在

[1] Koch C. The Quest for Consciousness: A Neurobiological Approach[M]. Greenwood Village, CO: Roberts & Company Publishers, 2004. 中译本:科赫.意识探秘:意识的神经生物学研究[M].顾凡及, 侯晓迪,译.上海:上海科学技术出版社,2012.

我以后的电子邮件中再讨论这些话题。在这里,我只想说,我对哈默洛夫(Stuart Hameroff)、彭罗斯(Roger Penrose)等人提出的意识量子理论持怀疑态度,因为在我看来,这只是一种猜测而缺乏经验证据,但也许我没有资格批评这一理论,因为我对量子力学知之甚少,我只在半个多世纪以前还是一名学生时听过课。

非常感谢你告诉我谷歌收购深心的消息。我还不知道这个消息。非常有趣,让我们看看接下来会发生些什么吧。

凡及

背景专栏 ‖ F5.1

"图灵测试"和"中文屋"思想实验

英国科学家图灵(Alan Mathison Turing)是数字电子计算机和人工智能的奠基人之一,也是一位密码专家。他在 1950 年提出了一条准则,以判断一台机器是不是有智能,或者说有思想。他的这条准则就被称为"图灵测试"(Turing test)。

在图灵的原始论文[1]里,他是这样说的:请想象一下,假定在两间不同的房间里各有一个男子和一个女子,通过电传打字设备在和外面的一位发问人进行对话。那位发问人必须搞清楚和他对话的究竟是男还是女,为此,他可以向他们提出任何电传打字设备能够传输的问题。那位男子在回答时总是试图让发问人以为他是那名女子,而那位女子则在回答时要发问人相信他确实是在和她对话。在会话进行时,如果用一台机器把那名男子替换下来,而使发问人还是以为他是在和那名女子对话,那么我们就说这台机器通过了这一测试,并且认为这台机器是有智能的。后来,人们就把这一假想的测试称为"图灵测试",并且把它简化为下列形式:在一间屋子里有一个人和一台机器,他们通过通信

---

[1] Turing A M. Computing machinery and intelligence[J]. Mind, 59: 433 - 460.

设备和屋子外面的一位发问者对话，发问者可以提任何问题。如果当机器和发问者对话时，发问者总是无法确定这时和他对话的是机器，那么我们就说这台机器通过了图灵测试，并且说这台机器是有智能的。

图灵的论文一发表就引起了各种专业的专家的极大兴趣，并且引起了广泛的争论。赞成他的人认为图灵测试给人们一直苦于无法严格定义的智能概念一个可操作的定义，并且为发明有智能的机器开辟了道路。这可能和当时心理学研究中的行为主义思潮有关，这种思潮认为以往在心理研究中占据主流地位的内省和听取受试者的主诉并进行解释的方法是不科学的，科学的方法是给予刺激并观察由此引起的行为。至于在脑子内部进行的智力活动则是无法研究的。脑成像技术的出现，特别是认知科学的诞生打破了行为主义的一统天下。人们对图灵测试也提出了种种质疑：首先，图灵测试远没有一开始所想的那么"简单"，所以要在实际上通过图灵测试判断一个机器是否是有智能的（即使姑且承认这一准则是对的），也难于操作，而且图灵测试过于人类中心主义，如果聪明过人，对话者也能判断出和他对话的不是人，那么这台机器究竟有没有智能呢？还是机器必须"装傻"才算有智能？其次，更为本质的是，图灵测试仅仅是根据行为来作出判断，但是智能是脑的内部性质，光从行为上加以判断是靠不住的。

对上述第二点的一个最著名的论据是由塞尔给出的。他提出了一个所谓"中文屋"的假想实验。他假定有一位完全不懂得中文的英国人关在一间小屋子里，外面的一位懂中文的人可以通过一条狭缝把一句用中文写成的问题写在一张纸上递进去。屋子里有一个规则库，里面的人把纸上的中文当做符号，并且根据规则库找出一系列的中文字写成一列，最后抄在纸上传出去。外面的人看了这张用中文写成的回答，可能会认为里面的人是懂中文的，但是如我们刚才所说，其实里面的那位对中文一无所知。这一假想实验说明光根据行为并不能判断对话者是否真正"懂得"。

虽然图灵测试受到越来越多的质疑，但是由于研究智能本质的困难性和重要性，只要这一问题没有得到解决，看来有关图灵测试的争论就会继续下去，而正是科学的争论推动了科学不断向前发展。

II-005　卡尔

摩尔定律;技术发展的新趋势;心智上传;意识
问题

2014－07－07

亲爱的凡及:

感谢你的电子邮件,也感谢你喜欢我的一些想法。你观察到人们很容易产生偏见是非常明智的,应该记住这一点。然而,我必须承认,尽管我很欣赏这种美德,但我常常无法在日常生活中坚持做到。我观察到我并不是唯一有这种问题的人,我怀疑这与我们心智架构中的基本原理有关。一个类似的,可能相关的现象是,我们在觉察自己的和他人的想法和行为中错误的能力是不对称的。为什么比起自己来更容易在别人身上发现错误呢?

我对此有一个假设,不过我想稍后再回来谈这个问题。请让我先来谈谈你所提出的摩尔定律的问题吧。

当然,对于一个由经验派生的规则是否可以无限外推的问题,你的质疑是正确的。即使只是出于数学和逻辑的原因,它也不能如此! 因此,这一"定律"最初的说法总有一天会不再成立。然而,这并不意味着计算机系统的性能改进很快就会放缓。相反,它甚至可能急剧增加并加速发展。请允许我介绍一些技术细节来解释上述听上去似乎自相矛盾的话。

当摩尔在 1965 年发表他的观察结果时,集成电路上的元件数量每两年翻一番,他预测这种速度可能会持续十年。这一预言已经实现,在研发出第一个硅芯片上的微处理器后,人们用每个芯片上的晶体管数量("晶体管数",transistor count)作为衡量芯片技术进步的标准已成为一种传统。自 20 世纪 70 年代中期以来,集成电路上的晶体管数量已从几千个增加到几十亿个。晶圆厂

（制造工厂）生产的半导体晶圆的特征尺寸（线宽）从 10,000 纳米降至 20 纳米左右，到 2020 年将进一步降低至 3 纳米。这个尺度即使从生物学领域的观点来看也是很小的，病毒的大小也要在 15 纳米到 400 纳米范围之内。3 纳米只有 30 个氢原子的大小，这一点可能会更好地说明这一点。所以这可能是关于尺寸在技术上做得到的极限，因为在这个水平上，由于量子隧道效应，导电部分之间的绝缘问题可能成了大问题。

　　然而，我不认会这种极限就无法突破，因为理论上常常谈到这类极限，但是工程师们总是能找到克服的方法。但即使 3 纳米是特征尺寸的极限，也并不意味着摩尔定律的终结。处理器的时钟速率还没有到头，这是决定计算机性能的另一个物理参数。这一参数告诉我们 CPU 每秒有多少个周期可以用来处理一堆任务。时钟速率从微处理器早期的不到一兆赫兹升至今天的几千兆赫兹。当我提到表示每秒周期数的物理单位"赫兹"时，我可以自豪地告诉你这一单位是以赫兹（Heinrich Hertz）的名字命名的，他于 1886 年就在卡尔斯鲁厄证实了存在电磁波。☺加大时钟速率是提高计算机速度的最简单和最便宜的方法。这非常象提高内燃机每分钟的转速是增加发动机马力的最便宜的方式一样。这就是为什么计算机游戏玩家常用这种办法（称为"超频"，overclocking）来调整他们的机器以达到最佳性能。然而，这也要付出代价（无论是对汽车还是计算机来说都是如此），这就是散热问题。实际上，热量问题是现代芯片技术的最大问题，也是生物在数据处理方面优于计算机的主要优势所在（维持相当凉快的 37 摄氏度）。集成电路（integrated circuit，简称"IC"）内部的所有导体都有电阻，所有通过它们的电流都会产生热量，必须把这些热量散发出去以防止 IC 损坏或熔化。过热非常容易损坏集成电路，因此无论在 PC 中或是 IT 中心的机组中，冷却系统都是关键设备。而且散热成本很高，特别是对那些巨大的数据中心来说就更是如此，例如互联网和发送这封电子邮件的中心就是如此。冷却计算机所需的能量可能超过了运行它们所需的能量。

　　当 IC 的表面很小并且 CPU 几乎赤热时，冷却很困难。虽然当 IC 的特征尺寸继续下降，热效率获得改善，这可能让工程师又增加几千兆赫兹，但这样的可能性非常有限。

多年以来,芯片工程师一直都以惊人的创造性寻找改进古老的冯·诺伊曼机器性能的方法,这种机器可以追溯到 20 世纪 40 年代后期。这正如汽车工程师一再找到方法来提高其内燃机的性能和效率一样,而内燃机的概念已经超过 100 年了。

所以我们的大多数现代计算机板上仍然是冯·诺伊曼机器,芯片仍然由硅晶片(或基于类似原理的砷化镓)制成。这不一定非得如此不可,因为对机器架构和使用的材料都可以有其他思路。工业上之所以坚持使用这些旧技术的原因在于,对熟悉的技术进行改进要比转向新平台更便宜。正如摩尔定律预测的那样,即使依旧采用硅和冯·诺伊曼架构,在较小规模上仍有很大的改进余地。就像我们在拥有多个内核的智能手机 ARM 处理器中看到的那样,下一步是将多个 CPU 集成在同一块芯片上。再下一步就是我们在片上系统(System on Chip,简称"SoC")概念中看到的那样,将更多设备集成到一块芯片中,并在一块芯片上构建复杂的数据处理中心。当你在同一块芯片上将数据从一个 IC 传输到另一个 IC,而不是通过导线在不同的 IC 之间传输数据时,你会获得更高的速度(并节能)。在 3 纳米的世界中,你可以在 SoC 上放置数量惊人的晶体管和功能单元。而且你不再限于使用晶体管,因为你还可以在硅片上构建机械、流体或气动结构。即使你已用尽了 3 纳米的平面世界中的潜力,你还可以进入三维世界利用第 3 个维度制作立体 SoC,这就是说把许多层 IC 彼此堆叠起来就又可以有更长的摩尔定律适用期。在芯片上、各层之间和外部设备之间使用光而不是电来进行通信是另一种提高速度和节能的可能途径。而且这一切都是在不离开我们所熟悉的硅世界的前提下就能做到的。

我们已经知道有速度更快、耗能更少的材料,而这我们还没有谈到像光计算机或量子计算这样的真正革命。

当然,你对摩尔定律只能在有限时间内成立的怀疑态度是对的。它不可能永远适用,因为以 2 为指数的增长效应将会在某一天耗尽一切。这非常像有关达希尔(Sissa ibn Dahir)的古老故事中所说,他曾与印度君主西拉姆(Shihram)下棋,并提出了一个看似微不足道的要求。如果他赢了,他要求在棋盘的第一个格子里给他 1 粒米,在第 2 个格子里给 2 粒,第 3 个给 4 粒,如此等等一直到

第 64 个格子。好吧,从那以后,不仅数学家,而且连君主们都知道了 2 的 64(当然要减去 1 ☺)次方是一个很大的数字。

对摩尔定律来说,只要计算一下如果把现代计算机中晶体管的数量翻倍,要翻多少次才能达到 10 的 80 次方幂就可以粗略地估计出其上限了,这个数也就是已知宇宙中所有原子的估计数。这个结果也就是直到宇宙中的所有原子都成了某台超级计算机的晶体管之前摩尔定律所预测的翻倍次数。

这个想法虽然听起来很疯狂,但其实早就有富于创造性的发明者想到了这一点。实际上,计算机的先驱楚泽(Konrad Zuse)[1]早在 1945 年就认为整个宇宙只不过是一台巨大的计算机。[2]

如果摩尔定律永远正确的话,那么总有一天宇宙中的每一个原子都得是晶体管,而一切也将就此告终,即使面对上述论点,摩尔定律的极端狂热分子也还不会放弃。他或她可能会提及埃弗里特(Hugh Everett)[3]对量子物理中多个世界的解释,并坚持认为所有这些世界都可以用于计算。☺

但是当我们从这样的宇宙学维度回归到我们不久的将来时,比如说未来 25 年,那么没有理由认为计算机就不能像我们在过去 50 年看到的那样以同样的速度提高其能力。然而,实际速度在更大程度上并不取决于技术上是否可行,而是取决于我们对这种增加能够承受的程度,例如我们愿意或能够为此付出多少。

摩尔定律在过去并非是对改进计算机技术可能速度的预测。这是工程师们向营销部门的同事作出承诺的路线图,按照这一路线图,他们可以以多快速度提供足够好(随之有更多功能)的新产品,让客户愿意购买它们。在 IBM 几乎垄断了数据中心市场的旧时代,全球数千个数据中心不得不承担开发下一代计算机所需要付出的代价。当微软和英特尔在 PC 革命中开始走到了 IBM 的前面时,这个基础扩大了,不久之后,数以亿计的用户为下一代晶圆厂买单。然

[1] Konrad Zuse(1910—1995),德国工程师和发明家,计算机先驱。他最大的成就就是在 1941 年发明了世界上第一台可编程的计算机。——译注
[2] https://en.wikipedia.org/wiki/Calculating_Space
[3] Hugh Everett III(1930—1982),美国物理学家。——译注

后是互联网和谷歌、脸书（Facebook）、百度等，以及他们所有的庞大的数据中心，这些数据中心使对计算机能力的需求呈爆炸性增长。这通常使这些数据中心成为其所在领域中对计算能力最大的消费者。最后，现在是智能手机的拥有者很快就达到了数十亿，并为此作出贡献。他们通常每隔一年，当制造商展示他们的新款时，就会更新手机。这种情况与摩尔定律同步进行，这并非巧合，而是因为它本来就是行业在其发展路线图中制定的计划。

在冷战和有名的太空项目的旧时代，是政府从其军事和研究预算中买单。今天，则是消费者为了更强大的计算机能力、人工智能以及有关的所有支持技术付钱。改进的速度之所以没有超出摩尔的预测，其原因是所需投资巨大。要开办一个新的半导体工厂需要数十亿美元。台积电（TSMC）[1]早在2010年就宣布将投资接近100亿美元的新型300毫米晶圆厂FAB15。

现在我们正在向450毫米晶圆和3纳米结构迈进，这需要更多的资金。顺便说一句，在我看来，台积电看起来要超过在晶圆技术方面的前领导者英特尔，如果他们首先进入3纳米领域，我不会感到惊讶。亚洲、特别是中国在美国选择使其成为现代计算机技术的工厂之后成了这一领域的重镇。有很长一段时间，像苹果这样的美国大品牌相信他们能够凭借市场营销和品牌知名度的强大力量来控制市场，就像IBM得以控制数据中心市场40年一样。

然而，这并非注定如此。令人惊讶的是，韩国和中国的制造商迅速地学会了不仅为美国品牌生产部件，还创建了自己的产品和品牌。这也适用于相关的软件技术。举例来说，很久以来百度就在搜索技术方面与谷歌相提并论，阿里巴巴与亚马逊相比也是如此。我相信在AI应用方面也会如此。我猜测新一代的IT将由消费者买单，如果这个猜测是正确的话，那么拥有庞大消费群的那些玩家就会占有很大的优势。

综合考虑这一切，你就可以懂得我以前提到过的安德森在其《创客：新工业革命》一书中向美国工业界敲响的警钟。他的信息很简单，这就是对于美国

---

[1] 台积电全称为台湾积体电路制造股份有限公司，成立于1987年，是全球最大的集成电路制造企业之一。在中国台湾，人们把集成电路称为积体电路。——译注

工业来说,让亚洲,特别是中国成为世界上最新一代计算机技术的工厂是个问题。

人工智能是下一轮从自动汽车到语音识别、机器人和工厂自动化等新技术的驱动力;没有哪一项别的应用比 AI 更需要强大的计算机能力了。因此,甚至可能需要以比摩尔定律更快的速度发展。

哦,我亲爱的凡及,你再一次用一个简单的问题让我偏离航程作了太长的旅行。☺☹但这不是你的过失,而是我自己的错,我只是希望你没有觉得那些技术问题过于枯燥。

尽管这封电子邮件已经很长了,但请允许我告诉你,我对你说的弗里曼关于数字计算机"是从对神经元如何工作的一种错误认识开始的"这句话有多赞赏。

我不认为我是第一个发现这个事实的人,但我特别高兴弗里曼同意这一点。一般说来,我觉得弗里曼的观点总是非常新鲜。你所引用的他认为计算机中的信息处理与头脑中发生的过程不同的观点也是如此。当我读到它时,我和你同样感到困惑不解——真是令人印象深刻! 我在他的著作《心智是如何在脑中产生的》一书的"意义和表征"(第 13 页)一章中,找到了更深入的解释。在这里,他得出的结论是,脑作为信息处理机器的整个想法是错误的和误导的。他(又一次)说,这个想法是由计算机科学家和逻辑学家引入的,而且由于这种错误的观点成了一种范式,整个心智研究产生了混乱。我读了相关章节好几遍才搞懂了他的意思。但我认为他是对的,至少他发现了这个观点是从外部引入的。我不确定他选择"意义"而非"信息"作为心智的货币是否是问题的最终解决方案。但无论如何,我可以说,你的朋友弗里曼在一个我认为基础非常牢固的领域中改变了我的观点。这种洞见卓识并不是经常有的,因此我要为这一卓越的想法再次表示感谢。很可惜弗里曼不在布莱克莫尔采访的研究人员之列。他的作品并不容易读,因为它充满了实质性的内容,而且他具有异乎寻常的独立性和原创性,并具有他自己的世界观。但是发现他的世界确实是一种喜悦,而且他肯定不属于那种大声吹嘘和装腔作势的人!

当你说你不明白"把心智下载到机器里去"究竟是什么意思,我相信库兹韦

尔的意思就是你可以在许多科幻电影中看到的那种。在那种电影里把人的心智复制到另一个人的记忆里或机器的存储器里并没有多难。所以他可能认为，所有这一切都可以被复制，其中也包括主观感受。我想他并不关心所谓的"无魂人问题"[1]（zombie problem），只要它的行为表现得像真人一模一样就行。我不得不承认，在我更多地了解到脑的结构及其与身体其他部分的紧密关系之前，我也相信过这种可能性。这个问题当然与你非常感兴趣的主观性和主观体验特性高度相关。

当我问你关于在布莱克莫尔采访过的研究人员中谁最接近你自己的观点时，我曾希望你会说是查默斯，我很高兴你也是这样说的。

这并不是因为在众多候选人中我最喜欢他的立场，而是因为我发现他的立场很难理解。因为你关心这些问题的时间比我长，我想通过向你学习，希望你能让我更容易理解他的立场。也许你可以从查默斯可能用来反驳丘奇兰夫妇[2]的批评的反对意见开始。特别是因为帕特丽夏的批评非常尖锐，她甚至否认查默斯把对意识问题的研究分成"困难问题"和"简单问题"的著名区分是有道理的。

丹尼特（Daniel Dennett）的论点也许也同样有趣，他认为根本就没有什么主观体验特性，并且似乎把自己看作是在这一问题上无所不知（know-all）。原则上我喜欢他的立场，因为这最接近我自己看待这一问题的、也许有点天真的立场。我读过他的一本书，虽然我喜欢他总的方法，但是令我感到不舒服的是他字里行间中表现出来的高度自信〔我是否应该把这叫做"库兹韦尔性"（Kurzweilness）呢？〕。并且似乎为了清楚表明他的高见，他总翻来覆去地用一堆话来阐述。他当然很聪明，但我更喜欢弗里曼的风格。但我在这里所说的，认真说起来，带有情绪性、非理性和主观性，也许我完全错了，并没有掌握精神

---

[1] Zombie 一译僵尸，不过在国家科技名词委员会生物物理专业组织的讨论会上，大多数专家认为僵尸在民间已有了固定的迷信意义，不宜采用，笔者提出今译并得到了与会专家的同意。这一术语是在哲学界，有人提出设想，有一种生物，从其行为上来看与正常人毫无二致，唯一的不同是他们没有意识，这被称为"无魂人"。后来克里克和科赫又把这一概念推广为正常人不需要意识参与的自动行为，并把负责这样行为的系统称为无魂人系统。——译注
[2] 妻子帕特丽夏·丘奇兰（Patricia Churchland）和丈夫保罗·丘奇兰（Paul Churchland）。——译注

实质,也没有弄懂主观体验特性究竟是什么。不过你可以告诉我!无论如何,我会尽力坚持原则不要成为自己偏见的受害者。☺

你一定会告诉我的!
祝好!

卡尔

II-006　凡及

心智上传;主观性是某种脑活动涌现出来的不可还原的基本涌现特性;恰当的问题是意识从什么样的脑活动中涌现出来的,以及涌现意识的充分必要条件是什么;科学家致欧盟公开信批评 HBP;美国 NIH 顾问委员会工作组提出对 BRAIN 倡议的建议

2014－08－01

亲爱的卡尔:

非常感谢你对摩尔定律有关问题的深入分析。特别是,谢谢你向我介绍了详细的技术背景,我就像大多数公众一样对此所知甚少。你向我说明了许多加快计算速度的方法,这是我以前所不知道的,其中有一些方法还只是在开始发展的阶段! 因此,虽然这个问题对你来说很简单,但对我和其他人来说却很复杂。你详细的解释正是我们所需要的!

读了你的论点之后,我不得不承认我以前的论点有点书呆子气。虽然我的论点原则上对无限时间尺度来说是正确的,但对在可预见的将来所能取得的技术进步的问题上,这样的论点可能毫无意义。从实际出发考虑,正如你在上一封信中所说:

"但是当我们从这样的宇宙学维度回归到我们不久的将来时,比如说未来的 25 年,那么没有理由认为计算机就不能像我们在过去 50 年看到的那样以同样的速度提高其能力。"

我必须从中吸取教训。

　　　意识之谜和心智上传的迷思　　一位德国工程师与一位中国科学家之间的对话

至于"把心智下载到机器里去"的问题,我认为主要问题在于我们甚至都不清楚心智究竟是什么。正如你在 2014 - 03 - 20 的信(Ⅱ - 003 卡尔)上所说:"你怎么可以在计算机上仿真你在现实世界所不了解的东西呢?"如果我们不知道心智是什么,我们怎么能把它复制到机器里去呢? 要是心智只不过是一种特殊的用语言表达出来的思想,那么下载它是有可能的,你甚至可以把这封信看作是我现在所想的一个拷贝——它是从我的心智中"下载"下来的! 然而,心智不仅仅只是记忆或思想,它不是静止的,而是不断变化的。它可能带有情绪,有些东西可能根本无法用语言来表达……在我看来,心智是在与内、外环境相互作用下的脑全局活动中涌现出来的,你怎么能把所有这些因素都复制到机器中去呢?

至于查默斯和丘奇兰夫妇、丹尼特之间的争论,我在上封信中所引用的查默斯的话最接近我的想法,虽然我不能说我会完全同意他的所有论点,而且我想他的想法也不是始终前后如一。因此我不会为他的每个论点进行辩护。当然,我并不期望我关于意识的所有观点都是正确的,正如你在 2014 - 01 - 15 的信(Ⅱ - 002 卡尔)上警告过我的话:"但是请不要期望过多,别忘了你正在研究的是多年以来都没有得到解决的事!"我想这句话不仅对我来说是对的,对查默斯、丘奇兰夫妇、丹尼特和其他人来说同样是对的。另外,我不得不承认,我没有读过他们的每一本书,我对他们的了解主要来自布莱克莫尔的采访,因此也许我会误解他们的某些论点。如果你发现我有错误,请不要客气予以指正。

当布莱克莫尔问丘奇兰夫妇为什么意识问题很特殊(困难)时,他们否认这个问题有什么特殊性。保罗说:"我发现现在当我们看待意识或者主观体验特性时感到某种迷惘,这并不是什么新的东西。"帕特丽夏补充说:"在任何科学理论的早期阶段,很多人都很惊讶。"他们认为如果人们对脑懂得多了,问题就会自然解决。他们认为,许多"谜题"在科学史上都已经得到了解决,这些难题在解决之前也显得非常困难,就像意识问题在今天显得很困难一样。从他们的观点来看,意识和脑活动只不过是同一个硬币的两面。在我看来,他们并没有驳倒查默斯的主要论点:意识是主观的,而在科学史上研究过的其他对象都是客观的。说意识和脑活动是同一回事并不那么令人信服。

我不得不承认,我对丹尼特的观点了解很少,我甚至没有读过他的《意识的解释》(*Consciousness Explained*)一书,我对他之所知也只来自布莱克莫尔的采访。从她采访所得的话中,我同意丹尼特的下列说法:

　　"我认为我们觉得意识之所以如此艰难的原因是,我们已经进化出一种自我感知的能力,这使我们能在某种程度上接触到自我,并给我们以主观体验,这给了我们从自我出发向外看世界的途径。而这一点是很难理解的。""让我们看看能不能以纯粹是物质的语言来给出发生下列现象的充分条件:存在着那种看起来好像存在着的东西(there to be something that it is like something to be),某种有内部世界的东西,某种有主观观点的东西。"[1]

　　然而,要找到这样的"充分条件"真的很难。就我所知,所有的科学家,包括埃德尔曼(Gerald M. Edelman)和托诺尼(Giulio Tononi)在内,找到的条件都只是必要的,而不是充分的! 我想有一天我应该读一下《意识的解释》,这样我就能更好地和你讨论他的观点了。

　　现在让我来讨论查默斯的观点。正如我上面所说,在我上一封信中提到的他的话给了我很大启发。我也同意他下面的话:"意识科学的核心是试图理解第一人称视角[2]"。但是,我不完全同意他的下列说法:

　　"这一领域的关键问题是,'我们如何能用科学所熟悉的客观过程来解释主观体验? 脑中相互作用的 1,000 亿个神经元一起如何会产生一种有意识心智的体验,包括其中所有各种奇妙的图像和声音?'"[3]

　　如果最后一个问题就是他所说的"困难问题",那么我不得不说,这个问题

[1] Blackmore S. Conversations on consciousness[M]. Oxford: Oxford University Press, 2005.
[2] 第一人称视角也就是主观性。——引者注
[3] Blackmore S. Conversations on consciousness[M]. Oxford: Oxford University Press, 2005.

提得不恰当。从这个问题听起来好像意识是一件东西，而脑活动是另一件东西，甚至是后者"产生了"前者！对我来说，前者是后者的涌现性质。正如一个分层非线性系统的所有其他涌现性质一样，你不能问系统中涌现特性是如何产生的。正确的问题是在什么条件下会涌现出这样的属性？因此，我同意在我上一封信中所引用的他的话。主观性是某种脑活动的一种不可还原的基本涌现特性。正如查默斯所说的意识是"一种基本特性"，但它只是某种脑活动的基本特性，而不是他所说的"世界"的基本特性。正确的问题是：意识是从什么样的脑活动中涌现出来的？这个问题真的很难。这个问题的最后答案应该给出某种充分必要条件，但现在还没有人知道。因此，虽然主观性是一种基本特性，但并不是每个系统都具有这样的性质，甚至也不是每种脑活动都有这样的特性！这是我受到查默斯和其他人（甚至包括一些他的对手）的启发所产生的观点。当然，正如我所引用的你的话那样，我并不期望我的观点能解决争端。我期待你的评论，甚至是批评。

也许你已经知道，上月 7 日 156 位神经科学家就有关人脑计划的问题发表了一封致欧盟委员会的公开信[1]。在信中，他们批评了 HBP 在研究方法上过于狭隘，从而导致无法实现其目标的很大风险。他们特别批评 HBP 取消掉有关神经科学中的整个一个子项目——认知神经科学子项目，这是 HBP 中唯一一个使用自上而下的方法的项目。他们呼吁，"HBP 已经偏离航道，欧盟委员会必须在和 HBP 续约之前对其科学和管理两个方面仔细审查。我们强烈质疑 HBP 在其目标和实施方面是否适合作为全欧合作的核心，以推进我们对脑的认识。"他们要求未来的评审应该是透明的、独立的、能反映出神经科学研究方法的多样性。在此基础上，他们要求 HBP 应该根据评审意见对其学术目标和管理进行大刀阔斧的改革，否则的话，他们呼吁欧盟委员会和成员国把现在给予 HBP 核心计划和伙伴计划的资助进行重新分配，把这些钱以广阔的神经科学为导向进行资助，以实现 HBP 的原定目标——认识脑功能及其对社会的影响。

---

[1] Open message to the European Commission concerning the Human Brain Project (http：//www. neurofuture. eu/).

万一欧盟委员会不能采用这些建议,他们将敦促科学家抵制 HBP。

这件事非常有趣也很严重!让我们看看下一步还会发生些什么事吧!

最美好的祝福。

凡及

附:在我 2013－10－12 的信(Ⅰ-011 凡及)中,我提到美国国立卫生研究院(National Institute of Health,简称"NIH")成立了一个主任顾问委员会(Advisory Committee to the Director,简称"ACD")的工作组,由神经科学家巴格曼(Cori Bargmann)和纽瑟姆(Bill Newsome)牵头。这个工作组将提出报告,向 NIH 主任提出美国 BRAIN 倡议的科学目标。现在我听说 ACD 的建议已经提交给了国立卫生研究院院长,并于 2014 年 6 月 5 日获得批准。据说,它建议在从 2016 年起的 10 年内给该项目 45 亿美元的资助。建议在头 5 年里,专注于开发新技术,而在后 5 年中,则将这些技术应用于脑研究。这个目标听起来不像 HBP 那样野心勃勃,也更为实际。然而,我还没有看到详细的相关资料,所以现在还不能发表更多意见。也许我们可以在稍后再讨论这个问题。

背景专栏ⅡF6.1

埃德尔曼和托诺尼有关意识产生条件的假说

在对什么条件下才会涌现出意识的问题上,诺奖得主、美国神经科学家埃德尔曼及其同事托诺尼所提出的一系列假说有着很大的影响。

埃德尔曼认为研究意识的正确做法是研究所有的意识过程有什么共同特性,意识在什么条件下才会产生,然后研究什么样的神经过程也具有类似的特性。由此提出假设,说明什么样的神经过程才对意识的产生有贡献。

埃德尔曼强调所有意识体验都存在两大特性:整体性(integration)和分化性(differentiation)。

所谓整体性就是说，每个意识状态都只有单一的"场景"（scene），不可能把这个场景分解成一些独立的成分。也就是说，意识体验是统一的和协调一致的，所谓"心无二用"就是这个意思。我们不可能在同一时刻执行两个不同的任务，除非其中的一个是高度自动的，不需要进入意识。我们在几分之一秒的时间里也不可能同时作出两个有意识的决定。当我们看一幅交变图（ambiguous figures）时（图Ⅱ F6.1），因缺乏必要的约束，同一幅图可有两种不同的解释。然而在某一瞬间，你只能感受到其中之一，而不可能同时"看到"这两种不同的情形。

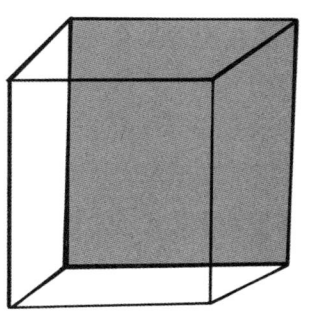

图Ⅱ F6.1　奈克尔立方体——一种交变图。灰色面是在最前还是在最后，两种可能性都存在。因此你既可以把上图看成是一个灰色面在前面的立方体，也可以看成是一个灰色面在背面的立方体，但是你绝不能同时看到这两种情形。

所谓分化性是指可以在极短的时间里体验到极大数量不同意识状态中的任何一个场景。呈现某一特定意识状态，意味着在所有可能的意识状态的清单（repertoire）中以极快速度选取其中之一。埃德尔曼特别强调，清单中的可能状态的数目是巨大的。意识到某一意识状态也就是排除了其他可能性，减少了不确定性，因此埃德尔曼也把分化性称为信息性（informativeness）。

托诺尼在他和埃德尔曼的合作基础上，于2004年提出了一种"整合信息理论"（integrated information theory）且不断加以修正。

托诺尼认为，要想形成主观的感知觉体验，脑必须把输入的感觉信号和储存在记忆里的信息整合起来，才能形成有关世界的协调一致的图景。问题是怎样把这种直观的想法精密化，为此他提出了有关意识的五条基本性质作为不证自明的公理，不过其核心依然是有关整体性和信息性的思想。这些公理包括：存在性（existence）、结构性（composition）、信息性、整体性、排他性（exclusion）。存在性是说，意识确实存在，至少"我"可以绝对肯定"我"自己是有意识的。结构性是说，意识的内容有一定结构，其中有许多不同方面。例如对眼前景物，我们可以同时意识到其中有许多不同颜色、不同形状、处于空间不同部位的内容等。信息性和整体性已经在上面讲过了。排他性是指意识无论就内容、时空尺

度来说都是确定的,不会同时有多个体验、或者以不同的速度展开内容多少有所不同的多个体验。

正是站在这些公理的基础之上,他们认为如果一个系统要有意识的话,那么这个系统就必须有和上述公理相应的性质,它应该是一个有极大量可能状态的统一整体,为此在有关脑区之间必须有交互作用。一旦这些脑区之间开始失连接,或者其组成部分失去特异性,意识就会消退,这便是在深睡、麻醉或者癫痫发作时的情形。意识的程度可用该系统超越其各组成部分所含信息量的信息量来度量,他们把这称为"整合信息"(integrated information),并用符号 Φ 来表示,以度量一个系统不能被还原或简约为其组成部分在互不相关时所具特性的程度。所以,有高 Φ 值的系统必定是由一些各具特异性的部分所组成而又高度整合在一起的系统。他们希望用这样一个量来度量意识的程度。确实,它也能解释一些问题,例如虽然小脑中的神经元数比大脑皮层还多,却没有意识,这是因为小脑各个模块之间缺少像大脑皮层各脑区之间的复杂双向联结,因而 Φ 值低。相反,丘脑-皮层系统内部有着大量双向的相互联结,因此对意识有重要贡献。当深睡和癫痫大发作时,大脑皮层各部分的活动高度同步,缺乏特异性和信息性,其 Φ 值也低,在这种情况下,即使丘脑-皮层系统也没有意识,或者说意识程度很低。

最近切鲁洛(M. A. Cerullo)对托诺尼的理论提出了批评,尤其对他把整合信息和意识等同起来的说法不以为然。他指出:"如果没有进一步的论据,不能认为信息排他性是不证自明的,也没有理由认为整合信息会产生(或者就是)意识(换句话说,整合信息对意识来说还不充分)。就算意识排他性是正确的,最多只能讲整合信息对意识是必要的,但并不充分。"[1]

笔者也认为他们的五条"公理"只给出了存在意识的必要条件,他们所定义的 Φ 值只适合刻画有意识程度的某些方面。上述指标完全没有涉及意识的主观性和私密性这样最关键的特性,也看不出主观性和私密性如何能从整体性和

----

[1] Cerullo M A. The Problem with Phi: A Critique of Integrated Information Theory[J]. PLoS Comput Biol, 2015, 11(9): e1004286.

分化性里涌现出来。如果他们提出的五条公理并不是存在意识的充分条件，那么即使某个系统有非零的 Φ 值，它也未必具有某种程度的意识。笔者还以为，期望通过制定一些定量指标来全面描述意识的所有方面，这是一个不可能完成的任务。当然，Φ 值可能给出了度量意识某方面的定量指标，这一理论尝试用一种新的、严格的、同时采用数学和经验的方法去探索心身问题的新途径，这还是值得人们关注和深入探讨的。

II-006 卡尔

心智上传;涌现现象;自然科学的历史就是一个
形而上学家不断失望的历史

2014 – 08 – 30

亲爱的凡及:

嗯,这是个有趣的消息。HBP 遭到攻击了,看来我们的朋友马克拉姆后面还得有麻烦呢。然而,粗看起来,这似乎更像是一种由于嫉妒或虚荣心受到了伤害所驱使的行为,而不是像你我那种对其总的概念所持有的严肃怀疑。我们知道马克拉姆在与其他人打交道时是不太通情达理的。但要协调好这样一个巨无霸计划确实需要高度的同理心,态度上也要温和有礼。我相信这些完全不是他的强项。好吧,我们就会看到他是如何反击这次攻击了。

但是,让我来谈谈你上次和先前的邮件中谈到的一些论点。现在我们正接近意识问题的核心,与此同时,我也开始更好地理解你的主要关注点是主观性和私密性。

我同意,要想理解意识即使不是不可能的话,也是非常困难的。这是因为对外部感觉的表征在每个脑中的"储存"方式都是不同的。我也同意,除了主观性之外,我们还面临一个棘手的问题,这就是你所强调过的"心智不只是记忆或思想,它并不是静止的"。你说对了。我们正在谈论的是一个连续的、高度动态的过程,从这个过程中仅摄取其中个别画面对我们来说并没有多大帮助。这是一个非常有力的论点,这是最天真的"心智下载"迷们所忽略了的!

最主要的问题是脑不能被看作是一个孤立的信息处理机器,因为它总是身处身体其他部分的生物化学环境之中,并不断接受到来自体内外的感觉信

号。我对此了解越多，就越清楚地认识到，试图下载心智就像想抓住彩虹并把其存放到一个盒子里一样虚幻。如果你真的想要拷贝你的心智，就必须逐个复制整个身体的原子。甚至这都还不够，因为为了抓住各种动态变化，例如体液的运动、动作电位和电磁波，你另外还需要有关每一原子的信息，甚至还可能需要每个电子在给定时刻的运动方向和速度。你需要实时获取所有这些数据！

要想用一台超级测量装置执行相关扫描，即便原则上可行的话，也只有当这种机器能提供同一时空中的所有物体的信息时才会有用。因此，这样的扫描机要么需要有和所扫描的物体一样多的传感器，并且还要以比光更快的速度工作；要么在以较慢的速度依次进行扫描的条件下，那就必须计算已扫描过的对象的位置，以确定当把它们组织在一起时所在的正确位置、方向和速度。这些就已经很不容易了，更不要去说海森堡的测不准原理，你这就可以认识到了要做这样的尝试只是一种幻想。然而作为一种思想实验，这样的想法可能还不错。至少在科幻电影中，这样做根本就不在话下，因为当太空船"企业号"上的船员利用"化成射线束"（beaming）到达另一颗行星时，他们使用的正是这种方法。

对人的身体进行扫描，去物质化并输送到另一个地方，到达之后重新再物质化（rematerialized），这一切都没问题。他们不仅老是这样行事，并且还轻松容易地就把某个人的拷贝储存了下来，这一点在某人遇害时是很有用的，因为人们可以用这种办法重建和受害者一样的完美副本，甚至包括其主观体验特性和其他一切。我说过库兹韦尔一定也读过我小时候读过的科幻作品，因为他仍然认为这样的事情很快就有可能实现。

通过去物质化，化成射线束避免了某个个体有多个副本的冲突，但这又引发了另一个远比上面提到的那些问题更为严重的问题。扫描一个对象到电子水平就已经很难了，但要把一个中等大小的个体化成射线束，这意味着要将 70 千克的物质在到达目的地并重新物质化之前转换成辐射。这意味着要释放相当于好几颗原子弹的能量。好吧，公平地说，库兹韦尔除心智上传之外并没有承诺过化成射线束。但如果他真的这样说了的话，我也不会感到惊讶。☺

但是让我回到那个思想实验和我想问你的关键问题上来。如果我们真的有一台像上面讲过的那样的扫描机器,这台机器使我们能对一个包括脑在内的人体做出一个完美的1:1的复制品,那么这个完全相同的副本也会有和真身相同的意识和主观体验特性吗?查默斯可能会说"不",而丹尼特一定会说"是的,当然"。你怎么想呢?

我站在"可能是"的这一边,不过我不确定我是否能同意丹尼特的进一步说法,他认为意识只是当脑复杂到一定程度不知何故自动产生出来的一种幻觉。

你指出过,意识是一种涌现现象,这种现象不容易从其所据的理化基质中推演出来。这我是同意的。这对于物质世界中的许多涌现现象来说都是对的,例如令人费解的各种相变就是如此。相变是指物质在某一个时刻前后突然表现出完全不同性质的现象。液态水冻结成冰可能是这方面最常见也是大家最熟悉的例子。虽然这种冻结现象在理论上很难解释,需要对量子物理学有深刻的理解,但最终还是可以通过理性的物理手段来加以解释。

我喜欢你所引用的 C.科赫那段话中有关涌现的内容。他沿袭诺贝尔奖获得者劳夫林(Robert Laughlin)[1]的传统,认为一般说来物理学家应该更多地关注涌现现象和相变,因为比起在这些神秘的奇点之间的可以用连续可微方程描写的"舒适区"来说,从涌现现象和相变中可以发现更有趣的现象。另一位诺贝尔奖获得者普里高津(Ilja Prigogine)强调在这些远离平衡态的惊人的和令人感兴趣的区域中,即使是最可靠的定律——热力学第二定律似乎在微观层面也不再成立。在物质的自组织世界中充满了这种令人困惑的涌现现象,但劳夫林和普里高津都认为不应该放弃理性物理的方法,退居到形而上学的云雾世界里,而后者正是许多人在提到涌现时常常会犯的错误。

当查默斯谈到意识(包括我们的主观感受和主观体验特性,暂且不管它究竟是什么吧)究竟是和物质结合在一起的呢,还是一个单独的实体时,我猜查默斯对涌现的解释要比你的更严格。我想他会坚持说,它与物质世界没有联系,而是另一个独立的实体。

---

[1] Robert Betts Laughlin(1950—),美国物理学家,1998 年诺贝尔物理学奖得主。——译注

在查默斯的这一立场问题上，我不喜欢之处是他的立场上的形而上学色彩，这让我想起了 19 世纪对"活力论"（vitalism）的讨论和 20 世纪对"活力"（élan vital）[1]的讨论。

自然科学的历史就是一个形而上学家不断失望的历史。每一次，当他们提出某种把无生命物质的世界与生命物质世界分隔开来的界限，认为自然科学的理性手段无法突破这一限制时，这种界限或早或晚总得后退。活力论相信有机世界是根本不同的，因为它包含了一些神秘的非物质要素。1828 年维勒（Friedrich Wöhler）[2]用无机物合成第一个有机分子——尿素，这给活力论以沉重一击。1959 年米勒（Stanley Miller）[3]和尤里（Harold Urey）[4]成功地创造出一个环境，氨基酸在其中能自发产生，这似乎是对传统活力论的最后一击。然而，这种思想也会间或回潮，而在和意识的关系问题上也会成为"主旋律"（Leitmotif）。丹尼特说，这是因为很多人觉得生活在一个只有物质的世界上很不舒服，有一种根深蒂固的愿望，希望在物质之外还应该有其他东西。就像"生命质"（essence of life），它使岩石和动物以及我们人类区分开来，并使我们与众不同。

在欧洲，这一思想与长期以来教会信仰体系主导下学究气大学的传统有着密切的关系。祭司们总是坚持除了物质之外还有其他的东西，比如就像完全无法用理性推理和自然科学方法研究的灵魂之类。

但我认为也会有类似的想法深深植根于亚洲尤其是中国的哲学之中。我不知道"气"中的"精气"（living force）概念和"活力论"思想之间的关系有多深，这些概念之间究竟有没有相互影响。但我相信你会知道得更深入一点。

我明白你并不同意查默斯所说的一切。但我为了批评方便起见，要建立某个和你立场接近的稻草人或形而上学假人，在此之前，我想请你在上述查默斯

---

[1] Élan vital 是法国哲学家 Henri Bergson1907 年在其著作《有创造力的进化》（*Creative Evolution*）一书中创造出来的一个术语，在英语中通常译为"vital force"，用以解释进化和个体发育。——译注

[2] Friedrich Wöhler（1800—1882），德国化学家。——译注

[3] Stanley Lloyd Miller（1930—2007），美国化学家。——译注

[4] Harold Clayton Urey（1893—1981），美国物理化学家，1934 年诺贝尔化学奖得主。——译注

和丹尼特的争论问题上回答我有关理性和形而上学（physics-metaphysics）的问题。也许你还有第三种看法，或者能给我某种更有帮助的观点。

不管怎样，我期待着知道你的意见！

就写到这里吧，祝好！

卡尔

心智上传;涌现特性与不可还原的基本特性;要
找到对任何主体都成立的意识涌现的充分必要条件
是非常困难的;主观性和私密性是意识研究困难的实
质所在

2014 - 09 - 28

亲爱的卡尔:

你绝对是对的,意识令我困惑的核心问题是它的主观性和私密性!

我认为我们取得了一点共识,这就是我们都认为心智上传几乎是不可能的,至少在实际上无法做到。

说到你的思想实验,我的观点是,如果你的扫描机器真的能做出包括脑在内的人体的一个完美的1:1的复制品(精细到每一个原子),又如果相同同位素的所有原子都是相同的(我不知道,但是我没有任何理由反对),那么我对你的问题"这个完全相同的复制品是否也有和真身完全一样的意识和主观体验特性?"的回答是"是"。我的逻辑是,意识是神经活动的一种涌现特性,如果所有的神经活动过程都相同,那么它们的涌现特性必定也相同。然而,这里仍然存在一个问题。正如我在上一封信中强调过的那样,心智是动态的,它是不断变化的,取决于主体与其环境之间的相互作用。因此,即使你的扫描机器可以造出一个完美的1:1的复制人体,直至其每一个原子,但是要使这个拷贝具有和真身完全相同的意识和主观体验特性,你必须将副本精确地放到原来的位置上(但要使不同的物体占据相同的空间位置是不可能的),否则你还要为这个副本完美地百分之一百地复制所有不断变化着的外部刺激,也就是你还得完美地百分之一百地复制其外部世界。因此,在实际上不可能有两个完全相同的心智,

或者将一个心智下载到另一个系统中去。

　　我必须承认，在我对涌现和相变的表达中有一些含糊之处。你正确地指出，涌现现象"不容易从其所据的理化基质中推演出来"，你在这里加了一个重要的词"容易"，这意味着某些涌现出来的现象是可以从元件性质和元件间的相互作用推演出来的。因此，说意识是神经活动的一种涌现特性并不意味着就不可能说明它是如何从神经活动中涌现出来的。当然，这也不意味着意识必然就能够用神经活动来解释。我想，查默斯以下说法可能还是正确的，不过需要稍作修改："意识是世界的一种不可还原的基本性质"。然而，它只是某种如清醒的人脑那样的复杂系统的属性，而不是一切事物的特性。我不认为意识是世界的不可还原的基本特性！根本的问题是什么样的系统才会有意识？如果我们知道某类物体有意识，比如活人，那么另一个根本问题就是给出神经活动的一个充分必要条件，在这种条件下会涌现出主观性。如果我们知道了这种条件，那么"困难问题"就会消失。然而，还没有人知道这样的条件。我也怀疑人们是否能对各种各样的主体都可以找到普遍适用的这种条件。主观性使所有这些研究极为困难，如果不是不可能的话，因为你无法证明一个行为复杂的系统是否是有意识的，不管它看起来多么像有意识，因为意识是主观的和私密的，你不可能分享它！也许它只不过是一个无魂人呢！

　　你把有关意识的争论和有关生命的争论进行比较的论点是非常有力的。查默斯和他的支持者认为，意识是由于其主观性而成为独一无二的，这是科学家们所从未见过的，所以科学在其他领域中取得的胜利不意味着意识之谜也一定会像其他谜题那样最终得到解决。你的论点是生命在科学史上也是独一无二的，但现在它已经被解决了。是的，"独特性"并不是解决不了问题的原因，你的这个论点我是同意的。然而，主观性却可能成为原因，也许主观性是不可还原的。归根到底最后必然会有一些不可还原的东西，就像数学中的公理一样。除了承认事实之外，你别无选择！由于主观性，你无法严格判断其他主体是否有意识。对人来说，因为所有人都有结构几乎相同的神经系统，也许我们可以把下面一条当作公理：如果其他人能像有意识的我那样行为，那么他或她一定是有意识的。高等动物有脑，也许我们可以采用类似的公理，承认黑猩猩也有

　　　　　意识之谜和心智上传的迷思　　　　一位德国工程师与一位中国科学家之间的对话

意识并非那么困难，甚至狗、猫和鹦鹉都是如此。但是低等动物呢？你相信虫子有意识吗？那么机器呢？由于意识的主观性和私密性，我们不可能知道任何其他主体的主观经验，我们只能观察它的行为。然而，正如塞尔在他著名的"中文屋"思想实验中所表明的那样，你不能仅仅根据一个主体的行为就判断它是否是有意识的，除非有一些公理，比如说，如果一个人的行为与我相似，那么他就是有意识的。你能把图灵测试当作公理吗？塞尔的"中文屋"思想实验似乎表明"不能"。即便如此，这样的"公理"是相当模糊的，"有与我相似的行为"究竟是什么意思呢？不同的国家有不同的传统，他们在相似的情况下可能表现不同，我是否可以否认在某种情况下和我有不同行为的外国人有意识？一个孩子确实有很多不同于我的行为，我能否认他有意识吗？诸如此类，回答显然都应该是否定的。

我认为，在目前，我们可以研究人类主体有意识的必要条件。近年来，人们对"意识的神经相关机制"（neuronal correlates of consciousness，简称"NCC"）[1]和"意识获取的印记"（signature of conscious access）[2]进行了许多有趣的研究，并朝着这一目标迈出了重要的一步。然而，这些研究大多集中在意识的某些方面——意识的某些特殊内容，或"意识获取"这样的意识的某个特定方面。至于意识获取的概念，我将在以后的信中讨论，否则这封信就会太长了！也许有一天，当有关上述研究的数据积累得足够丰富时，人们就会发现人类意识的充分必要条件，正如你和其他一些科学家所期望的那样，"世界之结"（knot of the world）[3]将获得解决。如果搞懂了人类主体的意识发生的充分必要条件，假设这些条件得到满足，那么意识就会自然地出现，"困难问题"也就可能自动

---

[1] 克里克和C.科赫认为为了产生某种特定的知觉，脑中至少需要某些特定部位的特定神经元集群以特定的方式活动，他们把此称为"意识的神经相关机制"（旧译"意识的神经相关物"，不过这一译名过于强调"物"，可能使人误解为仅指相关的神经元、神经回路或脑区。实际上按他们两位原来的意思，并不仅限于这些"物"，还和相关的活动模式等有关，故笔者建议改为今译，并得到了国家科学技术名词审定委员会有关专家的认同）。——译注

[2] 法国神经科学家德阿纳（S. Dehaene）的工作表明在不同的条件下，主体可能意识到刺激，也可能意识不到，他把意识到刺激称为"意识获取"。当刺激被意识到时，脑活动中表现出明显的特殊印记，这些印记可用特定的脑机制来加以解释。——译注

[3] 由于研究意识问题的困难性，叔本华把意识问题称为"世界之结"。——译注

消失。我不能说这是不可能的,但对我来说,不太可能在可预见的将来做到这一点。真正的问题在于你如何知道除了人之外的主体也有主观体验呢?要找到对任何主体都成立的意识涌现的充分必要条件是非常困难的!

我很惭愧地说,我从来也不知道"气"这个概念的真正含义,它是相当模糊的。在古代中国,人们认为"气"是万物的基本元素,它像气体那样到处流动,它也在身体中流动以保卫身体和给予精力。当然,现在我们知道体内并没有这种气。

以后再谈。

凡及

心智上传；意识和主观性；意识的神经相关机制；
依然有很大的可能性可以通过理性手段在自然科学
的范围里认识意识和主观性

2014 - 11 - 01

亲爱的凡及：

你有关意识之谜、主观性和私密性问题的说法都很吸引人，也很有启发性，我不得不承认你的论证很聪明。我又一次花了好几个小时去阅读和探索有关心智之谜各种立场的新、旧想法，以求有个概括性的了解。当然，我还没有完全想好，这里只是初步的结果。

我喜欢你对问题的讲究实际的解释而不是查默斯的立场。对我来说，他是想用一种众所周知的方法来捍卫自己的理论不受攻击。他把主观性定义为某种我们无法接触到的东西，所以任何试图检验这个理论的尝试都是徒劳的。当你这样做时，论证就会陷入循环。因为主观性是不可证明的，所以你不能证明主观性。对灵魂、气、Psi[1]或任何一种幻想出来的东西都可以这样做。

稍后我会回过头来再谈这个问题，让我先来谈谈你关于我有关完美复制包括脑在内的身体的思想实验的想法。

我很高兴听到你同意一个1∶1的身体拷贝也能复制意识和相关的主观体验特性的说法。然而，真要想创建这样一种拷贝很难，甚至是不可能的。

---

[1] Psi 是由生物学家 Berthold P. Wiesner 创造的一个术语，用以指不能由已知的物理或生物机制解释的超感觉知觉（extrasensory perception）中起作用的不明因素。这一术语来自希腊字 ψυχή（心灵、灵魂）的首字母 ψ（psi）。在西方流行文化中，"psi"已经越来越成为特殊的心灵、精神能力的同义语。——译注

你坚持说必须把身体的拷贝精确地放在真身的同一时间和空间位置上,因为心智是一种动态过程,并且从感官不断有信息流向脑。从技术上来讲,你这样的说法是绝对正确的。因为我们思考的是一个动态的问题,因此就要求不仅所有的原子都要在正确的位置上,而且还要在空间中有正确的速度和朝向。如果你想非常精确,你还得考虑到电子的自旋和所有各种亚原子粒子,以保证所有流动的体液、扩布着的动作电位以及相关的电磁场都要动态表示,并且一切都得处于正确的相位。

所以光是这些要求,就已经使得这种复制身体的机器在现实世界中是无法实际做到的,而你正确地指出我们还得把拷贝放到真身所在的同一个时空点上,这是一个更难得多的要求。如果只是为了可以继续讨论下去,而一定要使这个思想实验在原则上得以成立,那么我就只能求助于埃弗里特有关量子力学中的多宇宙思想了。为此我们就得创建一个得以实现拷贝的新宇宙。这并没有像初听起来那样疯狂,至少对策来说是如此,他以其在量子力学上的"去相干"(decoherence)概念而闻名于世,他提出了一种很好的论点,说明为什么这样的平行宇宙总是在产生。这是一个不寻常的观点,但如果你对量子理论的哥本哈根解释及其神秘的"互补性"(complementarity)感到有点不习惯,那么策可能会提供给你另一种选择。我不知道爱因斯坦、薛定谔和德布罗意(Louis de Broglie)会不会喜欢它。至少我是喜欢的,因为我真的和这三位现代物理学奠基人一样不喜欢这种模糊不清的"互补性"。他们总是拒绝成为"玻尔教会"的信徒,也非常不喜欢玻尔(Niels Bohr)及其追随者。然而要想像策那样从互补性中解放出来,你就不得不付出代价接受埃弗里特的多世界概念。是的,其中也包含了多个心智的思想。[1]

策的解释并不比互补性更荒谬,在我对此问题思考一会儿之后,我宁愿生活在一个策-埃弗里特的世界里,也不愿意生活在一个玻尔-海森堡的世界里。但这完全是另一回事了,我并不想施放烟幕来回避你的论点。所以我很高兴地同意你的有力论点,你赢了这一回合。☺

---

[1] https://en.wikipedia.org/wiki/H._Dieter_Zeh

一个更务实的问题可能是问脑的拷贝要精确和详细到什么程度才能拥有主观性和意识。从我们周围的物质世界中，我们可以知道物理对象在宏观层面上有很强鲁棒性[1]，尽管它们是由那些在微观层次上很难测量和定位的高度动态的物体所组成。月球作为一个绕地球运行的宏观物体，可以以难以置信的精度进行定位和测量，尽管它是由一些微粒组成的，当你想测量这些微粒时，它们都有不确定性。

另一方面，我们不得不问，当我们不那么准确时，我们会失去多少。主观感受和主观体验特性有多强鲁棒性而得以维持其恒常性？当你站起来、走路、跑步或做空翻时，它们有很大的变化吗？也许其中有些感受和主观体验特性很鲁棒，而另一些则没有那么鲁棒，至少你能发现这一点。你甚至能告诉我影响如何。这正如人们可以描述毒品的影响，毒品肯定可以改变知觉和意识。

你的有些主观体验特性可能对绝大多数机械变化或化学影响都有很强鲁棒性，也有另一些主观体验特性则没有那么稳定。我想要说的是，我们有可能通过检测你主观意识中的变化，至少知道你的主观意识中的某些部分。这并不容易，但在原则上并非不可能，如果研究人员付出足够的努力，那么我们也可能会看到有实验可以探索神秘事物。有时需要很长时间和巨大的努力才能找到一个理论上预测过的现象的经验证据。当爱因斯坦在 1916 年预测引力波时，他确信其存在，因为这是他的广义相对论的结果。但当时还没有技术可以检测极为微弱的相关信号，连爱因斯坦本人也怀疑是否有朝一日真能检测到引力波。我们仍然没有引力波存在的证据，但是理论告诉我们，我们必须寻找什么，我们应该期待有什么样的结果。仪器的分辨率正在不断提高，几乎没有一位物理学家不希望能很快就看到证据。[2]

我想问查默斯的问题是：要有什么样的实验结果，他才会放弃他的二元论

---

[1] 所谓"鲁棒性"，是指控制系统在一定（结构、大小）的参数摄动下，维持其他某些性能的特性。它也是在异常和危险情况下系统生存的关键。——译注

[2] 当卡尔写此信时，情况依然如此。但是 2015 年 9 月 14 日激光干涉引力波天文台（LIGO）首次探测到了双黑洞合并所产生的引力波。三位相关科学家荣获 2017 年诺贝尔物理学奖。到目前为止（2017 年 8 月）已经探测到了 6 个引力波事件。我国也有太极计划等重大科学项目进行引力波研究。

立场。如果他说,这是不可能发生的,因为主观性根据其定义就是看不到摸不着的,那么他的理论就毫无价值,而更像是宗教教条。用同样的方法,你也可以捍卫意识是由随便哪一种东西控制的理论,比如把这种东西说成是像荒谬的飞行意大利面条怪物(FSM)那样的东西也行。FSM(如图 II K7.1 所示)是意大利面条教(pastafarianism)[1]教会崇拜的神,该教会的成立是为了说明许多宗教的信仰体系的荒谬,特别是那些坚持智能设计的宗教。[2]

图 II K7.1 "飞行意大利面条怪物"

当然这个故事很有趣,但并非每个人听了都会发笑。我希望你会会心一笑。☺

所以我知道你并不会坚持在意识问题上的某种形而上学的立场,但我不知道查默斯会怎么说。

---

[1] Pastafarianism 是把 pasta(意大利面条)和 Rastafarianism(塔法里教)两个词组合成的,因此译为"意大利面条教",又称"飞行意大利面条怪物教会"。这一教会实际上是一种社会运动,反对在公立学校教授智能设计和创世说。"飞行意大利面条怪物"首先是由汉德森(Bobby Henderson)在 2005 年的一封带有讽刺性的公开信中提出来的,其目的是抗议堪萨斯州允许在公立学校的科学课程中教授智能设计。他在信中要求,在科学课中要和智能设计以及进化论一样用同样的时间教授飞行意大利面条怪物教教义。在他发表此公开信之后,飞行意大利面条怪物成了网红,并成为反对在公立学校教授智能设计的标志。——译注
[2] https://en.wikipedia.org/wiki/Flying_Spaghetti_Monster

我不确定丹尼特是否就一定正确,而和主观性有关的一切都只是伪问题。或许我们真的需要一门物理学和化学之外的新科学来认识意识和主观性。但在我愿意改弦易辙并选择形而上学之前,我宁愿竭尽所能用尽一切物理和理性的方法。

你提到"意识的神经相关机制"(NCC)这一新的研究领域,这对我非常有帮助,因为它表明确实有人正在研究这个神秘的现象。谢谢你告诉了我如何更好地支持我自己的立场——你真的很有公平竞赛的精神!实际上我并不知道早就已经有这个领域了。

他们似乎遵循着塞尔的建议,塞尔在接受布莱克莫尔的采访时也采取了一种和我类似的非常务实的立场。他说道,认为意识不可能做客观研究的想法是错误的,有可能对一个本体论意义上就是主观的问题进行客观的科学研究,只是这样的研究需要创意和努力罢了。我喜欢他对这个问题的回答,和你一样,他也不知道动物是否有意识和哪些动物可能有意识。他说我们不知道白蚁是否有意识,但"我猜它们是有意识的"。我也很喜欢他对专家们要怎样才能回答我们的问题的说法。他说道,如果我们发现脑中有某个特殊过程 xyz 负责意识(通过有可能使有特定脑损伤的患者恢复意识的方法),那么我们可以检查狗、猫和灵长类动物的脑中是否也有 xyz。我们可能会发现它们也有 xyz,因此它们一定也有意识。然后他说我们还必须更深入到种系发生的层次,最后也许会发现白蚁有 xyz,但是蜗牛则没有。除了还不清楚 xyz 究竟可能是什么这一问题之外,我觉得这是研究 NCC 的一个很好的计划。

另一种研究意识主观性方面的经验方法可能是通讯交流。虽然我们永远都不能确定你所看到的红色究竟是否就是我看到的红色,但我们可以学会接近别人的印象。我可以用我个人的经验来说明这个问题。我非常痴迷于蓝色,特别是一种像布加迪(Bugatti)赛车曾用过、后来凯旋(Triumph)TR 跑车(颜色稍鲜艳一点点)所用的称为"选美蓝"(pageant blue)的蓝色。我所有的朋友都知道我特别喜欢蓝色,阿迪(Adi)[1]当然所知最深。她完全清楚我喜欢的是什么

---

[1] 卡尔的夫人。——译注

样的蓝色,因为她能准确地告诉你,某种蓝色是不是太暗、太鲜艳、太淡,太偏紫,或者正好是卡尔最喜欢的蓝色。

我们都学习不仅仅通过语言,还通过共情(empathy)来理解或解释他人的主观感受。虽然这很难,也不是每个人都能做得一样好,但我们越有经验,做得就越好。我们越了解一个人,我们就越能评估他或她感觉如何,心情如何,以及某个人在某一情况下是否感觉良好或不舒服。有时候,我们甚至可以察觉到一个人的感受正好和他或她试图告诉我们的相反。

就像发射机-接收器的情况,通讯的时间越长,我们在语言和非语言交流中检测信号的能力就越强,这是因为随着时间的推移,索引表变得越来越完整,发生新的内容的可能性也变得越来越小。在语言交流中,一旦你开始并建立了一些你所理解的基本词汇,你就可以用这些基本词汇解释更复杂的术语了。这是一个在不断作尝试和错误的适应循环中进行猜测和逐步理解的自举过程(bootstrapping process)。这也是孩子学习语言的自然途径。他们并不是用专业语言学家抽提出来的抽象语法规则来学语言的。我们的学校常常试图通过规则和语法来教语言却失败了,这是又一个说明想对某个你所不懂的系统进行逆向工程失败的例子。当然语言学家们对语法逻辑的发现是很有意思的。但是他们的形式化知识就像人工神经元一样也是虚拟和人工的。这种形式化的逻辑原则与我们脑中的过程毫无关系。这就是为什么以通常学院式的方法,就像教数学一样地教语言往往事倍功半。孩子学会说话是通过全身心投入交流的情景之中,通过适应性的探索来达到的(就像弗里曼的蝾螈探索其气味环境一样),而不是从抽象的规则出发学说话的。

专注于"只管去做"(doing the thing)也能让你学会外语或更多地了解另一个人、动物或机器的内部状态。交互作用将逐步完善你对正在打交道的环境的知识。当然,你比较容易与一个有相近文化背景的人交流,而较难与一个文化背景很不一样的人交流。当你不得不需要沟通某种文化中不存在的术语时,你需要花更长的时间,但是你可以通过交换更多的细节来逐步接近。随着时间的推移和多次反复纠正错误,我们可以改进这种进程,直到我们达到可靠的沟通。大概没有比汉语和德语相差更大的两种语言了,也几乎没有哪两种文化比我们

相差更大的了。尽管如此，我们成功地运用了第三种语言，即英语，甚至对一些很复杂的问题成功地进行了交流，尽管我们俩都不能讲一口完美无缺的英语（至少我不行）。虽然在一位以英语为母语的人眼中看来，我们的英语在遣词用句方面并不总是完美的，但我认为我们已经学会了相当好地交换想法。在这样做的同时，我们也学到了一些关于对方主观感受的知识。我不知道这种知识的增长究竟有多大，但我认为它超过了零。

我在这里说的话，你可能已经意识到了，它是受弗里曼的观点启发而来的。他的主要观点是，与环境的持续互动是产生觉知和意识的关键。我敢肯定，他会同意塞尔和我有关意识可理解性的正面看法。当我环顾四周寻找他的新作品时，我发现他就在几个月前在伯克利研讨会上发表了一篇非常有趣的演讲，标题是《动作感知周期的演化和神经动力学：意识与量子场论何时、何处与如何相遇》("*The evolution and neurodynamics of the action perception cycle — When, where and how consciousness and quantum field theory enter*")[1]。

如果你还没有看到过，那么可以尝试去看一下，你很可能会喜欢它，特别是因为有他在 1999 年发表《心智是如何在脑中产生的》一书以后的新进展。基本上，他讨的仍然是同一主题，并以蝾螈嗅觉系统作为例子，所以你读起来会觉得很熟悉。我不确定增加量子场方面的问题是否确有必要，但我认为他已经很好地打理了他有关意识如何从与环境相互作用中演化出来的想法。

我特别喜欢弗里曼的想法，即在脑中我们涉及的是产生电磁场和振荡的神经网络，这些电磁场和振荡会扩布到脑的许多区域并表现为某种模拟系统的编码模式。这与那种信息储存在特定脑区、甚至是特定神经元的概念形成了鲜明的对比，后者就像把信息存储在数字计算机的硬盘上一样。弗里曼总是追随拉什利（Karl Lashley）[2]的大规模作用（mass action）的想法，以及他的老师普里布拉姆（Karl Pribram）[3]有关信息分散存储的想法，这更像全息图的原理。不

[1] https://archive.org/details/UC_Berkeley_FOM_2014_05_02_Walter_Freeman
[2] Karl Spencer Lashley（1890—1958），美国心理学家，以其对学习记忆的研究而闻名。他认为记忆分布在全脑，脑的各个部分对记忆的贡献是相同的。现在一般认为他对他的实验结果的解释有误。——译注
[3] Karl H. Pribram（1919—2015），美国心理学家。——译注

过现在他对这些问题表达得更为精确了。

你对他的工作要比我更为了解，如果你能看一看，并告诉我你对我认为弗里曼的理论有了新进展的看法是否正确，我将不胜感激。上面提到的网站里有录像，后面还有他的工作的相关介绍和他所有作品题目的清单。

我无法判断后来加上去的有关量子场的想法到底与意识有什么关系，这个问题是弗里曼在 2007 年与一位不那么知名的物理学家维蒂耶洛（Giuseppe Vitiello）合作写的一篇作品中提出来的。它似乎没有引起多大共鸣，不管怎么说吧，弗里曼似乎为这一领域中一些大人物所忽视，这一点是很明显的。我很惊讶他在坎德尔的圣经中连一次都没有被提到过。在我看来，他值得受到更多的关注。也许他太过于埋头苦干钻研一个难题，而不对弹奏"科学秀钢琴"（science-show-business-piano）上的库兹韦尔琴键（Kurzweil-keys）感兴趣。也许他根本就不在乎，但如果在此领域中有哪个人是"兴趣派"的话，那么这个人就是弗里曼。

但有时名气和声望并不一定取决于工作的内在质量。决定一个想法、理论或发明成败的常常是一些完全不同的因素。有时可能有某些完全非实质性的、根本就没什么关系的东西，却产生了巨大的影响。在接受布莱克莫尔采访，被问及"中文屋"思想实验于 1980 年发表后的影响时，塞尔就这个问题作了一个非常有意思的评论。他很有感触地说道，虽然这篇文章遭受了大量的批评，却也使他闻名世界，因为这篇文章成了现代科学出版物上被引用次数最多的文章之一，而"中文屋"一词几乎成了一个全球性的家喻户晓的名词。在他看来，文章激起的巨大愤怒，很显然是因为很多人认为这是在意识问题上对他们所持的基于还原论的世界观的一种攻击。但塞尔认为真正的问题并不在此。他认为，真正的问题是，他忽视了这样一个事实，即与此同时，这也对许多基金、职业生涯和预期中的资金流构成了威胁，因为原先许多人在可以重塑人的精神的许诺下花费了大量金钱。顺便说一句，他早在 HBP 诞生之前就说了这番话。☺

我不知道如何把这些在动物身上检测是否有意识的论点也用于机器。我们今天所拥有的人工智能还很原始，还谈不上这个问题。然而，显而易见的是，图灵测试本身太弱了，因为它仅限于语言。因此有些人早已开始寻找一种更复

杂的测试,称为全图灵测试(Total Turing Test,简称"TTT")[1],这可能是下一代机器所需要的。

　　总之,我想要说的就是,弗里曼、塞尔和卡尔认为,依然有很大的可能性可以通过理性手段在自然科学的范围里认识意识和主观性。

　　如果这还不能说服你,这里至少有些东西可以让你展颜一笑。这是科幻作家克拉克(Arthur C. Clarke)曾经提出的著名三定律[2]中的第一条:

　　"当一位杰出但年长的科学家说某事是可能的时候,他几乎肯定是正确的。当他说某事不可能的时候,他很可能是错的。"

　　来自卡尔斯鲁厄的晚安问候!

卡尔

---

[1] 全图灵测试是认知神经科学家哈纳德(Stevan Harnad)提出的一种图灵测试的变种,他对传统的图灵测试又加上两条额外要求,这就是提问者也要测试被试的知觉能力(这需要计算机视觉)以及操控物体的能力(需要机器人)。——译注
[2] https://en.wikipedia.org/wiki/Clarke%27s_three_laws

II-008 凡及

心智上传;私密性和共情;意识获取;意识印记;
神经全局工作空间假设;主观性;意识涌现的必要条
件和充分条件

2014 - 12 - 20

亲爱的卡尔:

我很高兴你能同意我关于意识主观性和私密性问题上的一些观点。当然,
正如你所说的,我并不指望我所有的观点都是正确的,只是提供一些想法供思
考和讨论。

至于你的问题,"脑的拷贝要精确和详细到什么程度才能拥有主观性和意识",
这让我想起你在 2013 年 9 月 7 日的信(I- 010 卡尔)中推荐的承现峻(Sebastian
Seung)的文章。[1]

我非常喜欢这篇文章,我甚至去订购了作者的另一本书《连接组:造就独
一无二的你》(Connectome:How the brain's wiring makes us who we are)
(2012)[2],作者在书中宣称你就是你的连接组。然而他并没有止于此,他更进
一步说道:

"脑中是否有什么和这一框架根本不相容之处呢?一个难点是神经元可以

---

[1] https://www.scientificamerican.com/article/massive-brain-simulators-seung-conntectome/#
[2] 有中译本:承现峻.连接组:造就独一无二的你[M].孙天齐,译.北京:清华大学出版社,2015.此译本翻译
得非常好。——译注

在突触范围之外相互作用。例如，神经递质分子可以从某一突触处逸出，并扩散开去而被更远处的神经元接收。这可能导致没有突触连接的神经元之间也有相互作用，或者甚至在彼此之间没有实际接触的神经元之间也可能有相互作用。因为这种相互作用是突触外的，所以它不包含在连接组的'接线图'中。也许有可能相当简单地对某些突触外相互作用进行仿真；但也有这样的可能性，即神经递质分子在神经元之间既狭窄又迂回曲折的空间中扩散需要很复杂的模型。

"如果突触外相互作用真的对大脑功能至关重要，那么可能就有必要拒绝'你就是你的连接组'的假设，可能还得依然用较宽的说法'你就是你的脑'，但这会更难用作上传的基础（把你的脑上传到计算机里去，这正是某些未来学家所主张的）。我们可能不能仅限于作连接组这样的抽象，还需要进一步深入到原子水平，你可以想象根据物理定律创建对脑中每个原子的计算机仿真，这将是对现实的一种非常逼真的仿真，远远超过基于连接组之上的仿真。

"问题在于这需要极其大量的方程，因为原子数太多。即使只想想这样做所需的巨大计算能力显得也很荒谬，并且除非你遥远的后代能在宇宙的时间尺度下一直存活，否则就完全是不可能的。目前，即使是那些很小的被称为分子的原子集合体也很难仿真。要想仿真脑中的所有原子几乎是无法想象的。有限的计算能力还不是唯一的障碍。获取初始化仿真所需的信息也很困难。这样做可能必须要测量脑中所有原子的位置和速度，这比连接组中的信息要多得多。目前还不清楚如何收集这些信息，或者如何在合理的时间内收集这些信息。"[1]

我认为他的观点和你的观点足以拒绝任何在实际上传输或下载/上传心智的可能性。

至于说到我们是否能够知道别人的感受的问题，我的猜测是既"是"又"否"。在这里，我认为也许我们应该把共情和确切地分享感受区分开来。我在前几封信中提到的所有论点都是关于如何确切地分享意识，由于上面讨论过的

---

[1] https://www.scientificamerican.com/article/massive-brain-simulators-seung-conntectome/#

那些原因,要想确切地分享他人的感受或主观体验特性看来可能性极小。其根本原因就在于没有两个脑是完全相同的。因此也就没有两个主体可以有完全相同的体验。想想看吧,蓝纹奶酪对于意大利人来说非常美味,但对于大多数中国人来说却很糟糕;相反,皮蛋对大多数中国人来说很美味,但对于西方人来说却很糟糕。人们不可能有和别人完全相同的感受!但是,这并不意味着他们不能大致分享他们的感受。一个中国人会认为意大利人吃皮蛋的感受必定类似于他吃一口蓝纹奶酪时的感受,因为两者都有厌恶中枢。当他们有相似的感受时,这些中枢必定是以类似的方式激活的。因此,共情是基于脑结构和活动之间的相似性,因为没有两个脑是完全相同的,共情就只能是近似的。也因此,我同意你的观点,即"我们至少可以知道你的主观意识中的某些部分",但决不能精确地知道你的体验。

那么你也可能会提出一种强烈的反对意见,因为宇宙中没有两个宏观物体是完全相同的。我为什么要强调不同主体不可能有完全相同的体验?我们是否应该将精确分享主观体验和共情作为一个连续体来考虑,认为它们之间没有不可逾越的鸿沟?对于第一个问题,我认为这仅仅是因为人们过去倾向于认为如果刺激相同,那么从信息处理观点来看,它们的神经表征应该也相同,然而事实上,相应的感知却可能很不相同。弗里曼强调,大脑的功能并不是表征刺激,而是要以其内在的体验来表达它的意义,因此意义对于主体来说是独一无二的。至于第二个问题,也许这是正确的,因为我们都承认体验是从脑活动中涌现出来的,那么我们就不得不承认,如果两个脑的活动是相同的话,那么它们的体验也必定是相同的,尽管这样的概率很小,即使不说是零也罢。要是这样的话,我们是不是可以说,如果两个脑的活动非常相似,那么它们的体验也应该非常类似呢?这听起来很合理,但我不太确定,因为脑是一个非线性系统,混沌可能在其活动中发挥重要作用,因此脑活动之间的相似性不能保证它们的体验也一定具有相同程度的相似性。

我赞赏你的态度,即"但在我愿意改弦易辙并选择形而上学之前,我宁愿竭尽所能用尽一切物理和理性的方法。"我的观点是主观性可能是大脑活动的一种不可还原的涌现特性,因此追问主观性如何从脑活动中涌现出来并不是一个

合适的问题,我们应该做的是努力寻找涌现意识的充分必要条件,首先是对人的意识这样做。如果我们真找到了这样的条件,那么当这些条件满足时,主观性就会自然地出现,而查默斯的"困难问题"也就消失了。只有在此之后,我们才有可能将这种研究扩展到其他主体,包括动物甚至人造物。然而,即使对于人的意识来说,要想找到涌现意识的充分必要条件也是困难的。关键难点仍然是老问题:你如何知道主体具有主观性?如果条件是充分条件的话,那么如果这种条件得到满足,主体就必定是有意识的,但是你怎么知道主体有意识呢?对于人类来说,我们可以把下面这一点看作为公理,即所有正常行为的人都是有意识的,尽管如何判断行为是否"正常"也是一个问题。但是,对其他主体来说怎么办呢?对于机器,我们可能会进行"全图灵测试",尽管到目前为止还没有机器甚至可以通过原来的图灵测试。至于对其他动物而言,显然我们不能采取同样的策略,因为动物的行为必然与人类有所不同!

至于说到塞尔的"xyz 理论",我有一个问题。让我们来进行一个思想实验:假如我们有一位脑干损伤患者,但有正常的丘脑和大脑皮层,并且假设我们找到了一种治好这种脑干损伤的方法,那么患者应该醒过来并再次具有意识,我们能说脑干是负责意识的吗?我不这么认为,否则,你会得出结论,所有的脊椎动物都有意识,因为所有的脊椎动物都有脑干。C.科赫从 NCC 中排除掉了这样的脑活动,并将其称为"前提因素"(enabling factors)以与"必要条件"区分开来。前提因素是一些前提。正常的血液供应也是意识的前提,但显然不是NCC 的一部分。如何把前提因素和必要条件区别开来也是一个问题。顺便说一下,今年有报道称,当屏状核受到高频电脉冲刺激时,受试者失去了意识,一旦刺激停止就恢复了,就好像屏状核是意识的开关似的。然而,除了正常的屏状核活动外,还不清楚同时还应该满足些什么其他条件。所以尽管屏状核的正常活动对意识来说似乎是必要的,但目前还不清楚这是否也是充分的。

因此,在我看来,现在研究意识的一种比较实际的途径是找出人有意识或表现出其某些方面的必要条件。同时,我们也应该注意看这些条件是否充分,尽管这很困难。

最近,我读了一本今年刚刚出版的新书:德阿纳的《意识与脑:解读脑如何

编码我们的思想》(*Consciousness and the Brain: Deciphering How the Brain Codes Our Thoughts*)[1],这确实是一本非常有趣的书。作者和他的同事们试图通过实验研究、理论假设和计算机仿真来解释意识如何从脑中涌现出来。他试图把这个问题变成一个实验问题。我不知道你是否已经读过这本书。让我试着总结一下他究竟做了些什么以及他的想法,并提出我的看法。

由于对意识仍然没有明确和普遍接受的定义,不同的人在不同的情况下对这个术语有不同的看法。为了避免混淆,德阿纳将他的研究集中在他所谓的"意识获取"(conscious access)(受试者意识到了他所注意的刺激并可以向其他人报告的现象)上,很少人会否认这是意识的一个重要方面,并且是通向更复杂形式的有意识体验的门户。我非常赞赏他的这种策略,正如克里克多年前指出过的那样:"人们并不试图构建某个旨在解释意识所有方面的无所不包的理论,这是典型的科学方法。……在战斗中,你通常不会全线出击。你寻找最薄弱的地方,然后集中全力攻击。"[2]

与行为主义者根本拒绝内省的态度相反,他们将受试者的报告作为有价值的原始数据,告诉实验者他们是否觉知到了刺激,而不是把内省当做研究方法。他们使用掩蔽[3]、双眼竞争[4]和其他方法表明,虽然刺激保持不变或几乎不变,但受试者的知觉却可能发生根本变化,例如从意识不到变成意识到,或正好相反,因此意识获取可以被视为唯一的变量,并可以通过实验对这一变量进行操纵。然后,他们就寻找当也只有当受试者对相应刺激有意识获取时才会出现的脑活动模式。他们将这些模式作为意识获取的标志,并称之为"意识印记"。他们发现有如下这些印记:(1)较低层次的脑活动增强,逐渐积聚力量并侵入到前额叶皮层和顶叶皮层的多个区域;(2)脑波晚成分中事件相关电位 P300

---

[1] 有中译本:迪昂.脑与意识:破解人类思维之谜[M].章熠,译.杭州:浙江教育出版社,2018.——译注

[2] Crick F H C. The Astonishing Hypothesis:the Scientific Search for the Soul[M]. New York:Charles Scribner's Sons, 1994.

[3] 如果单独给予刺激 A,主体能感知到这一刺激;假如给予刺激 A 的同时,或在其前后另外再给一个新的刺激 B,而使主体不再能感知到 A,那么就说刺激 B"掩蔽"了刺激 A。——译注

[4] 如果同时分别给左右两眼呈现两个完全不同的图像刺激 A 和 B,此时主体会轮流知觉到 A 和 B,但是不会同时完全知觉到两者。这种现象称为双眼竞争。——译注

突然增强;(3)高频振荡的晚期放大;(4)跨脑区域活动的同步化。到此为止一切都好,他们严格地表明了,他们的"意识印记"与意识获取密切相关,这告诉我们当受试者体验到某种意识状态时脑中发生了些什么。他总结说:

"在给刺激后大约300毫秒左右开始进入有意识状态,在此期间,脑的额区以自下而上的方式接受感觉输入,但是这些区域也以相反的方向自上而下发送大量投射到分布很广的许多区域。最终的结果是形成一个由许多同步活动的区域构成的脑网络,其各个方面为我们提供了许多意识印记。"

"当有有意识的知觉时,神经元群以协调的方式开始发放,首先是在一些局部的特定区域,然后蔓延到皮层的广大范围。最终,它们侵入到许多前额叶和顶叶脑区,同时与前面的感觉区保持紧密同步。正是在这个时候,突然形成了一个协调一致的脑网络,有意识觉知也似乎由此产生。"[1]

也许我们可以将他们的"印记"看作是人类有意识获取的必要条件。这是意识研究向前迈进了一大步!

基于意识印记,他们发现从无意识状态到有意识状态的过程就像是某种相变过程,作为其基础的脑活动必须超过某个阈限,然后自我增强,脑可能突然爆发出大规模活动模式,因此他们提出了一种"全局神经元工作空间假说"(global neuronal workspace hypothesis),试图以此解释意识的神经机制。作者说道:

"方案很简单:意识是一种在全脑范围里的信息共享。人脑中有高效的长距离网络,特别是在前额叶皮层,以选择相关信息并将其扩播到整个脑。意识是一种演化装置,它使我们能够注意某个信息并在这一扩播系统中保持活跃。一旦这个信息被意识到了,根据我们当时的目标,它可以被灵活地传送到其他区域。因此,我们就可以叫出它的名称,对此进行评估,记住它,或者用它来规

---

[1] Dehaene S. Consciousness and the Brain:Deciphering How the Brain Codes Our Thoughts[M]. New York: Penguin Books, 2014.

划未来。神经网络的计算机仿真表明,根据全局神经元工作空间假设能够精确地产生我们在脑的实验记录中所看到的那些印记。"[1]

他说:"当我们说我们觉知到了某种信息时,我们的意思就是说:信息进入了一个特定的存储区域,使其可以供脑的其他区域使用。"主观性消失不见了!作者在他的书的最后部分提到了对他的假设的一些批评,例如布洛克(Ned Block)说这个假设不能解释主观体验特性;查默斯说,它永远不能解释第一人称主观性之谜,或者意识的困难问题。尽管如此,他只是对这些批评作出了非常简短的回答:

"我的看法是,查默斯把标签搞颠倒了:正是他的'简单问题'才是困难的,而所谓的'困难问题'只是看上去困难,因为它是基于某种模糊的直觉之上。一旦我们的这种直觉受到认知神经科学和计算机仿真的检验,查默斯的'困难问题'就会消失。假设性的主观体验特性概念,纯粹的心理体验都和信息处理没有一点关系,这些都可以被看成是某种前科学时代的特殊概念,就像活力论一样。活力论是十九世纪的一种引人误入歧途的思想,它认为无论我们对生物体的化学机制研究得有多详细,我们永远都不能解释生命的独特性。现代分子生物学通过说明我们细胞内的分子机器构件如何形成某种自我复制的自动机,打破了这一信念。同样,意识科学也会使'困难问题'一步步退缩,直到消失不见……一旦当我们能够说清楚任何感觉信息如何进入我们的心智并得以报告出来,那么我们难于用言语表达的体验这一不可解的难题也就消失了。"[2]

我非常欣赏他关于意识获取印记的研究,他的全局神经元工作空间假设对于印记的解释似乎没问题,但我对他的假设也能够解释意识或者即使只是意识

[1] Dehaene S. Consciousness and the Brain:Deciphering How the Brain Codes Our Thoughts[M]. New York:Penguin Books,2014.

[2] Dehaene S. Consciousness and the Brain:Deciphering How the Brain Codes Our Thoughts[M]. New York:Penguin Books,2014.

获取本身持怀疑态度。我的主要观点如下：

（1）"意识获取印记"并不是"意识获取"本身，就像我的签名并不就是我自己一样。铁路列车发出的嘈杂的咔嗒声是列车运行时的一种印记，当且仅当列车正在运行时，你才可以听到这种声音，造成这种声音的机制可能与列车运行相关，但是，你无法用这种声音产生的机制来解释列车如何运行。当然，这种比较可能过于极端，但它确实表明你不能简单地用相应印记的机制来解释事物本身的机制。对印记的解释可能会对其所有者的解释给出一些提示，但并不能给出后者的确切解释。严格地说，他的研究只是表明，如果主体有意识获取，那么他/她的大脑中就有这样的印记，但反之并不成立。

（2）除了用受试者的主观报告来判断他们是否意识到了什么之外，德阿纳的工作并没有触及意识的主观性问题。他所有的实验和理论都是基于客观事实。他使用客观印记来取代主观的意识获取。因此，即使他的假设阐明了这些印记是如何起源于一些特殊的脑活动模式的机制，并且即使他的说法可以扩展到意识获取本身的机制，他的理论最多也就像关于立体视觉的双眼视差理论。后者确实解释了立体视觉在哪种情况下会出现，就像 3D 电影已经证明了的那样，但是它仍然不能解释我们如何能够具有这样的主观体验特性。就像巴里博士告诉萨克斯博士的那样。德阿纳博士无法解释为什么当巴里博士说她错了时是错的！当然，我并不是说他的工作根本就没有价值，类似于立体视觉的视差理论，他的工作可能是向意识获取涌现的必要条件迈出的一大步，但还不是充分条件。

（3）虽然他们的计算机仿真似乎支持了他的假设，基于这种假设建立起了一个神经网络模型，其行为模仿了这四个印记，发生了类似于相变的现象，然而，即使他自己也承认他的仿真"和有意识还相距遥远"。因此，他们的仿真可能证明了他的全局工作空间只是一个无魂人，正好与他自己认为这些印记会引起意识的观点相反！要想有意识获取似乎还需要有在他的假设之外的某种东西！

（4）然而，尽管他的相变概念或雪崩隐喻听起来很有趣，但我们也可以用这些隐喻来描述癫痫发作——一种绝对没有意识的状态。所以也许这些印记

只是必要的,但不足以产生意识获取。许多关于意识的理论研究都有类似的缺陷,他们把必要条件和充分条件混为一谈。

总之,他的印记和假设可能是产生意识获取的必要条件,但并非充分条件,他的仿真显示,即使是基于他的假设所建的网络也具有像印记那样的特征,这样的网络仍然远没有意识,因此必然有什么东西没有被考虑到!问题是:没考虑到的究竟是什么?主观性?我想是的,但还有其他什么吗?

如果人们能够找到足够多的必要条件,它们有可能也变得充分吗?当这些条件得到满足时,主观性能够突然涌现出来吗?我们还没有答案,但正如你强调的那样,人们应该努力尝试并且不要放弃。如果真的发生了这种情况,那么我们应该说查默斯的"困难问题"就消失了吗?让我再重复一次,我认为,问"主观性如何涌现出来"这样的问题是没有用的,当某些条件得到满足时它就会涌现出来,恰当的问题是:这些条件是什么?首先是意识的某些方面,例如意识获取,然后是一般的意识。首先是人的意识,然后是其他主体。你怎么想呢?

对不起,我的信太长了,不得不在这里停下来,你在上封信中提出的其他问题,我们稍后再讨论吧。

圣诞快乐,新年快乐!时间过得真快!

凡及

计算机仿真的局限性；主观性是有可能研究的

2015-02-19

亲爱的凡及：

也祝你新年快乐！

如果我指的是德国新年的话，那就太迟了，但使我高兴的是对你们中国农历新年来说，那么节日祝贺还正及时。无论如何，我祝愿你羊年大吉。

我很抱歉过了那么久才答复你非常有趣的来信，信的内容非常丰富，其中多数对我来说都是新的，并且充满了棘手的问题。我不想只坦率地回答你说"我不知道"，因此不得不做一些阅读和思考。不过，我仍然不确定我是否可以给你满意的答复。你对所谈的关于意识和主观性的问题已经思考多年了，也许你的有些问题没人能回答。但是我开始能更好地理解你的意思，我尽力而为吧。

阅读德阿纳的书《意识和脑：解读脑如何编码我们的思想》非常有趣。谢谢你的推荐！我选择了一本德译本，在我看来这本译本翻译得非常好，你作为资深翻译家可能会喜欢听到这个消息。

与你类似，我也不是对德阿纳所说的一切都同意，但在我看来，他即使是在走错了道的地方也依然对人有所启发。真正让我恼火的是，他也忽视了弗里曼，这不仅对你的朋友弗里曼来说，而且对德阿纳来说也很可惜。如果他遵循弗里曼的建议，不把脑当成一种信息处理机器，而是看作一种检测意义的系统，他本来可以获益匪浅。

我认为还有一个问题，就是你引用过的德阿纳的话："神经网络的计算机仿真表明，根据全局神经工作空间假设能够精确地产生我们在脑的实验记录中所看到的那些印记。"

这个说法让我感到怀疑，因为如果你知道要有什么样的结果，那么很容易进行计算机仿真，使其精确地显示你所期望的现象。你只需调整模型和参数的权重，直至看到你想要看的东西。这就像圣经的信徒们将圣经的特定文本输入到某个计算机程序里，这个程序就给出有关世界末日或泰坦尼克号沉没日期的神秘信息。事实上，如果你事先就知道你想要得到的消息，那么这样做就很简单。每个够格的程序员都可以通过编写一个算法来实现这种圣经代码奇迹，只要该算法的参数可以把文字和页面恰当地跳过或移动就行了。用这种方法你甚至可以从圣经中找到比如说北京烤鸭的配方，也可以从果蝇的 DNA 中找到德国国歌。诀窍其实很简单。你可以先想定你想要创建的句子，第一步先在你所选定的源文件中搜索所需的单词或字母。一旦你找到了它们，你所需要做的就是决定算法中要读取单词或字符的顺序。这会产生一个读取规则，该规则确定如何从一个页面跳转到另一个页面，从一行到另一行以及从一列到另一列跳转。这样，任何你所期望的结果都可以从任意来源读出。神奇之处并不在于算法，而是你仍然可以用这样的手段来制造头条新闻。☺

当然，我并不是说德阿纳是用欺骗手段来得出他的发现！我要说的是，当科学家使用计算机统计模型和仿真，在大量数据中发现他们正在寻找的东西时，我一般都不会感到惊讶，也不会留下深刻的印象。在这些系统中有太多的旋钮可以调整以适合自己的目的。我自己在计算机统计方面就工作了很长时间，因此深知人们禁不住想在这些神奇的机器上调整旋钮从而得出所想要的结果。

所以也应该要有其他人做类似于德阿纳的实验，因为这是科学的良好传统。

你所引用的布洛克和查默斯的批评原则上是正确的，但如果德阿纳理论中的其余部分能让我们更进一步地认识意识现象的话，我不会感到奇怪。

尤其是查默斯的极端言论，即这样的结果"永远不能解释此谜"，这加强了我在上封邮件中表达的怀疑。在我看来，这似乎是说无论你向他展示什么样的证据，他都不愿意放弃自己的立场，因为对他而言，主观性可能按其定义来说就是无法研究的。但是，也许你能告诉我说我错了。我应该感到高兴，德阿纳支持我的观点，即我们将逐步解决意识和主观性问题。虽然就这个预测而言，我同意他的观点，但正如我所说，我不喜欢他的信息处理研究方法和计算机仿真

　　　　意识之谜和心智上传的迷思　　　一位德国工程师与一位中国科学家之间的对话

会有很大帮助的想法。也许他只是因为是 HBP 的一员，至少在开始时是如此，这才加上了这些和计算机有关的材料，而要想成为马克拉姆俱乐部的一员，就得做点仿真。不过我不想对他这样一位显然可敬的研究人员不公平，而且无论如何，你可能比我更了解。

你的四点批评意见在我看来都是成立的。这是典型的凡及式的批评，很难反驳。特别是第一点批评在我看来至关重要，因为意识印记并非就是意识获取，这正如你正确地指出了的那样！

然而，我不确定你是否过于偏向查默斯那方面了，因为你也可能已经爱上了现在无法研究主观性的（也许以后也无法研究）这一观点。☺

这对我来说并没有什么问题，因为我不能排除查默斯和你正确的可能性。但我认为我们可以达成共识，即有可能取得进展，可以进行更合理的研究，并且德阿纳及其批评者都给出了有趣的提示，指出应该在哪些方向上取得进展。这和塞尔在接受布莱克莫尔的采访时所概述的研究方案很类似。

我可能不会放弃希望有一天 Siri[1]、Alexa[2] 或你的私人机器人会和你谈论他们的主观感受。万一真的发生了这种情况，你应该告诉他们打电话给我，让我知道我的胜利。☺

我很高兴你还发现承现峻关于连接组的想法非常有帮助。我不知道他的书《连接组：造就独一无二的你》，而只是从我发给你的那篇文章中知道他的立场。谢谢你告诉我！在此期间，我也看了一下这本书，看看他的立场是什么，这确实非常非常有用！但是我要到下一封信中才告诉你我的观感。

晚安，向上海致以最好的祝愿。

卡尔

---

[1] Siri 是一种智能个人助理，是苹果公司（Apple Inc.）许多操作系统的一部分。它使用语音查询和自然语言用户界面来回答问题，提出建议，并通过将请求委托给一组因特网服务来执行操作。——译注

[2] Alexa 是由亚马逊开发的智能个人助理。它能够进行语音交互，播放音乐，制作待办事项列表，设置闹钟，播放有声读物，以及提供天气、交通和其他实时信息。Alexa 还可以用于家庭自动化，控制多种智能设备。——译注

II-009　凡及

脑是创造意义的机器,而不只是信息处理系统;
私密性;共情;镜像神经元;欧盟人脑计划调解报告;
技术和科学

2015－04－01

亲爱的卡尔:

　　非常感谢你花了这么多的时间和精力来思考我提出的问题,甚至订购书籍来回答我的问题!我再也找不到另一个能够这样做的朋友了。

　　是的,我也为弗里曼的理论受到忽视而感到惊讶,只有少数人注意到了他的贡献,并且认为脑是创造意义的机器,而不只是信息处理系统。把脑完全当成某种信息处理系统仍然是神经科学领域的主流,特别是在计算神经科学领域。这可能是生物脑和技术制造系统之间的另一个重要区别吧。

　　弗里曼强调,脑是一种创造意向和意义的机器,是一种非线性动力系统,它只在小部分上受感觉刺激的影响,而在更大程度上受到自身正在进行的活动(包括其历史、注意力、行动和情绪)的影响。意向和意义先于意识。由于上述观点,刺激与意向、意义或知觉之间并不存在一一对应的关系。意义和知觉主要是在脑中产生的,它们是私密的。他说道:

　　"来自某个物体的感觉刺激的确会在脑中产生某种模式,但是当反复给予同一个刺激时,它不会在同一个脑中产生出完全相同的模式,更不用说在任何其他的脑中了。这是可以预料到的,因为不仅同一个对象对不同的人意味着不同的事物,即使是对同一个人,其意义也在不断地变化。我的结论是,不能把意义直接传递给脑,或在不同的脑之间相互传送,就像把信息和知识的表征传递

给机器,或是在机器之间相互传递那样。

……

"脑所能知道的一切都是在脑自身内部合成的,脑是以关于世界的假设以及他们对假设所作检验的结果(成功或失败以及失败的方式)做到这一点的。这是将每个人的主观体验特性与所有其他人的体验分开的唯我论区隔的神经生物学基础……"[1]

按照弗里曼的看法,私密性深深地植根于意向和意义创造之中,而这甚至还发生在意识之前。知觉不是感觉刺激的表征,它也取决于主体以前的经验,因此也是私密的。

他的论点也许可以帮助我回答你的大问题,即是否能知道他人的意识,尽管要想给出完全肯定的答案几乎是不可能的。事实上,我们在前面的信中已经讨论了很多关于意识的主观性和私密性的问题。在我看来,至少到现在为止,由于弗里曼在上文中解释过的原因,没有哪一种方法可以用来彻底获取他人的意识,我也不认为在可预见的将来这个问题将会得到解决。然而,与此同时,我并不否认人们可以有共情,可以大致了解他人的感受。人们甚至可能会说是镜像神经元负责共情。然而,即便如此,他们仍然无法解释镜像神经元如何在受试者有与其他人类似的感受的情况下会发放神经脉冲。正如我在前几封信中所讲的那样,如果真的发现了人有意识的某些充分必要条件(这也许是可能的,虽然很困难,尤其是如何知道他人是有意识的?也许只是基于常识判断?),那么也许没有人会再问人的脑活动会如何会产生意识的问题,这个问题也就自动消失不见了。由于人脑的结构彼此相似(但不完全相同),类似的脑活动可能会产生类似的感受,这可能是共情的基础。然而,问题仍然存在:一个人的脑活动如何触发其他人类似的脑活动?至于人造机器,也许我们就使用全图灵测试。尽管我们仍然不知道机器是否真的有意识,仅仅根据行为我们既无法排除这种可能性,也不能肯定这种可能性。虽然这样的判据对我来说不够满意,但我仍

[1] Freeman W J. How brains make up their minds[M]. New York: Columbia University Press, 1999.

然可以接受。但是,其他动物呢?我不知道。也许你可以提出一个判据来。

我不否认,在我看来,主观性的障碍可能永远无法被彻底克服,你永远无法准确地体验到别人的体验。但是,我承认部分或大致体验他人的感受是可能的。即便如此,你仍然无法解决"如何"的问题。你认为科学可以解决所有"如何"的问题吗?比如说,电子如何带有负电荷?也许对于那些非常基本的性质,我们不得不承认这是个事实,并把它们当作公理,而不是问"如何"的问题。

就像你说的一样,我不能排除我完全错了的可能性。我同意你所说的意识研究取得进展仍然是可能的,特别是在为人的意识寻找必要条件方面。我认为德阿纳在寻找意识印记方面的工作是朝这个方向迈出的一大步。

我不能排除有一天我的 Siri 会跟我谈论他们的主观感受的可能性,但是我仍然怀疑他们是否真的有意识,或者他们只是些无魂人。根据你的 TTT,我可能会承认他们是有意识的,但在我的心底,我可能……好吧,我认为我们都已经很清楚地解释了我们的观点,我想我们不容易在这个问题上达成共识。也许,让我们注意这个领域的进展,看看会发生什么,并再次考虑我们的论点,在以后再讨论这个话题吧。

我完全同意你对仿真局限性的评论,尤其是当有太多可以自由调整的参数时,在这种情况下你可以得到任何你想得到的结果。这种仿真很难解释任何事情。当然,如果几乎所有的参数都是由生物实验决定的,那将是另一回事。

现在让我把话题转到另一件非常有趣的事件上,你可能已经注意到了。就在上个月,HBP 的调解委员会(Mediation Committee,简称"MC")公布了一份调解报告[1]。该委员会负责调解 HBP 与 800 多名科学家之间的争议,这些科学家在去年 7 月向欧盟递交了一封公开信。

调解报告强烈支持了公开信中所提出的对 HBP 在目标、科学方法、领导管理方面的批评意见,指出该计划不仅管理不善,而且其仿真全脑的核心目标是不现实的,已经造成公众和科学界对 HBP 丧失信任。HBP 必须做出实质性的改革,以恢复信任。除了两名委员之外,调解报告得到其他委员的一致支持。

---

[1] http://www.neurofuture.eu/media/official_HBP_mediation_report.pdf

（在调解委员会全部 27 名成员中，有 10 名来自 HBP）

报告就 HBP 的科学内容和领导管理两方面分别提出了具体建议，我在这里只举其中最重要的内容。

在有关科学内容方面，HBP 的任务和目标要更加集中。仿真整个人脑难望在十年内取得成功，因此建议为脑仿真这一子计划另行确定研究内容。报告认为脑功能过于丰富，有关脑功能的现有知识尚很有限，不足以可靠地进行自下而上的仿真。HBP 应该将其目标重新定位到有限时间和有限资源条件下可望实现的某些具体目标上。报告建议 HBP 着重开展对神经信息学有用的方法和技术研究。HBP 应尽量利用已有的神经科学实验数据，不要企图自行填补结构和功能两方面所缺的一切，而应把精力集中到目标明确并为开发信息技术平台所必需的实验上。对于这些平台应通过跨学科合作研究来开发和检验，应该有认知神经科学家和系统神经科学家参与其中。认知神经科学和系统神经科学应重新整合到 HBP，作为至少有三四个方面的子项目。这些平台应针对具体问题进行开发，例如空间导向或者有目标的决策等。报告还指出，HBP 不应把有关非人灵长类动物的研究排除在外，因为这是从鼠脑到人脑的一个重要的中间环节。

报告认为以往 HBP 领导和欧盟的新闻发布，过分夸大了 HBP 的可能成就，使公众对 HBP 产生了不切实际的期望，比如在十年之内就能够认识脑功能，或是对神经退行性病变进行诊断与治疗。这种夸大使 HBP 失去科学信任。HBP 的名声归根结底要立足于令人信服的科学结果和为神经科学家广泛使用的信息技术平台。

在有关领导管理方面，调解报告的矛头几乎直指马克拉姆。报告说："负责的科学家……不仅是 HBP 中所有决策、执行和管理机构的一员，还主持着这些机构，并监督和支持这些机构的行政过程。此外，他还是所有顾问班子的成员，同时向这些班子递交报告。此外，他还指定管理团队的成员，并领导计划的实际管理。"[1]这种既当运动员又当裁判员的管理模式，自然怎么也说不过去。

报告要求经费分配透明，所建成的信息技术平台要易于为全体人员所用。

---

[1] http://www.neurofuture.eu/media/official_HBP_mediation_report.pdf

报告要求 HBP 的领导机构由马克拉姆所在的洛桑理工学院转移到对 HBP 最有贡献的一些单位所组成的新实体。

针对调解报告的严厉批评，HBP 不得不紧急采取了一些回应的措施。在领导管理方面，甚至在报告正式发表之前一个月左右，HBP 就不得不解散了大权独揽的三人执行小组，包括马克拉姆在内的三人都递交了辞呈。2015 年 3 月 18 日，HBP 理事会批准了调解报告。理事会成立了有许多国际科学组织负责人参与的领导工作小组，还成立了负责处理管理、评估科学事务和财务事宜的不同小组。看来 HBP 正开始进行大变革。

我很好奇，马克拉姆对这份报告会怎么想，他在今后的 HBP 中会扮演怎样的角色？是不是 HBP 理事会中的成员都赞同调解报告的意见？我想改革不会容易。

我同意调解报告的主要观点，虽然并不理想。事实上，在我看来对这样一个先天就有缺陷的计划，没有一种改革会是理想的。如果全面采纳调解报告的建议，HBP 的目标就要改为开发有助于神经科学家认识人脑及其疾患的新的信息技术（IT）和平台。这个目标虽然小得多，但依然十分宏大，它将和美国的"尖端创新神经技术脑研究倡议"（BRAIN 倡议）互为补充。前者集中研究 IT，而后者致力于研发观察、记录和成像神经回路活动的新技术。

事实上，技术发明在科学发展中起着非常巨大的作用，尽管在科学史上讲得最多的是采用新技术而有重大发现的科学家，而不是新技术的发明家。伽利略（Galileo Galilei）改进了利伯希（Hans Lippershey）发明的望远镜，并用它遥望星空，发现木星有 4 颗卫星绕它旋转。他的发现说明并不是所有的天体都像人们以前所相信的那样环绕地球旋转。至少有四颗星是绕另一颗行星转的。类似地，在神经科学史上，高尔基（Golgi）发明的染色法经卡哈尔（Santiago Ramón y Cajal）改进后观察了各种神经组织切片，结果使后者提出了神经元学说，奠定了神经科学的基础。电子管放大器的发明开创了电生理学的新时代，脑功能成像技术的发明使观察清醒受试者在作智力活动时的脑活动成为可能，推动了认知神经科学的产生和发展。因此，研发新技术对推动脑研究的重要性是无疑的，问题在于开发新技术如何和具体的重大脑研究紧密结合，而不是为技术而技术。否则 HBP 就成了一个纯粹的信息技术计划，而不是脑研究计划。如果

意识之谜和心智上传的迷思　　一位德国工程师与一位中国科学家之间的对话

这个计划又占用掉了脑研究所需要的经费，那么这对脑研究来说究竟是喜是忧还很难说。如果开发 IT 技术能够和脑研究中的重要课题紧密结合，如果新技术能极大地推进这种研究，而后者又能向 IT 技术提出新的要求，这就太好了。但是什么是当前应该集中力量进行研究，并有望在不远的将来可望取得突破的重大课题？我不知道现在有没有什么共识，也许这是神经科学界应该集中讨论的问题。不过这是不是 HBP 在现行架构下能做到的？我不知道。话说远了，这不是你我所能解决的问题，让我们拭目以待，观察今后的进展吧。信已经太长了，就此打住。

祝好！

凡及

背景专栏 Ⅱ F9.1

镜像神经元与共情

1991 年，意大利帕尔马大学的神经科学家里佐拉蒂（Giacomo Rizzolatti）教授领导的实验室发现猴子脑腹侧前运动皮层 F5 区中与手和嘴的运动有关的一小部分神经细胞，不仅当猴子取食时会产生神经脉冲，而且在看到其他猴子甚至工作人员在取食时也会发放。他们费了好几年时间才确证这一现象，并把这种细胞称为镜像细胞。这一发现第一次从神经生物学中找到根据，有可能说明当人们看到别人在做某个动作时，他们在内心里也在模仿同样的动作，甚至"理解"别人（猴）的"意向"。里佐拉蒂得出这一猜想的根据是，他们发现猴子脑中有的镜像细胞当猴子看到别人的手伸向某个目标时才有猛烈的神经脉冲发放，然而如果没有目标，只有伸手的动作，发放就很稀疏。然后在让猴子看到有目标以后，在其前面用一小块幕布把目标遮住，这时把手伸向幕布后，这个神经元还是有猛烈的发放，虽然这时猴子并没有看到目标，而只看到手的运动；但是如果先让猴子看到幕布之后一无所有，再把手伸向幕布后面，这个神经元就没有

发放多少。这意味着这个神经元能"懂得"对方的意图。这一发现成为近年来神经科学研究中最重要的发现之一。

后来人们发现在人脑中也存在镜像神经元系统。2003年法国神经科学家维克尔（Bruno Wicker）用功能磁共振成像技术发现：当受试者感到厌恶时，和他在看到别人厌恶的面部表情时，都引起了位于脑岛的同样的一些镜像神经元的活动。这可能意味着镜像神经元也是人们体验别人情绪的神经基础——"感同身受"（共情）的神经基础。进一步的研究发现：人的镜像神经元系统比猴子要更复杂、更灵活，它不仅能感同身受般地领会他人的感受和动作，而且还能领会他人的意图和情绪。对于后者，人们一直认为完全是一种像解决逻辑问题那样的推理过程，尽管在一些情况下推理过程确实也起作用，但是里佐拉蒂教授指出："我们是高度发达的社会性生物，我们的生存有赖于懂得别人的动作、意图和情绪……镜像神经元使我们得以一窥别人的内心，这不需要通过概念推理，而是通过直接模仿。这并不需要思考，而是直接感受。"

美国神经科学家雅科博尼（Marco Iacoboni）报道说，镜像神经元甚至能区分别人伸手拿杯是要饮水呢，还是整理桌面。如果杯子是放在一张除了茶具之外一无所有的桌面上，当人看到有人去取放在桌面上的一杯饮料时，有一些镜像神经元有发放；然而如果桌面很乱，那么看同一个动作时有发放的就是另一些镜像神经元，因为这时拿杯子很可能是要收拾了去清洗。

然而也并非所有科学家都认同对镜像细胞的上述研究，特别是有关其功能意义。一些科学家认为在运动区中强烈表现出镜像细胞特性的神经元数太少；有关镜像细胞的某些实验结果不尽一致；另一些科学家则认为对镜像细胞的直接记录绝大多数是在猴子身上进行的，其结果是否适用于人是个问题；对人镜像细胞的研究除了在癫痫病人手术前限定皮层位置所作的少量单细胞记录外，大多数通过脑成像等间接手段；还有科学家认为单个神经元不可能具有这样复杂的功能，神经元必定是作为复杂神经回路的一部分而起作用；最普遍的一个质疑意见是不应对镜像细胞的功能意义过度解读。2009年美国认知神经科学家希科克（Greg Hickok）从8个方面对镜像细胞理解动作的理论提出质疑。他认为理解动作在镜像细胞功能意义解读方面是最基本、也是最广为流传的，

如果对此都存疑,那么其他解读就更可疑。这些问题都还在激烈的争论之中。不过镜像细胞作为一类特殊的神经细胞,它们在主体看到其他动物做某种动作时会和主体自身在做同一动作时一样发放脉冲,其功能意义值得认真研究还是得到科学界的普遍肯定的。

背景专栏 II F9.2

弗里曼"脑是提取意义的机器"的实验证据

关于脑功能的一种传统的想法是:脑就是一种信息处理系统,外界刺激被忠实地"表达"在脑中。对嗅觉感受器的研究似乎表明,这种想法是对的。对同样的刺激,它们的反应是可重复的。这在他的意料之中,但是一旦进入脑,即使在第一站——嗅球上,情况就完全不同了。弗里曼把 64 根电极排成格阵,同时安置在嗅球表面,并由这些电极记录所在部位的局域场电位。这些电位表示电极下面神经元群体活动的程度。弗里曼通过建立条件反射对兔子进行训练,让兔子学会识别某种气味。在实验之前,有一段时间不给兔子水喝,然后进行实验。他用了两种气味(例如香蕉水和酪酸)作为刺激,香蕉水是条件刺激,而酪酸则是无关刺激。给水则是无条件刺激,并引起用舌头舔水的反应。在多次给予香蕉水刺激同时给水以后,即使光给香蕉水而不给水也能引起舔水的反应,这就表明此时兔子已经学会识别香蕉水了。反之,由于在给酪酸作为刺激的同时,从来也不给水作为奖励,因此兔子对它只有嗅的反应,而不会舔。另外还用空气作为对照。每次实验都记录 6 秒钟的 64 导脑电,每导脑电都包括一段对照期和一段试验期,试验期中吸进什么气味是随机安排的。他发现,所记录到的脑电非常不规则,即使在同样的实验条件下重复记录,每次记录所得的波形也各不相同。而且在吸进熟识的气味时,脑电突然变得规则一些,其幅度和频率也变高。频率落在 20~80 Hz 左右(也就是 γ 波),形成所谓的"簇发发放"。结果发现在同一次记录中同时记录到的这 64 导脑电中的 γ 波的载波波形都是

相同的,只是幅度不同。如果把这些波形的平均幅度标在脑区表面,用等值线把相同幅度的点联结起来,画成等高线图,就可发现脑电 γ 成分幅度的空间分布模式(表现为等高线图)在吸同一种气味时是可以重复再现的,尽管每次的载波波形都不一样。这就是说,关于某种气味的嗅觉信息就携带在脑电 γ 成分幅度的空间分布模式之中。单个嗅觉神经元不能辨别特定的气味,只有一大群神经元的共同活动才能识别。所以,他在后来说道:"简而言之,感觉到一种有气味的物体只需要有少量神经元的网络,而要知觉到一种气味则需要嗅球中所有的神经元。"

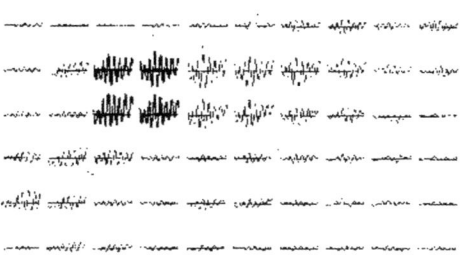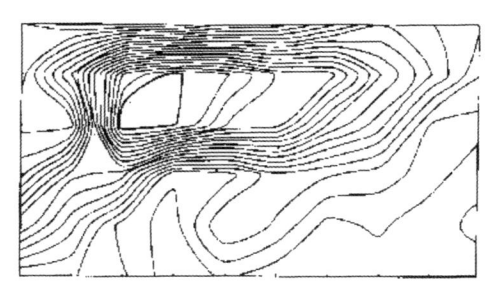

图ⅡF9.1　兔识别气味时的脑电及其平均幅度等高线图

(左图)当兔子在识别一种气味时,从其嗅觉皮层上同时记录到的 60 导脑电中的 γ 波;(右图)按照左图中各导脑电的平均幅度画在脑区表面所得的等高线图。(感谢弗里曼教授允许引用他的图)

令人奇怪的是,如果把奖励在两种气味之间加以切换,也就是说,把原来的条件刺激变成无关刺激,而原来的无关刺激变成条件刺激。这时虽然这两种气味本身并没有什么变化,它们所引起的脑电调幅空间模式却完全改变了。同样,当在实验计划中添加进某种新的气味时,所有早先已经存在的调幅模式都会发生变化。甚至当按次序用几种气味对兔子进行训练,然后再回到第一种气味进行训练时,出现的也不再是原来的模式,而是一种新的模式。因此,只要气味环境有了变化,实验对象能够识别的所有气味的调幅模式都要跟着发生变化。弗里曼由此得出结论:"嗅觉的调幅模式并不和刺激直接相关,而是和刺激的含义相关。"知觉不仅仅和外界刺激有关,还和过去的经验等脑的内部活动有关;知觉也不能用单个神经元的性质来加以解释,而要涉及大量神经元群体的协同活动。不光对嗅觉知觉来说是这样,对其他知觉也是如此。

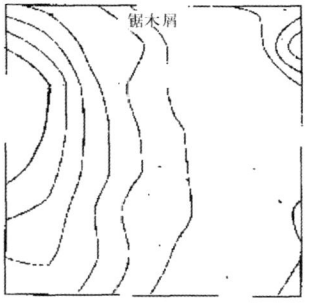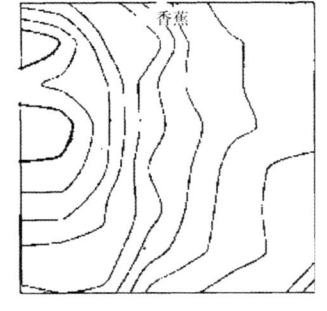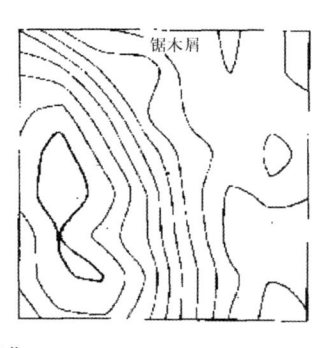

图ⅡF9.2　兔嗅球脑电调幅模式的动态变化

（左图）当兔子学会识别锯末味以后，在嗅球上记录到的嗅锯末味时的脑电调幅模式；（中图）改用香蕉味作为条件刺激进行训练而学会识别以后，在嗅球上记录到的嗅香蕉味时的脑电调幅模式；（右图）重新再用锯末味作为条件刺激进行训练以后，在嗅球上记录到的嗅锯末味时的脑电调幅模式。注意，虽然刺激和左图一样都是锯末味，但是脑电的调幅空间模式发生了根本的变化。（感谢弗里曼教授允许引用他的图）

正是在这些实验的基础上，他提出了一种学说，认为脑并不只是某种信息处理系统，而是一种创造意义的系统。信息表达只发生在自下而上的初级阶段。他认为，主体根据自己的目的，创造出有关外界环境的某种假设，并按此假设采取行动，根据这种行动的结果和自己对这种行动的感觉监视，证实或否定这种假设，从而对假设进行更新，开始新一轮意向性的行动——知觉周期［intentional action-perception（A－P）cycle］。意义就体现在这种目的或意向性之中，而不一定需要意识。他认为，脑中这样复杂的过程是传统的线性因果论所解决不了的，而需要用循环因果论(circular causality)来理解。

II-009　卡尔

*私密性;技术进步正在以指数速度向前发展,而
生物学和认识脑的进展只是线性的;连接组;心智上
传;交流*

2015－05－14

亲爱的凡及:

　　谢谢你出色的来信和美言。这样说吧,我必须告诉你,读你的那些见解和
问题对我来说是很大的乐趣,在这里面总是蕴含着如此多的智慧和新的观点。
如果没有你,我绝不会花费精力深究你引起我注意的所有那些令人激动的事
情,或者表述一些我已考虑了相当一段时间的想法。比起那些在头脑里放了段
时间而还没那么清楚的思想来说,要想把想法写下来就迫使人们用一种更加具
体和精确的方式来表达。

　　当然这很耗时也需要付出一定努力。但是当你这样做的时候,经过反复考
虑和产生一系列思想,并且当你觉得你更好地抓住了某件事,开始把握这件事
的问题所在时,我相信这是人可以体验到的最有回报感的事情之一。作为一名
经验丰富的研究人员,你一定比我更经常地感受到这种兴奋心情,你肯定知道
我在说什么。当然,我离通晓这一领域还相差很远,但我可以感受到知识在增
加,也更能把握自己在这个领域中的方向。如果我不再像以前那样无知,那么
这得归功于我们之间的对话,因此也要归功于你,我亲爱的朋友。

　　我自己的学习经历和弗里曼强调意义而非信息作为脑中交流的主要形式,
都让我思考脑中这种"模式形成"是如何工作的,这使我们能够从感觉输入中提
取出意义并将其与意向联系起来。很明显,这只能以非常主观的方式发生,因
为每个脑都是通过不同的渠道搜集输入,并且先后次序也不一样。这样的系统

都要根据其先前的状态对新输入作出解释并给予意义。虽然在某些情况下（例如同卵双胞胎）两个系统的初始状态可能相当类似，但是输入序列中很小的差别也可能很快就在与此相关的意义方面造成实质性的差别。结果这些系统很快就会产生非常不同的意向。由神经元之间的突触连接构成的"意义滤波器"（meaning filters）一定得彼此依存，并且可以像阵列中的功能构建块一样堆积或链接起来。每个新输入都必须嵌入到现有的神经网络中，因此创建了系统的一个新状态，意义也就从知觉中提取出来，并根据期望和意向进行解释。这种网络在每个脑中必定是非常不同的，因为当身体穿越时空时，所有（外部和内部）感觉输入、体验和潜在意向的序列都是独特的。在出生后 10 分钟，即使同卵双胞胎也必定有完全不同的突触联系。实际上，他们的脑永远不可能完全相同，这是因为 DNA 中可以编码的信息量远远不足以确定已经存在于"裸"脑中的达到天文学数字的突触联系。

你对弗里曼关于脑如何处理知觉、意义、期望和意向的立场给出了一个绝妙和清晰的解释。说真的，这是我所见过的最好和最清晰的解释！

我花了相当长的一段时间来掌握弗里曼的见解的本质，要是我能早些得到你的帮助就好了。但是也许理解过程需要多次反复，就像拉小提琴一样不断练习。

弗里曼早在他 1975 年的书《神经系统中的大规模作用》（*Mass action in the nervous system*）一书中就发表了他关于脑振荡场的想法。四十年来，他一直在研究这一现象，并且仍然着迷于它。这是一种惊人的奉献精神和表现。当然，这并不意味着他的理论就一定是正确的。但是，他确实不断地在取得进展，积累了极其丰富的经验，这是当他谈到脑时，你一下子就可以感觉得到的。对我而言，在迄今为止我读过其作品的所有作者中，他是最切题的一位。

我完全同意你所讲的他对私密性的解释：

"他认为，私密性深深植根于意向和意义的产生过程中，这甚至还在意识之前；知觉并不是感觉刺激的表征，它也取决于主体以前的经验，因此也是私密的。"

我不确定弗里曼是否更趋同于我的乐观看法，即相信我们可以通过自然科学研究意识。但如果你和我都同意，基于上述定义的主观性使得我们每个人都是独一无二的，那么我将乐于同意。我也同意你说我们在这个问题上的立场都已说得很清楚了。

当谈到对意识的进一步研究时，在当下我愿意采取一种务实的态度，也就是你讲到"常识"时所表达的意见。我不认为我们能够解决所有"如何"的问题，但我相信工程师会再次胜过自然科学家，比起生物学家认识脑功能来说（更不要去说像意识这样复杂的事物），工程师会以快得多的速度创造出更多的智能机器。我的猜测是像那些发明了 Siri 和 Alexa 的工程师们将会一往无前，并建立更好的 AI 系统。机器人会移动并表现出越来越像人的样子，会与我们交谈，使我们产生它们是有智能的幻觉，甚至还会觉得它们也有共情。比起库兹韦尔所预言的期限来说，大概需要更长的时间才能使机器人成为真正的无魂人，但这种无魂人可能不会像许多人所害怕的那样吓人。

如果一个人形机器人表现很好、友善、能帮助人、充满智慧、办事周到还有幽默感，那么我要问的问题是人们是否还会不要这样的无魂人，而宁愿要一些攻击成性、抱有敌意、愚蠢而又脾气暴躁的真人？

我同意德阿纳这样的人有很好的想法，但我的猜测是技术进步正在以指数式速度向前发展，而生物学和认识脑的进展只是线性的。

我不确定像对待工厂那样做计划和进行组织，投资数十亿美元的项目是否会改变这种状况。我甚至担心这会使情况变得更糟。

但是在过去几周里，我忙于弥补自己的知识空白，所以我没有足够的时间来考虑你早先就提到过的 BRAIN 倡议，也没有来得及考虑关于马克拉姆 HBP 帝国的宫廷革命。这两件事确实都非常有趣和重要，我们应该详细讨论。但请让我先回过头来告诉你我在最后一封信的结尾处所做的承诺，即告诉你我从承现峻那里学到了什么。

在读到承现峻的书之前，我以为由美国国立卫生研究院于 2010 年发起的美国人类连接组计划是奥巴马政府在 2013 年以 BRAIN 计划（尖端创新神经技术脑研究）名义宣布的大项目的前身。但是，比起 HBP 所获得的大量资助来

说,我一直奇怪为什么这个 NIH 连接组只获得相对较少(3 000 万美元)的资助。

现在我从承现峻的书中得知,我们在这里谈论到两种不同的连接组。以下是他对 NIH 连接组计划(第 181 页)的介绍:

"大多数人没有理解到这个计划只是关于脑区之间的连接组,而与神经连接组无关。"

哎呀,我不得不承认我也是这群"大多数人"中之一。☹

他的解释非常有用。当医生和神经病学家谈论连接组时,他们指的是脑功能区之间的联系。就像布罗卡-韦尼克-语言模型中所讲的那种联系,在那个模型中,负责语言理解的是一个脑区[韦尼克(Wernicke)区],它联结到了另一个负责生成语言的脑区(布罗卡区)。在发现那些功能区的日子里,把这类联系称为连接组还是有相当道理的。当时用于检测和成像的工具及其分辨率仅允许识别相对较大的区域。

当承现峻在你引用的那段话里谈到连接组时,他谈论的是最低层的神经元和分子,就像我们在前面的讨论中所做的那样。这是相当不同的!

他对我们迄今为止所讨论的图景上添加的复杂性很重要。他说神经元不单是通过突触连接,这确是事实。神经元的确可以受到各种分子(例如神经递质)的影响,这些分子可以漂浮在神经元之间任何有空隙的地方。我非常赞赏他的一点是,如果最后证明突触外相互作用确实对脑功能是至关重要的话,他也愿意质疑"你是你的连接组"这一范式。

我认为这是非常可能的,这是因为,正如我已经提到过的,首先神经元也是一种细胞,在其膜上密集分布着大量的跨膜蛋白质,所有这些蛋白质实际上都是些开关,随时准备开启或终止各种过程。尼古丁、咖啡因、酒精和许多其他药物都可以影响连接完好的生物神经网络,大家都知道这些化学物质如何影响我们的知觉、觉知和意识。这可能是通过干扰突触间隙并因此干扰了连接组而发生的,但也可能是干扰了神经元的其他细胞机制。但无论如何,很明显,在神经网络模型中把神经元当成像晶体管那样的器件,这只存在于逻辑学家和计算机

科学家的幻想之中,而不存在于真正的脑中。与许多玩抽象逻辑神经元模型并在计算机上仿真它们的人不同,承现峻掌握了其间的本质差异,他也深知这对心智上传和"炼金术士们"给我们有关脑的其他天花乱坠的承诺都是不利的。当然,这与在计算机上可以运行基于理想化神经元的神经网络的应用程序,并得出很棒的结果无关,其中的许多方面甚至比生物脑更好。

所以结果是,虽然脑系统由于考虑到连接组而使其复杂性极大地增加了,但仍然不足以阐明脑的方方面面并搞清楚个性。再次,我们不得不说,"不,你不是你的连接组",就像我们不得不说"不,你不是你的 DNA"和稍后"不,你不是你的蛋白质组"一样。

无论如何,我很高兴你找到了一位对心智上传持怀疑态度的专家,并得出与我们两人相同的结论。这种观点虽然还不那么流行,对库兹韦尔一类人来说是一个坏消息,但可能却是对的。

所以当你想要研究清楚意识和主观体验特性时,试图百分之百地复制大脑并不是一个有效的选择。但也许这根本就没有必要。也许我们应该使用更粗大的建筑模块,而不是使用非常精细的纳米工具在分子水平上认识脑功能。

神经生物学似乎正在进行一场竞赛,寻求有更高时间和空间分辨率的精细仪器。但即使对最好、放大倍数最高的显微镜来说,问题也在于要找到合适的聚焦水平。当你用肉眼看细胞时,你看不到太多细节,但当你使用电子显微镜时,你可能看到的细节又太细,因此和细胞本身反倒关系不大。这就像你拍摄人脸的快照并放大到皮肤毛孔的水平。当你看这样的快照时,你可能会得到关于这个人的皮肤的非常有趣的信息,但是你却无法认出这个人是谁。

现在我们已经有仪器可以研究到每个原子,那么对研究人员要想研究的效应来说,其可选择的粗粒度(granularity)范围就太大了。如果你没有一个理论告诉你需要看什么,那么你就可能无所适从,只能寄希望于碰运气在大海里捞针了。

前一段时间,生物学家拉泽布尼克(Yuri Lazebnik)写了一篇题为《生物学家能够修理无线电吗?——或我从细胞凋亡上学到了什么》(*Can a Biologist Fix a Radio? — or, What I Learned while Studying Apoptosis*)的发人深思的文章

（这篇文章在互联网上很容易找到），他的博士论文做的正是有关细胞凋亡（也就是"细胞死亡"，这可能是认识癌症的一个重要现象）。[1]

他问的问题是：生物学家（他本也可以说神经生物学家）是否可以用他们研究人细胞的相同方法，以他们所有的仪器去认识和修理一台晶体管收音机？好吧，正如你可能已经猜到了的那样，答案是否定的。他给出了很好的例子，说明为什么生物学家甚至没法处理一个远比细胞（这他们也没搞懂）要简单得多的信息处理设备。我不敢肯定他的建议（在生命科学中建立类似于数学那样的更形式化的方法，和使用更精确的研究语言的方法）是否能改变这种情况，但他的分析充满了智慧。它读起来就像是对现代生命科学的费恩曼式的讽刺，他怀疑现代生命科学除了论文外几乎没有产出，但却需要越来越多的钱。

拉泽布尼克还对这些越来越细的实验所产生的大量数据究竟有什么用提出质疑，并说我们已经到了这样一个阶段："这可以用下面的悖论来总结，即我们知道的事实越多，我们对我们所研究的过程的认识反而越少"。[2] 而当说到我前面提到的应该到哪里去找"预期中的金矿"问题时，他引用了一则中国谚语："在黑暗的房间里很难找到一只黑猫，特别是如果根本就没有猫。"

我不确定这是否真是一句中国谚语，因为西方人倾向于把所有此类深刻而又诙谐的智慧贴上中国标签，但你可以告诉我。无论如何，它很好地描述了我们在神经生物学方面所面临的窘境，以及要想在脑中找到心智和意识的起源的尝试都不那么成功。

也许我们真的是在错误的地方或错误的层面上寻找。如果弗里曼是正确

[1] Lazebnik Y. Can a Biologist Fix a Radio？：or，What I Learned while Studying Apoptosis[J]. Cancer Cell，2002，2：179－182.受到拉泽布尼克问题的启发，Jonas 和 Kording 分析了一种人为的更为复杂的情况，这就是用神经科学的方法来分析数字微处理器，并得出很明智的结果：Jonas E，Kording K P. Could a Neuroscientist Understand a Microprocessor？[J]. PLoS Comput Biol，2017，13（1）：e1005268. doi：10.1371/journal.pcbi.1005268

[2] 20世纪90年代瑞典裔美国听觉科学家穆勒（Aage Møller）也说过类似的话，这是因为在以前人们曾经认为听觉器官就是一个傅里叶分析器，它把输入的声音分成不同的频率成分，并为不同的神经通路传送。人们曾经认为已经懂得了这一过程。但是正如我们在"Ⅰ－006凡及－2013－03－15"一信中说明的那样，现在知道事情远没有那么简单。所以穆勒感叹说我们现在对听觉系统的认识比我们曾经以为已经认识到了的要少。——译注

的,而且我们主要面临的是振荡场的话,那么当生物脑是模拟滤波系统时,即使创建更精细的连接组图谱或是在数字计算机上仿真逻辑门网络就都没有用。

这一切听起来都很令人沮丧或甚至几近绝望。但在另一方面,尽管宇宙中没有两个脑是相同的,而事实上,纳米尺度上的所有事物都是令人难以置信的复杂,那么两个脑如何还能够交换想法,而且不仅仅是通过言语交流理性的信息,也可以通过调节声调或通过触摸来交流共情。甚至一个中国人的脑和另一个德国人的脑,通过电子邮件这样局限性很大的手段并用第三种语言猜测他们的脑的结构,其效果也非常出色。虽然这些猜测可能大多是错误的,但是这两个脑中的两个有意识的部分都非常有信心地认为,这样做对这两个差别显然很大的脑都产生类似的乐趣!

你同意我的话吗? ☺

我对这种沟通之谜的兴趣也许和你对意识和主观性之谜的兴趣一样历时久远,如果你愿意的话,我们可以更详细地讨论这个问题。

这也必然要说到你问到过的关于意识测试标准问题和语言这一令人困惑的现象。但这是一个全新的领域,在脑功能的层次结构上至少要高出一个或者好几个层次,因此让我就在这里结束此信吧。

晚安,对上海致以最好的祝愿。

卡尔

内心活动;图灵测试;美国 BRAIN 倡议工作组的
建议

2015－07－01

亲爱的卡尔:

　　再次感谢你的美言,虽然我感到羞愧,因为我远不像你所说的那样聪明。相反,你的话倒正好说出了我对你的信件的感受。特别是,"如果没有你,我绝不会花费精力深究你引起我注意的所有那些令人激动的事情,或者表述一些我已考虑了相当一段时间的想法。比起那些在头脑里放了段时间而还没那么清楚的思想来说,要想把想法写下来就迫使人们用一种更加具体和精确的方式来表达。"和朋友讨论,特别是和一位聪明的朋友讨论,是磨炼人头脑的最好的磨刀石,许多困惑不解之处会突然变得清晰。在讨论期间,人会突然悟出从未想到过的想法。结果就成了"人可以体验到的最有回报感的事情之一"了。

　　我同意你对弗里曼的评价。我应该不好意思地承认,我只是在最近才开始理解他有关意义和意向的出色想法,虽然我认识他已超过 25 年。我认为他的想法是超前的,甚至到了今天也没有多少人认识到它的重要性。人们仍然只把脑看作一种信息处理系统,仅此而已。我也同意,虽然我们不能断定他的理论一定正确,但他启发我们以一种新的角度思考脑功能,这深深植根于他半个多世纪以来对嗅觉系统的研究。

　　我也应该不好意思地承认我曾经低估了技术进步。在经过我们长时间的通信后,我不得不承认你的下列信念一定是对的:"工程师会再次胜过自然科学家,比起生物学家认识脑功能来说(更不要去说像意识这样复杂的事物),工程师会以快得多的速度创造出更多的智能机器。"我曾经高估了复制生物机制在

开发新技术中的作用。当然,从生物体中获得灵感对于开发新技术来说很重要,然而,模仿只在某些时候起作用,但并非总是如此。你的猜测"技术进步正在以指数式速度向前发展,而生物学和认识脑的进展只是线性的"很可能是正确的,至少从定性上来说是如此。如果一定要认识清楚生物机制之后才能发展新技术,那么技术与神经学进步速度之间的显著差异就无从解释。

我不能排除总有一天能构建出一个通得过 TTT 的人形机器人的可能性。根据我们同意的公理,我可能会承认,他可能是有意识的,但是在我的心底里,我可能仍然怀疑他是否真的有意识,因为我不能分享他的体验,我还知道他的材料、结构和我完全不同。我不确定他是否有内心世界,也就是类似于我们的感受、体验、思想、意义、意向或意识之类。或许他只是从行为上看起来像,听起来像,或甚至摸起来像人一样,但是却根本没有这种内心的精神活动。其实只不过是一个无魂人!尽管看起来不可能有生物无魂人,但技术上的无魂人也许是可能的。

另外,关于 TTT,甚至是关于图灵测试本身,我还有一个疑问,说你不能判断对象主体是不是人究竟是什么意思。一般来说,我想其意思是说要是主体不犯任何正常人不会犯的错误,那么我们就判断他为人。然而,要是主体太聪明了,他回答得太快了,太完美了,没有一个正常人能做到。在这种情况下,我想大多数人会怀疑这个主体还是否是人。我们经常把一个天才说成不像人。因此,在我看来,图灵测试可能是判断主体是否有智能的一个很好的标准,但不是判断主体是否有意识的标准。即使是判断智能,这也只是一个充分的条件,但不是必要的。为什么一个有智能主体的行为就得像普通人一样呢?你怎么看?

我很遗憾地承认我从未听说过拉泽布尼克。☺根据你对他的描述,他一定非常聪明,但我从来没有听说过中文中有"在黑暗的房间里很难找到一只黑猫,特别是如果根本就没有猫。"这样的谚语,也许你的解释是正确的。

你所说的沟通之谜必定是最困难的问题之一。虽然心智的私密性似乎无法被否定,但不同的脑仍然可以在彼此之间有效地相互沟通,就像你强烈主张的一样。脑又是如何做到这一点的呢?一个脑如何才能拥有另一个脑的共情呢?里佐拉蒂、拉马钱德兰和雅科博尼把这归之于镜像神经元,然而,希科克和

意识之谜和心智上传的迷思　　一位德国工程师与一位中国科学家之间的对话

其他人则对此表示怀疑。似乎还没有一个普遍都接受的共情理论，尽管共情现象似乎不容否认。一定有某些负责它的神经回路，至于是否就是镜像神经元或是镜像神经系统，我不知道。

你的信提出了很多重要的问题，这些问题不容易回答，其中一些我从未想过。这些问题确实很有启发性，我必须更多地阅读和思考。

现在，我必须把我的话题转到美国的 BRAIN 倡议的进展上来，尽管我很抱歉讲得太迟了，几乎迟了近一年。从另一方面来说，延迟让我有更多时间去思考这个计划。

如你所知，BRAIN 虽然脱胎于"脑活动图计划"（Brain Activity Map Project，简称"BAM"），但是这一倡议是由奥巴马政府在没有广泛征求神经科学家意见的情况下提出来的。许多神经科学家发现奥巴马政府给 BRAIN 倡议所提出的目标远超 BAM，虽然吸引公众眼球和激动人心，但是却不切实际，难于实现。2013 年 5 月 6 日，美国国家科学基金会（NSF）等为此召开了一次讨论会，但是众说纷纭、莫衷一是。会议的组织者神经生物学家韦登（Van Wedeen）总结说："（与会者的）信念是要有一次大飞跃，但究竟是什么大飞跃、跃向何方，大家争论不休。"后来不得不要求参加者提出一页纸的建议，描述自己所认为的神经科学家所面临的关键问题。虽然由此还不能得到共识，不过在我看来，这样做是很有意义的，至少打破了长官拍脑袋和个别专家定方向的路线，也可以算是对应该研究什么在顶尖的神经科学家中所做的一次普查，这对制订一个合理的计划是很重要的。最后，由洛克菲勒大学的神经科学家巴格曼和纽瑟姆领导的 BRAIN 倡议工作组（有 15 个成员，号称"梦之队"），最终确立了这一计划的目标、具体内容、时间表、标志性成果和经费预算等方面。有趣的是，巴格曼曾经对 BRAIN 提出过尖锐的批评。工作组把其建议提交给美国国立卫生研究院（NIH）顾问委员会（ACD）。其任务首先是要提出在第一年（2014 年）的计划，在 2013 年 9 月把此计划上交给 NIH；之后再制订长期的实施计划，并于 2014 年 6 月上交。

工作组首先召开了四次学术讨论会，以广泛听取神经科学界的意见。这些讨论会的主题集中在下列四大领域：分子技术、大规模记录技术、计算神经科学

和理论神经科学、人脑研究。每次讨论会都有十几个特邀报告,每场报告结束以后,报告人就和"梦之队"闭门讨论。巴格曼声称这些讨论充分保护隐私,因此发言人得以自由地甚至批判性地评论不同的实验方法。尽管如此,由于神经科学范围广阔,还是有许多神经科学家,其中包括分子神经科学家、细胞神经科学家、发育神经科学家和临床神经科学家,觉得他们的领域没有得到应有的重视。

2013 年 9 月 16 日,工作组向柯林斯递交了有关 NIH 在 2014 财政年度应优先资助领域的过渡期报告。2014 年 6 月 5 日,又向柯林斯送交了有关 BRAIN 的长期计划报告并得到了批准。

无论是过渡期报告还是长期报告,在其提出的任务中都包括:对神经细胞作详尽的分类;绘制不同尺度的脑结构图谱;在神经回路层次上同时记录其所有神经细胞的活动;开发操控回路神经元活动的各种工具;把神经元活动和行为联系起来;把理论研究、建模、数据分析、计算和实验结合起来进行研究;开发正常和病理状态下的人脑数据共享平台;改进和开发为完成上述任务所需要的技术。他们提出在前 5 年里集中于技术开发,后 5 年则把这些技术应用到脑研究中去。当然两个阶段并非泾渭分明,而只是重点有所不同罢了。

在我看来,从一些科学家提出 BAM 到奥巴马政府提出 BRAIN 倡议,再到组织"梦之队"征集神经科学领域中广大顶尖科学家的意见,反复讨论论证,科学家基本上坚持了实事求是的原则,没有屈从于长官意志,也没有只是为了从政府取得巨额经费而提出根本就没有实现可能的"宏伟"目标。经过反复讨论,无论从目标还是技术路线方面,都更趋于实际可行。

事实上,美国的艾伦脑科学研究所(Allen Institute for Brain Science)和HBP 在神经细胞分类上已做了很多工作;从 2009 年启动的美国人类连接组计划(Human Connectome Project,简称"HCP")也在不同层次上,特别是在宏观层次对脑内白质的连接图谱做了不少工作,尽管离搞清楚突触层次上人脑内所有神经元相互之间的连接图谱距离尚远。在同时记录和操控大量神经细胞活动方面,他们聚焦在回路层次。比起 HBP 或是奥巴马总统在演说中所

提出的愿景，其目标要有限得多，他们的时间表也要合理得多。10 年内实现整个果蝇脑的活动成像[1]，15 年做到对小鼠脑的成像，只有在做到了这两步之后才逐渐向灵长类动物前进，并且也没有对此作出时间上的承诺。

当然，为争取政治家和公众的支持，该计划也打出了诊断、防止和治疗人的脑疾患的旗号，不过他们并没有给出时间表。我想这也无可厚非，只要不给公众虚幻的许诺就行。事实上，柯林斯等人就强调指出："我们必须十分小心，勿轻易许诺在这方面取得立竿见影的结果。"他们又表示："描绘脑要比描绘基因组复杂得多，这里并没有线性的序列需要解码，也不存在明显的终点。但是从这一有关工具开发、伦理考虑和伙伴关系等的早期努力中所吸取的经验教训，将有助于我们进行一场新的、甚至更令人望而生畏的艰难征程。"

总之，在我看来，美国科学家在对待脑计划方面的表现值得称许，他们对计划的目标和内容事先进行了充分的讨论，把问题聚焦到在有限时间和有限资源条件下可望突破的有限目标。当然不可能每个人对该计划的所有内容都有相同的看法，我也不例外。例如对于一旦能同时记录人脑中所有神经细胞的活动就能揭开脑之谜的说法，不要说技术上的可行性，即使技术上能做到同时记录，这种说法也有点像说如果我们能同时记录全世界每个人的活动，人类社会的一切问题都将迎刃而解一样。对于同时记录某个回路中所有神经细胞的活动能为阐明该回路功能的机制做出多大贡献，我不知道，让我们拭目以待吧。

　　信太长了，就此打住。
　　祝好！

凡及

---

[1] 2018 年 7 月 19 日《细胞》(Cell)杂志报道了美国科学家已经将果蝇脑切成 7 000 个切片，由此得到 2 100 万个像，并进行 3 维重建，从中可以看到每一个神经元，也可以追踪其中任意两个神经元之间的联结。目前已得出了果蝇脑中很小一部分(蘑菇体)的线路图，但是要想得出整个线路图尚需时日，而要想由此解释果蝇的种种行为则依然是个挑战。此外，这一工作还只是解剖结构上的图谱，而非神经元的活动图谱。详情可参阅 http://www.sciencemag.org/news/2018/07/tour-de-force-researchers-image-entire-fly-brain-minute-detail?utm_campaign=news_daily_2018-07-19&et_rid=260067369&et_cid=2192551。

背景专栏 II F10.1

NIH 顾问委员会 BRAIN 倡议工作组的研究方案

过渡期报告

结合奥巴马总统和柯林斯院长所提出的目标,顾问委员会工作组认为,分析回路中神经元的相互作用最为合适,并有可能取得革命性的进展。要想真正认识这些回路,就要求对其中的各种细胞——加以识别,并研究其特性,确定它们彼此之间的突触连接,观察在动物活动时这些回路活动的动态模式,对这些模式加以干扰以检验其意义。此外,也需要认识在回路中信息处理所用的算法;也要研究脑作为一个整体,在其中相互作用着的各个回路彼此之间所用的算法。

他们提出应该坚持下列 6 个基本原则。

(1)选取适当的实验系统和模型。虽然目标是要认识人脑,但是许多方法和设想先要在动物模型上进行试验。

(2)进行跨学科合作。最有希望的途径是进行跨领域研究,把实验和理论、生物学和工程、工具研发和实验应用、人脑研究和动物模型结合起来。

(3)兼顾空间尺度和时间尺度。对脑的全面认识要跨越许多空间和时间层次。

(4)建立起共享数据的平台。

(5)新方法要通过制造者和实验者之间的反复交流加以严格检验,一旦通过检验,则需要大力推广。

(6)注意神经科学研究的伦理方面。

按照该报告,NIH2014 财政年度的资助应该集中在下面 9 个领域,并把有关细胞、回路、脑和行为的研究整合在一起。

(1)建立脑细胞类型的分类细则。确定神经系统中的所有细胞类型,研发在体记录、标记和操控这些分类精细的细胞的方法。

(2)建立脑的结构图谱。现在越来越有可能建立局部回路中相互连接的

神经元的图谱,以及分布在许多脑系统中神经元之间的联结图谱。这样就可以认识神经结构和功能之间的关系。

（3）研发新的、大规模记录神经网络活动的技术。要应对从整个神经网络中记录动态神经活动的挑战,从长远来说,甚至记录所有的脑区。为此必须改进现有方法和研发新工具。

（4）研发一套操控神经回路的工具。通过直接激活和抑制神经集群,神经科学就可以从仅仅观察现象进展到研究因果关系。

（5）把神经活动和行为联系起来。巧妙地应用虚拟现实、机器学习、记录装置微型化,有可能大大推进我们对认知和行为的神经活动机制的认识。

（6）把理论、建模、统计和计算与神经科学实验结合起来。这是认识复杂的非线性的脑功能所必需的,而不能只靠直觉。新类型数据的飞速增加,要求新的数据分析方法和解释。这就要求实验工作者和统计学家、物理学家、数学家、工程师和计算机专家通力合作。

（7）阐明人脑成像技术的机制。进一步改善空间分辨率和时间分辨率。

（8）建立起一种机制,使得能把人的数据汇集在一起以进行科学研究。特别是临床数据,这要求临床医师、工程师和科学家通力合作,因此要求建立起一种新的机制,以最有效地搜集这些无价的信息,并为脑失常病人造福。

（9）传播知识和训练。如对计算神经科学和各种新技术举办夏季训练班。

长期计划报告

报告把 BRAIN 的目标定为绘制出脑回路图谱,测量这些回路中电和化学活动的变化模式,并认识这些活动的相互作用如何产生我们人类独有的认知和行为能力。同时,把下列 7 项作为优先目标。

（1）识别不同的脑细胞类型,并对它们进行实验研究,以确定它们在正常和病理情况下的作用。

（2）绘制具有各种分辨率的神经回路图谱,其分辨率从突触开始直到全脑。

（3）通过研发和应用大规模监测神经活动的改进方法,建立脑功能的动态

图景。

（4）应用精确改变神经回路动态特性的介入工具，把脑活动和行为联系起来，阐明因果关系。

（5）通过研究新的理论工具和数据分析工具，建立认识内心过程的生物学机制的概念基础，确定在其中起作用的基本原则。

（6）研发认识人脑和治疗其失常的创新技术，建立和支持把脑研究整合在一起的网络。

（7）把上述各点中用到的新技术和新概念整合起来，以揭示神经活动的动态模式如何转换成健康时和有病时的认知、情绪、知觉和动作。

为此必须坚持下列 7 大原则。

（1）对人的研究和对动物模型的研究并重。

（2）提倡多学科合作研究。

（3）兼顾空间尺度和时间尺度。

（4）建立保存和共享数据的平台。

（5）验证和推广各种技术。

（6）重视神经科学研究的伦理问题。

（7）建立对 NIH、纳税人、基础神经科学家、转化神经科学家和临床神经科学家负责的机制。由于 BRAIN 倡议具有多学科性，且在 NIH 内外有多个单位参与，须建立某种监督机制，保证 BRAIN 的经费资助得当，符合公众与科学界的利益。

在经费方面，他们建议在头 5 年每年 4 亿美元，以后每年 5 亿美元。

只有开放的头脑才能从讨论中受益;计算机与脑;图灵测试;欧盟人脑计划;美国 BRAIN 倡议

2015-08-17

亲爱的凡及:

谢谢你的来信,这封信一如既往充满了有趣的想法和意见。

但在谈到科学问题之前,我必须告诉你,我的女儿杰尔卡(Jelka)和她的丈夫于尔根(Jürgen)于 7 月 7 日产下了一个男婴。他的名字叫亚历山大(Alexander),他在一个充满雷电的夜晚中表现得相当出色。母子表现良好,阿迪和我正要成为出色的祖父母。当然,我们确信他是最最好的孙子,现在我有了良机去观察年轻的脑是如何发育的,以及意识是如何产生的。如果我有了什么有意思的新见解,我会让你知道的。☺

虽然我不知道你对我们的通信究竟是怎么想的,但我很高兴知道我们对我们交流思想的经历似乎都很满意。虽然我们永远不会确切地知道其他人心中究竟发生了什么,但我的假设是,这种主观愉悦在你我心中必定非常相似。☺

我非常喜欢你的话"讨论,……是磨炼你的头脑的最好的磨刀石"。但只有开放的头脑才能享受这一过程并从中受益。在对立的两个学究式学派的追随者之间的讨论通常毫无价值,他们都深信只有自己才掌握真理。这种辩论的目的通常只是为了取胜,并战胜另一派或另一种思想。这类讨论经常会流于形式。对立双方不会互相倾听,所讲的论点都是事先就可以想得到的,因此这些讨论大多都枯燥乏味。传统的正统学派常常以类似的态度攻击新思想。普朗克(Max Planck)自传中有对这种行为的非常深刻的见解。当普朗克年轻时奠定量子物理学的基础时,他为同行的固执和以自我为中心的盲目所苦。因此,

他得出的结论是，一个错误的理论决不会因为其倡导者意识到他们的错误而死亡并被推翻。他有点痛苦地说道，一个理论只会在它的拥护者都死后才会死去。他自己的经历可能促使他，在自己成为一位有影响力的物理学教授之后，竭尽所能地宣传一位当时还没有名气的年轻的爱因斯坦的革命性思想，后者在当时也受到物理学权威们类似的抗拒。

有时候我们被误导了，并不是因为我们是某个特定学派的忠实信徒，而是因为我们像其他人一样不假思索地使用了某些术语、概念或观点。仅仅是因为大家都认为除此观点之外别无其他，而且这些概念也没有任何问题。在爱因斯坦之前的时代里，有关空间和时间的作用就是物理学史上的一个极好的例子，而把脑看成为信息处理装置则是现代神经科学中的一个例子。我感到惭愧，直到最近，甚至也在我们的通信过程中，不假思索地用了这种信息处理的观点，就好像它是理所当然的一样。但事实并非如此，我也和大多数人一样连想都没有想，只是在我们讨论的过程中，我才意识到这是一个非常具有误导性的想象出来的视角，这种观点是逻辑学家和数学家将其引入到他们自己对此不甚清楚的生物学问题中去的。

在这种情况下，我必须告诉你我的两个令人愉快和令人鼓舞的最新发现，尽管其实我可以并且应该早就发现了。第一个发现是，现代计算机体系架构之父冯·诺伊曼自己早就已经意识到，把计算机比作脑的概念中有些根本错误的东西。当 1955 年他被邀请去为享有盛誉的西利曼讲座（Silliman lectures）作讲演时，他选择了"计算机与脑"（The Computer & The Brain）这个主题。由于沉疴缠身，他未能举办讲座，他的手稿只在死后才于 1957 年发表。

尽管 1958 年以同名出版的这本书只是一个片段，但它在展示冯·诺伊曼的想法、对此问题的深入分析以及对与其相关问题的理解这些方面都令人印象深刻。阅读此书是一种乐趣，人们可能会希望许多现代神经科学家也能够以如此清晰的方式表达自己的想法，就像冯·诺伊曼甚至在临终时所能做到的一样。

我不知道你是否阅读过《计算机与脑》，这本书只有 80 页，但内容丰富。如果你没有读过，那么我要强烈推荐给你，无论如何，我很想听听你的想法。对我来说，很明显，冯·诺伊曼已经认识到脑并不是按算法进行计算的机器，也不是

数字设备,而是模拟设备。他对这种模拟设备所可能达到的准确性作了估计,然而脑却依旧有如此惊人的表现,这些都非常非常有趣！他对脑可能的并行功能的猜测也是如此。他的令人惊讶之处还在于他的下述见解:当时有关脑的整个理论肯定存在根本性错误。他非常清楚自己设计的以他的名字命名的计算机架构与脑如何工作毫无关系,而且当时也没有人知道如何解决这个难题。在我看来,他也充分地认识到他最初尝试构建一个受脑启发的计算机失败了,这种机器基于麦卡洛克和匹茨关于神经元功能的想法,以及香农(Shannon)关于其逻辑和信息处理基础的想法。冯·诺伊曼也许可能再试一下进行纠正,也许他曾经希望这样做过。但他没有这样再做尝试的机会,结果现代计算机技术是基于逻辑运算的中央时间同步(central-time-synchronized)数字机器这种相当不恰当和非常低效的原理发展起来的,而并不是按高度并行的模拟原理发展起来的。技术发明就像生物进化一样,有时也会进入死胡同,纠正这些错误需要很长时间。蒸汽机和内燃机能效差、副作用大是大家都知道的例子。奥斯特瓦尔德(Wilhelm Ostwald)赞同燃料电池和电动机,而反对热机和火力发电机,他早在1894年就已经对那些错误趋势非常愤怒！但是人类花费了120多年的时间,消耗掉了大部分化石能源储量,导致空气污染以及其他问题,这才得出热机并不是最好的技术的见解。我们使用冯·诺伊曼计算机已经大约70年了,这也是一种浪费能量的怪物,问题是我们是否需要再等50年才能学会如何做得更好。

无论如何,当涉及脑功能时,冯·诺伊曼和弗里曼有几乎相同的见解,这导致了我的第二个发现。你可能听说过马尔斯伯格(Christoph von der Malsburg)的名字,他是一位物理学家,后来成了神经科学家,他经常被引用为视觉皮层信号处理的先驱。他还以对认知理论中所谓的"绑定问题"[1]的贡献而闻名。[2]

我知道他的名字,也知道他从事面部识别问题。但我不知道他与弗里曼的观点有很大的共同之处,也不知道他对流行的关于脑的信息处理范式也同样持

---

[1] 视觉系统在不同的脑区抽提对象不同的特征(例如轮廓、颜色、运动方向等),脑如何根据这些散布各处的不同特征组织成一个统一协调的整体,以及把视野中不同对象的特征分别组织成不同整体,这两个问题就被称为绑定问题。——译注

[2] https://en.wikipedia.org/wiki/Christoph_von_der_Malsburg

否定态度。

他也认为,控制我们脑的主要是模拟现象,而不是数字现象。我相信你比我更了解他,并且更好地知道他在神经科学中的作用。我对这个发现感到很高兴,因为我高兴地看到弗里曼在其观点方面并不孤单。对我来说,他们对脑功能的两个非常具体和明确的领域进行了多年的实验研究后,最后都深入到内心问题并不奇怪。弗里曼研究了嗅觉系统,而马尔斯伯格则研究视知觉系统。他们一定知道对方,并有自己的评价,因为我查到他们互相引用。这对你来说可能并不新鲜,你可以告诉我更多关于这方面的信息,也可以告诉我认为两者在研究方法上有类似之处的想法是否有错。

当你说你高估了复制生物机制的作用时,你可能属于大多数人。然而,有些人的观点甚至更加鲜明,并且说大自然在进化过程中所做的一切,其神奇再怎么说都不为过。其中之一就是美国神经生物学家林登(David Linden),他因其著作《偶然产生的心智:脑进化如何给予我们爱、记忆、梦想和上帝》(*The Accidental Mind: How Brain Evolution Has Given Us Love, Memory, Dreams, and God*)而闻名。他的观点非常有独创性,而且完全不属主流。《新闻周刊》(*Newsweek*)记者贝格利(Sharon Begley)很好地总结了他的观点:

"把新的部分加到老的部分之上,脑就像是把八轨盒式放音机改装成的iPod一样。爬行动物的遗产之一是当我们的眼睛扫过视野时,它们会产生微小的跳跃。在眼睛跳跃过程中到达脑的信息是模糊的,所以视觉皮层看到的是被跳跃切割后生成的神经等价物。然而脑由此却创造了一种连贯的知觉,填补了跳动输入的空白。你看到的是连续的、逐渐变化。正如经常在拼凑起来的系统中看到的那样,系统的改进版中常常用到旧的组件,在现在所讲的这种情况下,就是把一个在视觉上运行良好的系统运用到更高级的认知中去。由一系列片段的输入中构造出一个完整的故事,例如当问失忆症患者昨天做了什么时,他们会从记忆碎片中构造出一个故事。"[1]

---

[1] http://www.newsweek.com/human-brain-marvel-or-mess-97675

这和弗里曼的观点类似,它也支持了下列见解:一旦你按照某种差劲的原则制造了某个有机体(或机器),那就很难纠正这种生物学(和机器)上的错误,你可能被迫为了弥补以前概念上的缺点,而对一个先天不良的概念层层修补。

在许多这样的情况下,你宁愿从头开始,和大自然比较起来,这样做对工程师来说要容易得多,但是尽管这样,工程师们也往往会在很长一段时间里坚持不适当的观念。

你有关全图灵测试是否合适,以及机器人是否可能因为太聪明而通不过测试的问题都很有意思。作为一个例子,我立即想起伦德贝利(Gene Lundberry)的《星际迷航》(Star Trek)科幻系列中的太空船"企业号"上的中尉指挥官,著名的人形机器人 Data。正如你所说,他常常太精确和准确,例如他给出的时间估计("我们将在两年 157 天 9 小时 28 分钟 37 秒……后到达")或者从百科全书中逐字引证,普通人永远都不会做同样的事。是人就会有点模糊和不那么精确,所以你的问题确实很聪明,我之前没有这样想过。在这里有一种不对称的地方,超级智能总是可以假装成不那么聪明,而反过来则是不可能的。图灵作为一位逻辑学家必定首先考虑过机器是否有可能接近人类能力的问题。

另一方面的问题是共情和幽默,这些都是 Data 虚拟生活中所面临的真正挑战。他知道他的人工脑[被称为"正电子脑"(positronic brain)以表明与人脑之间的根本差别]架构的这一缺点,并试图作出改进,所以他去数据库,读所有的笑话和幽默故事,并观看所有历史上记录下来的喜剧。因此尽管他明白在舞台喜剧录像中一定发生了什么有趣的事情,因为观众都笑了,但是他不明白这是为了什么,非常可惜他也不能懂得真正的幽默感。他想鹦鹉学舌讲笑话来模仿幽默,却可悲地失败了,因为没有一个人笑。当他最终偶然讲到了一些滑稽的事情时,他不明白为什么每个人都会笑。早在 20 世纪 80 年代,罗登贝里(Roddenberry)显然对建造真正的类人人工智能所面临的困难有着深刻的理解。可能在这个问题上他比图灵认识得还要更清楚。幽默,特别是讽刺确实非常棘手。讽刺需要更深层次的知觉和意识,这是孩子学会的最后一件事。而当我环顾四周时,我的感觉是,许多成年人都还从未学会这一点。☺

你提出了一个关于共情和镜像神经元的有趣问题。究竟是镜像神经元呢,

还是镜像神经元系统？我想是系统，而在弗里曼的理论中，最好还要在大范围里寻找振荡。弗里曼认为人类是唯一具有节奏并且可以分享节奏的物种。对他来说，分享节奏和舞蹈是产生语言的起点。在这里，我找到了一篇格尔德（Natalie Geld）写的有关这个问题的有趣文章。[1]

但在我们转向我所爱好的"语言"问题之前，让我来看看有关你讲到的欧洲人脑计划（HBP）危机的有趣新闻，及其美国竞争对手 BRAIN 倡议在研究方向上所作的更改。

HBP 有不寻常的事情发生了。当你去年告诉我有关欧洲脑研究人员反对 HBP 的组织和管理方式的强烈抗议时，我说过看起来马克拉姆面临麻烦了。是这样，实际上他也确实遇到了非常大的麻烦，因为组织调解过程本身就已经是很大的羞辱了。但报告的内容现在等于是在谴责骄傲的发起人和组织者。以这种方式受到批评和公开羞辱对科学家来说是非常不寻常的，因此感觉一定很糟。长远来看，我无法想象出还有什么他能继续待在这个项目中领导位置的解决方案。这非常难，并且清楚地表明了在 HBP 研究人员的圈内一定存在严重问题。我们从他和 IBM 的莫德哈的恶斗中知道，在处理与竞争对手的关系方面，马克拉姆并不是一位最能易位思考和通情达理的人。这个计划的组织方和资助方让他成了国王，他们必定也知道这一点，他们甘冒此风险，可能是因为有时候一位强势的领导者对这样一个巨兽般的项目有好处。当然，那些没有得到他们的份额，觉得有人偷走了他们所感兴趣的领域的政府资金的研究人员也深感嫉妒。但是，当除了两人之外的几乎所有调解委员会的成员（也包括那些参加 HPB 的调解委员会成员）都批准了这份报告时，这清楚地表明了，马克拉姆一定是以非外交手段做出过分的反应，结果使太多的人转而反对他。对一个十亿欧元项目的建筑师和领导者来说，这种经历可能已经让人感到很不愉快了，因此这个项目的结果如何可能就显得更加重要了。这个名声很大的计划的面子丢大了，因为你从报告的科学部分中引用的内容已经清楚地表明，现在每个人都认为目标太过野心勃勃，而且所作承诺也夸大了。建议重新调整研究目标

---

[1] http://mbscience.org/mbsci/walter-freeman-the-dance-of-consciousness/

和领域只能说明最初的想法是错的。

好吧，这一切都不好，而且令人沮丧。尽管我从来就不曾喜欢过这个想法，并总是相信通过在计算机上仿真一个我们不理解的生物器官来认识其功能并不是一个好策略，但我对这种发展深感遗憾。特别是因为这场恶斗的输家是科学进步，而赢家则是官僚主义。我记得在我的最初的某封电子邮件中，我说过我宁愿给一千位博士生每人一百万美元，也不愿意把十亿欧元的宝押在某个研究巨兽上。在一门尚处于起步阶段的学科中，只要我们对脑中发生的事情还只有某种模糊的概念，像"登月计划"那样细致入微地规划的大型研究项目很可能到头来只能是一场细致入微地计划好的大失败。

至少你应该感到高兴，因为调解委员会对这种想法的陈述恰恰是你在最初考察了 HBP 的想法和它的满口许诺后就批评了的！你从一开始就对这种想法持批评意见，而且在很久之前在你预判研究计划难以奏效的文章中说到了点子上。你说皇帝没有穿新衣，现在——事情就是这样——每个人都知道了。我再次为你感到骄傲！

我想知道你的同事现在会说些什么，特别是那些质疑你有权利或能力来批评这样著名的西方科学家的想法的人。☺

感谢你向我介绍对有关美国 BRAIN 倡议所作的调查。阅读所有这些报告和评论工作量很大，我非常感谢你的勤奋和好意，使我得以分享你的发现。

迄今为止，我对有关这个计划的材料还只读了个大概，你的总结很有帮助，尤其是当你将其与欧盟 HBP 进行比较时。你对这第二个十亿美元的脑计划持更为正面的看法，我想你是对的。我知道组织者和科学家已经从 HBP 发生的事情中吸取了教训。虽然后者的问题只是最近才公开发布，但我确信美国研究人员必定知道许多问题，特别是那些没有参与该项目的人的不满。神经科学家在全球的圈子很小，我敢打赌，在美国和世界其他地方早就听到了来自欧洲的雷声。因此不难理解，美国的管理人员（不管是政府方面还是研究人员方面）尽全力防止他们的欧洲同事在准备项目、制定研究计划以及建立管理体制时所犯下的错误。当然，研究人员也足够聪明，降低了他们承诺要跳过的横杆的高度。在我看来，他们降低得也未免太多了一些，但当然这是可以理解的。我本来希

望会有一个更具竞争性和更少点学究气的途径,并且希望能更接近解决现实世界中的实际问题。

我发现很有意思的一点是,BRAIN 倡议与私人基金会、艾伦脑科学研究所之间也有联系。其首席科学家 C.科赫是一位非常有声望的人,让我们希望他也有恰当的社交技能来处理好和他的同行科学家之间的关系。我记得你曾把他的一本或多本书翻译成中文,我说得对吗?所以你可能认识他,并告诉我更多有关他的思维方式。

无论如何,在此领域中正在发生很多事情,我们有很多有趣的事情需要观察。

晚安,向上海致以最好的祝愿。

卡尔

　　6 年前我们开始对话纯属偶然，我们探索性的讨论既没有计划也没有明确的目标。当时凡及在 HBP 这一著名而雄心勃勃的概念中偶然发现了一些自相矛盾之处，但他不能确定这种不安是否合理。起初我们只是像汽车修理工在类似情况下所做的那样，打开车盖好奇地进行检查，共同努力想了解究竟发生了些什么。

　　回顾以往，我们自己都对所发现的东西以及关于脑、心智、人工智能和大科学项目的讨论所走过的道路感到惊讶。

　　正如我们在序中指出过的那样，本书更像是一场随意漫游，从一个领域转悠到另一个领域，并在需要的地方深入下去。我们只是受到好奇心的驱使，并且当我们希望对事情理解得更为精确时或者当我们需要填补我们的知识空白时，我们花费了更多的精力。我们常常喜欢对知识追根溯源，包括追溯到我们两人极为不同的文化背景的历史中去。虽然我们这样的做法显得混沌无序，但我们感到，通过持续而且有时引起争议的辩论，我们获得了在更系统的方法下很难获得的见解。

　　回想起来，我们自己都对以前在某些事物、理论和概念上的看法以及在今天再看这些问题时有多么不同而感到惊讶。我们注意到，我们的观点逐渐发生了变化，随着事情发展而变得清晰起来。

　　例如卡尔认识到，看待脑还有更有希望的观点，而不是通常地把脑当作是一种信息处理设备。

　　在我们开始通信的时候，卡尔是（现在仍然是）一位神经科学的业余爱好者，他很高兴像凡及这样一位经验丰富的专家愿意与他讨论与最新研究进展有关的微妙问题。凡及则是（现在也仍然是）信息技术的门外汉，虽然他对这一技

术在近年来的迅速发展抱有浓厚的兴趣，但是他很难找到一位像卡尔那样同时熟悉脑科学的信息技术专家可以向之请教。专家很少喜欢与外界讨论他们的业务，特别是当他们的意见可能受到质疑时。他们中没有多少人会耐烦别人像儿童问出来的那种基本问题，并以小孩都可以理解的方式表达他们的想法。

我们两人来自不同的领域，从我们的专业和文化背景来说，再也找不到更不一样的人了。但是，我们都充满了好奇心，也都崇尚理性思维，并且都有一个特点，就是对什么事都要质疑一番，特别是当问题涉及公认的学术见解和有人声称在科学或技术上取得了突破时。

虽然下面的这番话我们早已在序里说过一次了，但是由于其重要性，我们想在这里再说一遍：

我们都喜欢从孩子的视角来看问题，他们会提出简单的问题，以了解真相。有时孩子可以看到皇帝的新衣并不像所说的那样华丽。但是我们也不想过于夸大，因为说我们就是著名童话故事《皇帝的新衣》中那个勇敢说出看不到别人"看到"的东西的孩子，就未免太自以为是了。

然而就 HBP 而言，卡尔坚持认为，从很早开始，当其他一些人还在赞扬它的时候，凡及就认识到这个令人印象深刻的计划存在缺陷。

我们在早期的信件中花费了大量的精力来说明并使自己确信在 HBP 的概念中有多处错误，我们不应该对此计划期待过高。由此开始了我们的通信，它成为探索脑和心智及其与人工智能和计算机技术的可能联系的许多基本方面的良好试验田。

今天，在这个项目的名声在公众面前已严重受损之后，这种批评很常见，而我们过去的批评在一些人看来似乎有点像在打"落水狗"。也许现在一般性的批评甚至过多了，因为在我们看来，HBP 概念中也确实有一些有趣的部分值得再作尝试。

针对内容的激烈辩论曾经是常用的科学争论风格，就像体育比赛中的情况一样。但是今天这种非常成功的方法似乎已经过时，所以如果有人这样做了，

就几乎必须为此道歉。也可能这种方法在欧洲更容易接受一些，而在中国则有可能被视为粗鲁和不恰当。

且不管我们在读者中引起了何种感受，重要的是要记住，在自然科学的黄金时代所取得的许多惊人进步可能都要归功于这种严酷的方法。只要读读书刊就可以看到，19世纪和20世纪之交科学家之间的辩论并不是基于客客气气的共识和你好我好大家好，而往往是很激烈的争论。通常朋友们竭尽所能为了辨明真相和知识而激烈争论，但这只是为了驳倒对手的理论而不是诋毁他的人格。当然，这样的争吵也可能有伤害并留下伤痕，但他们这样费力是为了找到更好的解决方案并为科学事业而战。

无论如何，我们（特别是卡尔）想要说清楚，如果我们在批评某种理论或观点时用了重话或隐喻，那么这只能以这种传统的体育精神来理解，我们的话只是针对这种论点，而并不是要抹黑某个人。

※　※　※

在卡尔收到的凡及的第一封电子邮件里充满了他无法回答的问题，其中有些问题他也不懂。但他为这些话题所深深吸引，他也赞叹他所遇到的头脑的严肃性，这个头脑总想弄懂并完全搞清楚。他也对其同伴尖锐的论点和勇气留下了深刻的印象，甚至对大权威也敢于提出尖锐的问题。即使在今天，卡尔也还不清楚这种非同寻常的执着和坚持不让步，直到一切都尽可能清晰和透明，究竟是凡及的天性呢，还是他在把许多当代神经科学巨著翻译成中文过程中获得的技巧。

无论如何，对于卡尔来说，凡及在与许多一流作者的讨论中磨炼了自己的技能，他会问一些难以回答的问题。当凡及把我们的英文信件翻译成中文时，凡及迫使作者澄清他想说的话，卡尔对此深有体会。凡及一次又一次地写信来确认他是否已经清楚了一段话的真实含义，有时甚至只是一个词。这个过程并不总是令人愉快和轻松的，因为它可能表明问题并不出在译者的理解有误，而有可能是由于对思想的表达不清楚，或者更糟糕的是这一想法本身就有问题。

凡及则非常欣赏卡尔的博学、认真和有问必答，为了回答问题，他甚至专门

订购书籍。凡及很难找到像卡尔这样一位朋友，愿意费时费力地思考、回答并和他讨论使他感到困惑的问题。凡及知道他找到了一位很好的朋友和老师。

这种不断澄清我们想说的话并真正理解彼此观点的过程，还要考虑到我们主观上以及文化上的差异，不仅其本身就是一种享受，而且还与我们共同感兴趣的主题息息相关。

在我们的讨论中，我们一再谈到是否有可能将一段话的内容（指的是它的意思）从一种语言准确且完整地翻译为另一种语言的问题，以及如何才能做到这一点。我们一次又一次地回过头来讨论这些问题，这不仅是因为凡及是一位经验丰富的翻译人员，他经常为这个问题而苦苦挣扎，而且也因为这是一个非常合适的实证测试平台。翻译问题与将内容从一个脑转移到另一个脑或机器所遇到的困难，在原则上是类似的。我们也花了很多功夫来讨论意识中的主观性和私密性以及由此如何产生意义。我们在通信中也一再谈到这些内容是否可以完全还原为物理过程，并且可以通过客观的理性研究来阐明。当然这和当前流行的下列讨论是相关的：借助于现代人工智能，我们是否可以制造出具有类似人类意识的机器。这一基本问题是我们在信中处理的两大谜题之间的自然联系。一个是脑的自然智能，即便在好几代最聪明的研究人员花费了巨大努力试图揭开这一秘密之后，我们对它的了解仍然很少。另一个是运行在计算机上的人造算法，也就是人工智能，计算机是完全不同于脑的机器，并且其能力以惊人的速度在不断提高。

实际上，人工智能在许多方面已经优于人类，但是人们往往忽视这一点，只是因为讨论集中在那些机器还不能做的事情上。

我们讨论的关键问题是：自主学习机器在所有智能领域（包括创造力和其他许多被认为是真正的人类的能力，如情感和幽默）中能否和人类一样，甚至超越人类？如果能的话，他们最终能够发展出意识和主观性吗？

这是我们仍然持有不同立场之处，虽然我们两人都在对方论点的影响之下有所改变。在我们开始讨论之前，卡尔对主观性问题、意识可能在什么条件下出现以及如何出现等方面几乎没有想过。对他来说，它似乎更像是一个可能会消失的伪命题，就像物理学中旧的"生命力"（élan vital）问题消失了一样。许多

　　　　意识之谜和心智上传的迷思　　一位德国工程师与一位中国科学家之间的对话

工程师认为，当人工智能在高速计算机上编制出足够多的算法以最终执行我们心智所有的各种智能功能时，意识可能就会以某种方式涌现出来。当凡及坚持认为这不是一个显而易见的问题时，卡尔最初怀疑他是否会退回到了一种形而上学的立场上，从而偏离了讨论的理性基础。凡及不太确定卡尔是否抓住了让他思考多年的问题的核心。另一方面，卡尔认为凡及可能低估了硬件仍在发展的速度，并可能相信摩尔定律即将结束的流行观点。

我们共同对许多问题进行思考，试图对科学和技术中许多相关领域的发展有一个概括性的了解。通过反复交换意见，我们终于设法较好地理解了问题的所在，以及对方的立场。对意识这个困难的问题，凡及并没有后退到形而上学，而是提出了支持他立场的有力论据，这给卡尔留下了深刻的印象。

我们最终同意如果能百分之百地复制身体和脑的话，那么这一拷贝也可能包含主观性的内心状态，甚至是神秘的主观体验特性。但是凡及成功说明了这样的拷贝实际上是几乎不可能的。另一方面，卡尔不得不认识到，要想建立起能发展出意识和主观性的强大的通用人工智能系统，可能比他希望的要更困难。凡及也不得不承认他低估了 AI 的发展速度。

所以如果在今天问凡及，他可能仍然怀疑是否能实现强人工智能（在人工智能有意识的意义下），而卡尔仍然会选择一个肯定的答案。尽管我们从来没有在我们的信件中明确表达过这一点，但我们都有一种感觉，我们已经设法在对方的脑海中种下了一颗怀疑的种子。

无论如何，我们都同意，心智上传几乎是不可能的，即使是建造出人工无魂人，也要比库兹韦尔的信众所希望的花费长得多的时间。更不要说所谓的近在眼前的奇点了。我们也同意，要在机器、身体和脑之间建立神经联系比许多人认为的困难得多。

我们一直在思考的另一个问题是：工程师为什么在有关计算机和人工智能软件的概念上会取得如此巨大的成功，尽管它们都是基于显然非常不准确的脑模型之上？

如果工程师使他们的机器更像脑（brain-like），是不是对实现他们的目标会有帮助呢？借助 AI 概念和工程师、数学家和逻辑学家的更多帮助，是否能启发

神经科学家更好地理解脑和心智呢？用更大更快的计算机进行仿真并由此改善神经脑模型对神经科学家理解脑和心智是否会有好处呢？

对于像脑这样高度动态的功能系统，我们甚至在物理层面都还很不了解，更不要去说更高的功能层次。我们都认为进行计算机仿真没有多大意义，然而试图这样做可能有助于改善所需的工具。

我们也最终同意，试图在硅片上通过逆向工程建立一种生物脑并没有太大希望。卡尔早就提出，工程师不会从脑研究的结果中获益太多，走他们自己的路而不用理会生物模型会有更好的结果。一开始，这在凡及看起来是一个很奇怪的观点，但后来他觉得这一观点并非那么荒谬了。

我们仍然无法回答许多令我们困惑的问题，甚至无法确定我们有共识之处是否就一定对。当然，我们并没有声称我们已经获得了任何新的科学见解。

※　　※　　※

关键在于我们两人在更好地认识这个领域方面都取得了进步。我们的这一段经历对我们来说都很有收获，但愿也能启发他人以类似的方式去探索这个领域。读者可以决定哪种立场较适合他们，也有些人可能对我们两人的观点都不同意。

出版我们之间通信的目的并非是要宣传某些理论，而是提出一种获取知识和探索个人以前未知领域的方法。

作为对那些对这种主观探索感兴趣的人的一种启发，我们要再次强调我们认为对我们最有帮助和最有启发性的观点。

在讨论所有这些问题时，有一个人给了我们最大的帮助，他给了我们一个非常有意义的理论参考框架。他就是凡及的朋友弗里曼，他在我们的通信过程中不幸逝世。我们推崇他为该领域最杰出的思想家和研究人员之一。可惜的是，尽管他很有名，但在我们看来在主流神经科学界他并没有得到应有的认可。

对卡尔来说，凡及向他介绍并不那么容易搞懂的弗里曼的工作是他在这场对话中最具挑战性和有益的经历。几年前，当他第一次面对神经系统时，他将大脑视为一种信息处理机器。他曾经相信过心智上传的可能性，冯·诺伊曼、

香农和图灵建立起来的以逻辑和数学为核心的范式似乎是破解这个谜团的正确方法。然而,他对这个问题考虑得越深入,对这种方法的数字本质以及脑作为计算装置的想法越使他感到不安。

弗里曼是卡尔遇到的第一位既对脑有极为深入的了解,又分享了这种不安的人。他甚至更进一步把"信息处理"原则称为是对脑的根本误解。通过我们的讨论,甚至凡及也对弗里曼的框架的意义有了新的认识,尽管他之前曾将弗里曼的著作翻译成中文。

试图理解弗里曼的观点,特别是他认为我们的脑并不计算信息,而是在持续的相互作用过程和"循环因果性"回路中提取意义,这一观点对我们两个人都有着至关重要的作用。这个观点可能并不尽善尽美,也可能并不适用于所有脑区和功能,当然也不会是这场辩论的结论。

我们在其他一些问题上仍然有不同的看法,例如主观性的本质,意识是否会从物质中涌现出来以及如何涌现,机器是否也可能有意识,我们对查默斯和丹尼特等作者也有不同的评价。但是我们都同意弗里曼工作的价值这一点远比我们之间的其他分歧更为重要。

对我们而言,弗里曼早已在《意识、意向性和因果关系》(*Consciousness, Intentionality, and Causality*)(1999)中提出了他的观点,他的观点可以作为一种有希望的理论框架,这一框架本可以在 HBP 和 BRAIN 倡议等研究计划中得到应用和检验。

我们的脑不是一种数字信息处理机器,也不是按照逻辑建立起来的,也没有进行数值计算,而是一种模拟装置,可以从噪声中提取出意义。这种观点对急于在更快的计算机上求解偏微分方程的数学家和程序员来说可能不是很吸引人,但我们希望它能成为寻找魔法城堡不同入口的年轻无偏见探险家一个有启发性的起点。

只要我们能激励几名年轻学生或许还有一两名有经验的研究人员尝试这种方式,那么我们的工作就已经值得了。

除了对脑、心智和人工智能之谜的质疑之外,我们还讨论了很多有关科学方法论以及如何在知识方面取得进展的最佳方法。尽管我们广泛地批评了一

些研究的做法，但我们当然不会声称我们有权告诉别人如何去做科学研究。

关键在于科学本身应该服从理性思维，很明显，在寻求破解科学谜团方面有些方法和研究做法对揭开科学之谜比起其他的来说更有希望。首先，我们在自然科学中拥有的实质性知识宝藏在很大程度上是由前几代人积累起来的。当我们观察知识增长曲线时，令人惊讶的是，在19世纪和20世纪之交的这个阶段，这个增长曲线是多么的陡峭。而且，更令人惊讶的是，为我们的科学世界观做好准备的群体是多么小，而付出的成本又是多么的微不足道。

如果把它比作一栋房子，我们就会看到地下室和一楼是由几个工匠在短时间内建成的。今天，整群工匠都在施工现场工作，但进度似乎已经放缓，甚至小小装修的成本也比造最初几层楼要高。

有人认为，近代科学界在过去几年产出的知识超越了历史上积累起来的所有知识。如果以发表的论文数量作为衡量的标准，那么这话可能是对的。但是，如果我们使用可以被称为"创造性"的标准，也就是看看我们做到了哪些先人无法做到的事情，那情况又将如何呢？这是一个非常平常而务实的观点，但可能是一个非常有用的观点。

我们说过，如果按上述标准来衡量，那么物理学家特别是工程师的表现优于生命科学家，并且工程技术似乎是以指数式速度发展，而神经科学和医学的发展则可能只是按线性发展。

我们一直有点吹毛求疵，我们说过，我们对许多研究计划的产出和称为"有趣发现"的结果不满意，而病人还在那儿等待治疗。也许我们这样说对于那些在这些领域努力工作的许多人来说太过分了，并且可能会发现我们的批评不公平和令人沮丧。特别是对于参与大型科学计划的人来说，情况就更是如此，他们没有多少自由选择的余地。但是，我们并非专指哪些特定的科学家，而只是描述看到的情况，当然我们也理解在唾手可得的果实已经被我们的先辈摘走之后，剩下的问题有多么的困难。凡及还觉得大型科学计划如果从其本质上来说是工程技术性的，如果已有扎实的科学理论基础并以解决问题明确的重要任务为目标，那么这样的计划还是重要和可行的。多学科开发新的研究工具和大规模采集脑的基本数据也许对脑研究来说是重要的，并且需要巨大的投资和团队

　　　意识之谜和心智上传的迷思　　　一位德国工程师与一位中国科学家之间的对话

攻关,这可能是大科学最适合的场所。但这并不能解决脑研究中的所有问题,特别是那些极度需要创造性和洞见的问题。

※　※　※

我们的意思并不是说我们没有看到很多成就,而是说有些东西言过其实了,这似乎助长了整个领域的某种浮夸风。现有的学术奖励制度只看出版物和引用数据,让这种情况更加恶化。这助长了科学和研究中普遍将叶子装扮成果实的倾向,因为围绕几个谜题发表一些"有趣的发现"可能比冒着风险尝试破解困难的谜题对前程来说要更保险一点。

对这个问题我们并没有一个简单的解决方案,但是因此就连提都不敢提也是不对的。

我们很清楚批评容易做起来难。虽然我们手里并没有什么神奇的解决方案,但我们不想避而不答"我们该怎么办?"的问题。

当我们总结这场对话并为写跋做准备时,凡及提到卡尔在早先的信中半开玩笑的提议,他宁愿给一千个聪明的学生每人一百万,而不愿把10亿欧元投入到像HBP这样的"研究怪兽"中去,并提出了以下问题:

我同意你所说的关于医学的现状,但是在我们讨论之后,我的脑海里始终存在一个问题,那就是:我们应该在医学上做什么?削减对这些研究的财政支持?我并不认为应该这样做。所以,虽然你的质疑是正确的,过去的方式似乎并不奏效,"大科学"似乎也没有解决太多问题,那么我们该怎么做呢?照技术界那样行事?由于性质不同,那也不太可能,那么我们还有什么途径可以选择呢?把钱分给许多博士后?也许吧,但他们又应该怎么做呢?

凡及提出的问题让我们再次陷入思考和讨论。

卡尔仍然认为,即使是这样一种看似疯狂的想法,就像把希望寄托在一些年轻的不受拘束和心无旁骛的学生的创造力,在某些情况下可能是一种有效的选择。但是,对于所有的大科学来说,这当然不是一种现实的替代方案,而可能

只是对某个小实验的选项。这是因为大科学是一个已经建立的价值数十亿美元的行业，不可能在一夜间就发生巨大的革命性变化。在医药行业就更是如此，无论如何，这个行业有许许多多规定，有很多法律限制和伦理界线，这些都是有充分理由建立起来的。正如我们从普朗克 100 多年前的抱怨中所能看到的那样，所有这一切还要加上科学长期以来形成的惯性。

所以当我们试图对"我们应该做什么？"这样一个问题给出一个实用的答案时，我们不用去考虑战舰舰长或者大科学舰队司令官的意见，因为他们早就知道我们所描述的所有问题。我们提出的建议可能只会增加他们的沮丧，因为他们看到这样大的结构即使要在路线上稍做改变，也没有多少回旋余地。

所以从他们那里，我们至多只能期望他们不要把我们当成天真的浪漫主义者。我们的对话表明，对于那些充满决心和自信心的思想家们来说，我们确实都有点浪漫，他们永远关心某个问题，并且在问题得到解决之前不会放弃。

但我们当然也知道现代科学是一项团队运动，需要不同类型的人来建立一个成功的团队。除了个人的技能和雄心之外，团队精神也是决定性的。在科学史上，表现为把激情集中在某个特定问题上的团队精神非常有帮助，这一点也很重要。

因此，当我们提出建议时，我们想到的是充满热情的个人和小团队，就像小型研究快艇的船长及其船员，尤其是那些年轻的学生，他们早就感觉到自己有很大潜力，并且正在寻找真正的挑战，这些人可能从一点鼓励中获益。

对他们来说，非常简单而不足为奇的建议是：

● 努力工作，坚持钻研令你着迷的问题（就像弗里曼所做的那样），这和备战奥运的运动员很相像。

● 努力尽早成为某一领域中充满热情的专家，并真正精通。

● 不要错把高分和得奖当作真正的洞察力或实际掌握（这意味着要避免费恩曼所形容的巴西大学中的那种学究式错误）。

● 寻找/生产真正的果实，不要欺骗自己，误把树叶当果实。

● 不要只是为了发表而将无足轻重的结果当成"有趣的发现"。

● 让你的同行了解你，而且还要让其他有关的人也理解你，让他们知道你

的能力,并与你合作以破解一个问题明确的谜团。

好吧,我们并不认为我们心目中的大多数人确实需要这样的建议,因为他们早就已经这样做了,而且他们也太忙并全神贯注于他们所做的事情,没有时间来理会我们的建议。

但是也可能还有人仍然没有拿定主意该走什么道路,如果他们倾向于我们所主张的道路,那么我们的看法对他们也许有点好处,并从中受到鼓舞。

卡尔把他起草的跋给他女儿和一些年轻朋友看,想知道我们的建议会让读者们怎么看,结果得到了一些很有意思的反响。他们认为我们不应该只关注优秀发明家和发现者那样的天才,而且也要关注那些天赋上稍差一点,但是也能作出重要贡献的人。我们大多数人都不是天才,如果想让我们像天才那样行事也没有用。然而,天赋不太高的人也可以通过很好地补偿天才型人才的缺陷,而在团队中作出非常有益的贡献。这是一个很有道理的论点,为把科学发明转化为产品而创立团队和公司时尤其是如此,社交能力往往比科学卓越更为重要。我们提到过许多科学是团队运动,此时合理调配资源对成功起到决定性的作用,我们应该汲取这一教训。

※　　※　　※

当我们回顾自然科学的黄金时代时,还有另一个更普遍的观点值得注意。这就是学者开始进行研究的年龄。一般说来,比起 100 年前,现在开始研究的年龄要大得多,只是因为完成学业需要花费的时间很长。问题是在所有情况下是否真的有必要花这么长的时间。

有一些令人鼓舞的例子,说明并不一定需要如此。100 年前,两位 19 岁的年轻人受到学术界同行的鼓励去重塑物理学世界,他们很快就取得了成功,并使他们的教授不知所措。在他们成就背后的驱动力是好奇心和雄心壮志,以及对理解并破解迷人谜团的渴望。当泡利(Wolfgang Pauli)和海森堡(Werner Heisenberg)中学毕业进入大学时,他们已经知道了当时所有有关相对论和量子理论的知识。这并非因为他们是超级聪明的神童,而仅仅是因为他们长久以来一直为一个特定的问题着迷。两人是幸运的,他们的教授索末菲(Arnold

Sommerfeld）赏识他们的热情和才能，为他们作了特别安排以满足他们的学业要求。并不是每位同事都喜欢这样做，但是索末菲的勇敢举措得到了回报，两位年轻人都在 24 岁时完成了主要工作，并且都获得了诺贝尔奖。

我们今天在世界各地都有这样充满激情的年轻人。他们醉心一件事到了废寝忘食的地步。他们越是痴迷于某件事，就越容易忽视其他一切，因为这些事对他们的主要兴趣没多大帮助，使他们觉得枯燥乏味。传统的基于高分或应试技能的入学标准让他们名落孙山。所以问题是：我们应该如何吸引这些人并将他们引入"竞技场"？

当然，19 岁的年轻人不可能破解所有的科学谜团。但科学与其他类型高难度运动没有太大的区别，我们经常发现运动员在 30 岁时的表现就已经下降了。

当我们回顾历史时，我们会看到有很多突破是由极其年轻的研究者作出的，特别是在数学方面，那里的果实-树叶比总是很高。数学统计的一条基本原理是"最小二乘法"。它和现在我们所称的"曲线拟合"都是由 18 岁的高斯（Carl Friedrich Gauss）于 1795 年发明的。它是构建现代 AI 深度学习算法的核心原理。

当然，我们并不是说要一位 19 岁的黑客神经病学家在车库里为他们的邻居做脑部手术，对于这样的医疗手术，有充分的理由需要建立专业规则和管理部门来保护患者。

但我们要说的是，其他一些非常聪明的年轻人类似地在车库里就电子学和软件做出疯狂之举，并在过去几十年中创造出全新的行业。他们中的许多人都是辍学的人，他们觉得在现实世界中创造奇迹比完成学业更有吸引力。

正是这种情况才使我们有了现在所经历的技术巨变。

在我们的通信中，卡尔没有使用过"贸易战"这个词，但最后你可以在字里行间感觉到这一点。与此同时，在美国对中国宣布将在 2030 年成为世界主要人工智能创新中心的雄心作出激烈反应之后，类似的用语在西方成了头条新闻，大家也都在这样谈论。

我们很惊讶地看到我们感兴趣的这一领域的技术方面成了有关全球工业领导地位辩论的中心议题。在中国宣布其人工智能计划之后，世界上又发生了

许多变化。

当我们开始分析理解脑、智能和人工智能的作用时，这还只是一个有一定知识的人以及一小群专家关注的主题。今天其中有关人工智能的部分已经成了媒体中的一个主要话题，从科学界到经济界和政界，到处都在讨论。2016 年有人曾预测人工智能是一种炒作泡沫，很快就会进入下一个冬天。但事实却恰恰相反。人工智能成为有史以来炒得最热的技术，而且在中国宣布其人工智能计划和美国做出激烈反应之后，人们对人工智能的兴趣与日俱增。

对我们来说，感觉上就像是一场海啸，涌入了我们小心翼翼划着的友谊小船的小河中。

现在每个人似乎都对 AI 有看法，似乎也都明白是怎么回事。当然，我们并不因为我们比某些在今天谈论这个问题和写文章的人可能考虑得稍长远一些，就声称我们对该领域正在发生什么以及接下来会发生什么有更好的认识。特别是凡及，他依然只是一个对此领域抱有强烈兴趣而好质疑的"门外汉"。

一些记者和分析师对该领域的要点做了很好的总结，突出了关键点和所面临的挑战，我们对此印象深刻。这再次说明，即使是借助像老式的互联网及其陈旧的搜索引擎那样的老版本 AI，在获得有关新领域和复杂事物的见解方面也大有帮助。

而且我们也从这个领域的日益普及中获益，因为许多聪明人已经对这个话题作了进一步的解释，这帮助我们获得了新的见解。

所以读者可能会听到我们谈论最近才知道的事情，现在这些事情几乎在一夜之间就变得广为人知了。我们认为，我们的思考并没有太偏离正轨，并希望它可能会让一些人对人工智能问题有一个更加深入和清醒的认识。

这场人工智能热的主要后果，或许也是会引起巨变的后果，是现在有更多的钱涌入这一领域。早在脑研究计划时，人们可能就有这样一种印象，即没有任何一个工业化国家的政府可以不启动这样一个计划。在 AI 这一领域就更是如此，并且奖券更高。在五年以前，10 亿美元对一个研究计划来说似乎已经是一笔惊人的数目，而现在，单是一家中国的初创公司寒武纪投在运行 AI 应用程序的硬件芯片上的资金就达到 10 亿美元（这还是我们当时得到的数字，到了

2018 年 6 月我们发现投入已经上升到了 25 亿美元）。而且还有不计其数的这类创业公司正像雨后春笋一样到处出现。

这种冲击的副作用是，懂得如何建立人工智能系统的人，尤其是那些同时也懂神经科学的人非常抢手。世界各地的工业领袖都试图到处挖掘这样为数不多的专家，特别是从大学和研究机构挖人。这对专家来说固然是件好事，但对这些机构来说却是一个问题，因为他们往往无法在财务上进行竞争。但与此同时，它也为学术界、工业界和建立初创企业的投资者之间的新型合作方式以及各种研究合作和研究机构的新形式提供了机会。看到众多有创造性的新星正在涌现以及官僚主义的障碍正在快速被克服，这真是令人惊叹。中国似乎在这方面特别有创意和行动力。

因此，上面讲到的一些与纯学术研究有关的问题，也可能以这种方式得到解决。如果有一部分研究将从大学转向产业，我们丝毫不会感到惊讶，我们将看到过去几十年中 IT 行业那样的成长和发展，相关的基础研究将转移到像贝尔实验室或 IBM 那样的大型公司的实验室中进行，其中也产生出诺贝尔奖获得者。

由于激情和团队精神在这种研究项目中所起的重要作用，我们认为至少在早期阶段，我们所钟爱的小型快艇比大科学战舰更有优势。

但是这就已经是猜测性的了，要预测接下来会看到什么就更纯属猜测了。现在除了变化更快之外，我们还没有看到有比我们在信中讨论过的更多的问题。

有很多东西我们没有谈到，或者只是简单地提了一下。其中之一是人工智能可能产生的社会影响这样的大问题，从可能引起的失业潮到个人数据的隐私权问题。

我们没有深入讨论这些问题，这并不是因为我们没有看到这些问题。我们没有这样做，只是因为在这些讨论中对这些危险有太多的猜测和喧嚣，以至于我们不知道该再讲些什么真知灼见，我们也不想比别人提出更为阴郁的负面猜测。

然而，有一个方面我们想再讲几句。

我们经常提到冯·诺伊曼，他肯定是 20 世纪真正的天才之一，并且对现代科学的贡献比大多数人都多。其中之一是他在 1944 年与摩根斯坦（Oskar Morgenstern）一起发表的博弈论。这一理论几乎与以他的名字命名的计算机架构一样成功。尽管这令人印象深刻，但也有一些缺点。其中最主要的一点是它只研究总额固定的博弈，其中一个玩家的胜利就是另一个玩家的损失。这种观点不仅在冷战时代的将军中非常流行，而且还成了一种范式。

另一位科学特立独行者和数学天才纳什（John Nash）证明，在经济学中，为了改善福祉，合作策略比尝试击败竞争对手更有前途。

而这些合作策略肯定适用于获得科技知识的过程中，特别是当涉及一场值得为之一战的斗争，那就是对类似阿尔茨海默病、帕金森病等危害人类健康的疾病的攻关。

因此，中国科学家和德国工程师一致认为，我们最好在友好竞争中做到这一点，而不是错误地认为我们正在玩零和游戏。如果这个想法不容易在人们的头脑中传播，那么希望下一代人工神经网络能够为自己得出同样的结论。

扫一扫，获取更多有关
脑研究与 AI 的信息

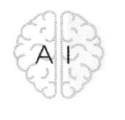

　　本书是我第一次翻译一本自己合著的书。尽管翻译自己写的部分比较容易，我当然知道自己想说些什么，但是，本书的近三分之二是由卡尔写的，由于他的思想深刻、知识面广，这使我在翻译时很不容易，尤其是当话题涉及先进技术、西方哲学和除神经科学之外的科学史时更是如此。幸运的是，卡尔非常耐心地回答我的问题，甚至重新撰写某个段落，以便我能理解。这使翻译得以尽可能地忠实于原文。因此，我想对我亲爱的朋友表示衷心的感谢。没有他的帮助，我就无法把本书翻译成现在的样子。这让我想起目前机器翻译的一个问题，它不理解文本的含义！即使是一个很差的翻译人员也总可以知道自己在原文中哪些地方没有看懂。如果他或她是认真的话，就会去查阅字典、百科全书、背景资料，甚至询问原作者或其他专家，直到弄懂为止。然而，现在的机器翻译从来都不会提出任何问题，它甚至把错字也照翻不误，就好像它不是一个错字一样！

　　正如在序言中所说的，本书主要是为了引起读者的思考，而不是告诉他们结论。作者不想假装知道一切，并给读者正确的答案，他们也不想假装成为预言未来的先知。他们只是想通过理性思考告诉读者他们的想法，并希望读者也能类似地进行思考。我写过几本关于脑和心智的科普书籍，我的一位朋友曾问我如何能保证我所说的所有内容都是正确的。我的回答是："不，我不能保证，我也不认为任何作者能作这样的保证，我所能做的只是保证我所谈论的所有材料都有一定的根据，你可以实证地加以检验或否定。"本书也是如此。更进一步说，两位作者对一些话题都仍然有不同的看法，就像卡尔在跋中所总结的一样。每个人都有自己的理由，但无法说服对方。对于某些话题，我们可能会在认同的程度上有所不同。例如，两位作者都认为对整个人脑进行仿真是不现实的，

甚至可能是毫无意义的，然而，如果大多数参数都在生理范围内，那么对神经回路等的仿真还是有意义的。认识脑不能光依靠大科学工程，但是当问题涉及昂贵的设备、有坚实理论基础的大规模数据收集、不同领域的专家之间的广泛合作，大科学工程可能仍然是必要的。问题是这种项目的内容、组织形式以及如何在战舰和小型快艇之间取得适当的平衡。我们两人都同意目前技术正呈指数式增长，而脑科学和医学则呈线性增长。但是其原因是否就是因为前者兴趣派占了主导地位，而后者则是规矩派一统天下？在译者看来在这两个领域中都存在着兴趣派和规矩派的斗争，虽然他们在这两个领域中的比重不同。他们发展速度之差主要是由其学科性质决定的，虽然这两派的斗争也起作用。这些都仍然是没有解决的问题，值得读者思考。

不管怎么说，写这本书和与卡尔的讨论改变了我的许多想法，甚至改变了思考问题的方式。我希望有类似想法的读者在读了本书之后也会有同样的感受。翻译让我有机会重新审查整个过程并整理我的想法。现在我希望以新的视角看待这些尚无定论的问题，我可能可以看得更清楚一点，对这些领域快速发展中出现的问题更敏感一点。尽管这些问题最终还是没有解决，甚至也不能在可预见的将来得到解决，这也正是这一领域之所以如此迷人之处。

仔细的读者也许会发现，本书的中文部分比英文部分多了一些内容。其实这些内容只是为了帮助国内读者更好地理解书中的某些内容所提供的背景材料，其中包括译注、专栏（其实专栏只是一种扩大型的译注）和插图。对于熟悉这些背景材料的读者来说，当然无需去阅读它们。由于这些材料并未译成英文让卡尔过目，因此在这些材料中如果有什么错误的话，那么责任应该完全由译者来负。

顾凡及记于复旦大学

2019 年 6 月 30 日

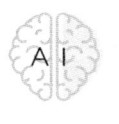

致 谢

　　世界上最可宝贵的并非金钱，而是时间。我们的家人、朋友和同事以其关心和合作帮助了我们。我们非常感谢他们投入的大量时间，使本书得以最后出版。

　　我们都以有一位睿智的妻子为傲，这是我们的福气。我们的妻子在 50 多年来一直在不知疲倦地减少我们生活中的"熵"。我们也都有幸有一个女儿，她们既是我们的朋友，也是非常有价值的顾问。因此我们首先并且永远要感谢马月美，阿迪（Adi），顾以蒙，杰尔卡·塞茨（Jelka Seitz）和于尔根·塞茨（Jürgen Seitz）。特别要感谢阿迪一遍又一遍地整理和编辑我们的草稿。我们也要感谢杰尔卡和她的丈夫于尔根。关于由我们的调研中可以得出些什么结论以及如何实现的问题，他们和我们进行了令人兴奋的讨论和争辩。亚历山大·塞茨（Alexander Seitz），作为他外祖父的一位现年 4 岁的老师，可能并没有注意到他在心智和意识如何发育的问题上给了他外祖父多大的帮助。

　　关于因果关系，我们要感谢我们共同的朋友汉斯·布劳恩教授（"德国汉斯"）。正是他启动了我们之间的联系和友谊，因此也就有了这本书。有些混沌学专家说，上海蝴蝶翅膀的扇动不会导致加勒比地区的龙卷风。但是杰出的混沌学专家汉斯已经表明，通过只是将一封来自上海的电子邮件从马尔堡转发到卡尔斯鲁厄，可以让两个人工作六年。这件事值得蝴蝶、龙卷风和混沌学专家思考！

　　我们也要特别感谢"德国汉斯"的一位亲密朋友和同事。这就是凌瀚思（Hans Lilienström）教授（"瑞典汉斯"），他也是凡及的老朋友和极有价值的顾问。

　　得自凡及的朋友江渊声（Nelson Y. S. Kiang）教授的支持和建议特别有帮助，我们也要感谢他慨允我们引用他在私人通信中所说的话。

还有许多其他朋友和同事在和我们的讨论过程中帮助我们解答了一些具体问题，或扩展了我们的研究。我们不能一一列举他们的名字，而只能提到一些其支持对我们特别有鼓舞的人。

在中国方面我们要感谢陈宜张教授、寿天德教授、梁培基教授、王培教授、唐孝威教授、郭爱克教授、杨雄里教授、洪波教授、吴思教授、俞洪波教授、林龙年教授、童勤业教授、曹建庭教授、孙哲博士、顾金涛博士、潘颖女士和宋蔓女士。

在欧美国家，我们要感谢哈霍夫（Dietmar Harhoff）、郝斯克（Gert Hauske）、恰伊赫（Ehrenfried Zschech）、贝特格（Matthias Bethge）、吉塞尔（Dirk Kanngiesser）、拉古纳夫妇（Susanne and Rafael Laguna）、霍夫曼（Ingo Hofmann）、多甘（Serdar Dogan）、汉森（Ben Hansen）、博克（Andreas Bogk）、乌尔曼（Bernd Ulmann）、赫费尔（Hendrik Höfer）、霍尔策（Mirko Holzer）、M.帕莱塔（Matthäus Paletta）、赖特（James Wright）和凯（Leslie Kay）。

陈宜张、梁培基、王培、唐孝威、顾金涛、宋蔓、布劳恩、哈霍夫、郝斯克、恰伊赫、贝特格审阅了全稿，并提出了许多宝贵的意见，这使我们得以在付印之前减少错误和写得不清楚之处。我们非常感激他们付出巨大的努力为本书撰写评论和书评，这些评论是对我们信件的某些部分的介绍。使我们感到不好意思的是他们通常对我们过于宽容和慷慨，我们相信他们或许能够把我们自己想表达的意思讲得更为清楚。

我们要感谢哈特曼（Patrick Hartmann）、A.帕莱塔（Agnieszka Paletta）和布鲁姆菲尔德（John Broomfield）纠正了我们的中式英语/德式英语中的一些主要错误，同时又不失去我们电子邮件中原来的语气。

特别感谢上海教育出版社的编辑黄伟先生和沈明玥女士，感谢他们巨大的支持和对改进本书的许多有益的想法，例如列出多种目录的建议，以及使本书同时兼具中英文版的奇妙形式。他们还邀请了一位画家陈楚桥先生为中文部分绘制插图，并对插图的内容提出建议。当然，我们也要在此向陈先生表示感谢。

我们不知道凡及的朋友弗里曼（Walter Freeman）是否会喜欢我们要说的话，但是我们都觉得有必要通过将本书献给这位杰出的研究者来表达我们的感激、钦佩和敬仰，从他那里我们得到了如此多的启发和洞见。